Delay Tolerant Networks

Protocols and Applications

Edited by
Athanasios Vasilakos
Yan Zhang
Thrasyvoulos V. Spyropoulos

CRC Press
Taylor & Francis Group
Boca Raton London New York

CRC Press is an imprint of the
Taylor & Francis Group, an **informa** business

AN AUERBACH BOOK

CRC Press
Taylor & Francis Group
6000 Broken Sound Parkway NW, Suite 300
Boca Raton, FL 33487-2742

First issued in paperback 2019

ISBN-13: 978-1-4398-1108-5 (hbk)
ISBN-13: 978-0-367-38220-9 (pbk)

Visit the Taylor & Francis Web site at
http://www.taylorandfrancis.com

and the CRC Press Web site at
http://www.crcpress.com

Contents

Preface   v

List of Contributors   ix

1 Delay Tolerant Networking   1
   Maode Ma, Chao Lu, and Hui Li

2 DTN Routing: Taxonomy and Design   31
   Thrasyvoulos V. Spyropoulos, Rao Naveed Bin Rais, Thierry Turletti,
   Katia Obraczka, and Athanasios Vasilakos

3 Energy-Aware Routing Protocol for Delay Tolerant Networks 69
   Seung-Keun Yoon and Zygmunt J. Haas

4 A Routing-Compatible Credit-Based Incentive Scheme
   for DTNs   101
   Haojin Zhu, Xiaodong Lin, Rongxing Lu, Yanfei Fan, and Xuemin
   (Sherman) Shen

5 R-P2P: a Data-Centric Middleware for Delay Tolerant
   Applications   127
   Corrado Moiso, Antonio Manzalini, Francesco De Pellegrini,
   Iacopo Carreras, Daniele Miorandi, and Athanasios Vasilakos

6 Mobile Peer-to-Peer Systems over Delay Tolerant Networks 159
   Angela Sara Cacciapuoti, Marcello Caleffi, and Luigi Paura

7 Delay Tolerant Monitoring of Mobility-Assisted WSN   189
   Abdelmajid Khelil, Faisal Karim Shaikh, Azad Ali, Neeraj Suri,
   and Christian Reinl

8 Message Dissemination in Vehicular Networks   223
   Shabbir Ahmed and Salil S. Kanhere

**9 Delay Tolerant Networking (DTN) Protocols for Space
    Communications**   **261**
    Ruhai Wang, Xuan Wu, Tiaotiao Wang, and Tarik Taleb

10 DTN and Satellite Communications   **283**
    Carlo Caini and Rosario Firrincieli

Index   **319**

Preface

The standardization of the IP protocol and its mapping into network-specific link-layer data frames at each router as required supports interoperability using a packet-switched model of service. Although often not explicitly stated, a number of key assumptions are made regarding the overall performance characteristics of the underlying links in order to achieve smooth operation:

- an end-to-end path exists between a data source and its peers,

- the maximum round-trip time between any node pairs in the network is not excessive, and

- the end-to-end packet drop probability is small.

Unfortunately, a class of Delay Tolerant Networks (DTN), which may violate one or more of the assumptions, is becoming important and may not be well served by the current end-to-end TCP/IP model.

DTN arise primarily as a result of various forms of host and router mobility, but may also come into being as a result of disconnection due to power management or interference. Examples of such networks include:

• **Terrestrial Mobile Networks**: In many cases, these networks may become unexpectedly partitioned due to node mobility or RF interference. In other cases, the network may never have an end-to-end path and may be expected to be partitioned in a periodic and predictable manner. For example, imagine a commuter bus acting as a store and forward message switch with only limited RF communication capability. As it travels from place to place, it provides a form of message switching service to its nearby clients to communicate with distant parties it will visit in the future.

• **Exotic Media Networks**: Exotic communication media includes near-Earth satellite communications, very long-distance radio links (e.g., deep space RF communications with light propagation delays in the seconds or minutes), communication using acoustic modulation in air or water, and some free-space optical communications. These systems may be subject to high latencies with predictable interruption (e.g., due to planetary dynamics or the passing of a scheduled ship), may suffer outage due to environmental conditions (e.g., weather), or may provide a predictably available store-and-forward network service that is only occasionally available (e.g., low-earth orbiting satellites that "pass by" one or more times each day).

• **Ad-Hoc Networks**: These systems may be expected to operate in hostile environments where mobile nodes, environmental factors, or intentional jamming may be cause for disconnection. In addition, data traffic on these networks may have to compete for bandwidth with other services at higher priority. As an example, data traffic may have to unexpectedly wait several seconds or more while high-priority voice traffic is carried on the same underlying links. Such systems also may also have especially strong infrastructure protection requirements.

• **Sensor Networks**: These networks are frequently characterized by extremely limited end-node power, memory, and CPU capability. In addition, they are envisioned to exist at tremendous scale, with possibly thousands or millions of nodes per network. Communication within these networks is often scheduled to conserve power, and sets of nodes are frequently named (or addressed) only in aggregate. They are often interfaced to other networks by way of one or more "proxy" nodes that provide protocol translation capabilities. Given the large accumulated experience and number of systems compatible with the TCP/IP protocols, it is natural to apply the highly successful Internet architectural concepts to these new or unusual types of networks. While such an application is conceivable, the effects of very significant link delay, non-existence of end-to-end routing paths, and lack of continuous power or large memory at end nodes present substantial operational and performance challenges to such an approach.

In an effort to adapt the Internet to unusual environments, one class of approaches attempts to engineer problem links to appear more similar to the types of links for which TCP/IP was designed. In effect, these approaches, which we term "link repair approaches," "fool" the Internet protocols into believing they are operating over a comparatively well-performing physical infrastructure. They strive to maintain the end-to-end reliability and fate sharing model of the Internet, and generally require the use of IP in all participating routers and end nodes.

Another common approach to deal with challenged networks is to attach them to the edge of the Internet only by means of a special proxy agent. This provides access to and from challenged networks from the Internet, but does not provide a general way to use such networks for data transit. Without supporting transit, the full capabilities of these networks will go underutilized. Indeed, supporting transit is often of particular interest because remotely-deployed conventional networks (e.g., Intranets) may only be accessible through challenged intermediate networks.

The book is written for a broad audience of researchers and practitioners. The book could be a useful reference material for graduate students and senior undergraduate students in courses of networking, wireless, and mobile communications. Furthermore, because the book is targeted towards the latest developments in the aforementioned domains, as well as towards the visions for our middle- and long-term future, it is expected to be useful also to senior researchers and technology analysts who want a glimpse on the future of networking.

List of Contributors

Shabbir Ahmed
School of Computer Science and
    Engineering
The University of New South Wales
Sydney, Australia

Azad Ali
Dependable Embedded Systems and
    Software Group
Darmstadt University of Technology
Darmstadt, Germany

Angela Sara Cacciapuoti
Department of Biomedical,
    Electronics, and
    Telecommunications Engineering
    (DIBET)
University of Naples *Federico II*
Naples, Italy

Carlo Caini
Department of Electronics Computer
    Science and Systems
University of Bologna
Bologna, Italy

Marcello Caleffi
Department of Biomedical,
    Electronics, and
    Telecommunications Engineering
    (DIBET)
University of Naples *Federico II*
Naples, Italy

Iacopo Carreras
CREATE-NET
Trento, Italy

Francesco De Pellegrini
CREATE-NET
Trento, Italy

Yanfei Fan
University of Waterloo
Waterloo, Ontario, Canada

Rosario Firrincieli
Department of Electronics,
    Computer Science, and Systems
University of Bologna
Bologna, Italy

Zygmunt Haas
The School of Electrical and
    Computer Engineering
Cornell University
Ithaca, New York

Salil S. Kanhere
School of Computer Science and
    Engineering
The University of New South Wales
Sydney, Australia

Abdelmajid Khelil
Dependable Embedded Systems and
    Software Group
Darmstadt University of Technology
Darmstadt, Germany

Hui Li
School of Communication
    Engineering
Xidian University
Xi'an, Shaanxi, People's Republic of
    China

Xiaodong Lin
University of Ontario Institute of
    Technology
Oshawa, Ontario, Canada

Chao Lu
School of Communication
    Engineering
Xidian University
Xi'an, Shaanxi, People's Republic of
    China

Rongxing Lu
Broadband Communications
    Research (BBCR) Group
Department of Electrical and
    Computer Engineering
University of Waterloo
Waterloo, Ontario, Canada

Maode Ma
School of Electrical and Electronic
    Engineering
Nanyang Technological University
Singapore

Antonio Manzalini
Telecom Italia Future Center
Torino, Italy

Daniele Miorandi
CREATE-NET
Trento, Italy

Corrado Moiso
Telecom Italia Future Center
Torino, Italy

Katia Obraczka
Computer Engineering Department
Jack Baskin School of Engineering
University of California, Santa Cruz
Santa Cruz, California

Luigi Paura
Department of Biomedical,
    Electronics, and
    Telecommunications Engineering
    (DIBET)
University of Naples *Federico II*
Naples, Italy

Rao Naveed Bin Rais
CIIT-Lahore
Pakistan

Christian Reinl
Simulation, Systems Optimization,
    and Robotics Group
Darmstadt University of Technology
Darmstadt, Germany

Faisal Karim Shaikh
Dependable Embedded Systems and
    Software Group
Darmstadt University of Technology
Darmstadt, Germany

Xuemin (Sherman) Shen
Department of Electrical and
    Computer Engineering
University of Waterloo
Waterloo, Ontario, Canada

Thrasyvoulos V. Spyropoulos
Institute Eurecom
France

Neeraj Suri
Dependable Embedded Systems and
    Software Group
Darmstadt University of Technology
Darmstadt, Germany

Tarik Taleb
Network Research Division
NEC Europe Ltd.
Heidelberg, Germany

Thierry Turletti
Planète Group, INRIA
Sophia Antipolis, France

Athanasios Vasilakos
University of Western Macedonia
Kozani, Greece

Ruhai Wang
Phillip M. Drayer Department of
    Electrical Engineering
Lamar University
Beaumont, Texas

Tiaotiao Wang
Lyle School of Engineering (EE)
Southern Methodist University
Dallas, Texas

Xuan Wu
Phillip M. Drayer Department of
    Electrical Engineering
Lamar University
Beaumont, Texas

Seung-Keun Yoon
The School of Electrical and
    Computer Engineering
Cornell University
Ithaca, New York

Haojin Zhu
Trusted Digital Technology Lab
Department of Computer Science &
    Engineering
Shanghai Jiao Tong University
Shanghai, Peoples Republic of China

Chapter 1

Delay Tolerant Networking

Maode Ma, Chao Lu, and Hui Li

1.1 Introduction...2
    1.1.1 History of Delay Tolerant Networking.....................2
    1.1.2 A Delay Tolerant Network................................2
    1.1.3 Requirements on DTNs...................................3
1.2 The Architecture...4
    1.2.1 Overlay Architecture....................................4
    1.2.2 Store and Forward Message Switching......................5
    1.2.3 Routing and Forwarding..................................6
    1.2.4 Fragmentation and Reassembly............................6
    1.2.5 Custody Transfer.......................................7
1.3 The Bundle Protocol..9
    1.3.1 Bundle Service...9
          1.3.1.1 Terms...10
          1.3.1.2 Service Offered by Bundle Protocol Agent.......11
    1.3.2 Bundle Format...11
          1.3.2.1 Self-Delimiting Numeric Values.................11
          1.3.2.2 Endpoint IDs in Detail.........................12
          1.3.2.3 Formats of Bundle Blocks......................12
    1.3.3 Bundle Processing......................................14
          1.3.3.1 Bundle Creation at Source......................15
          1.3.3.2 Transmission by Source.........................15
          1.3.3.3 First-Hop Processing and Forwarding............17
          1.3.3.4 Second-Hop Processing and Forwarding..........18
          1.3.3.5 Bundle Reception by Destination................18
1.4 Routing Schemes in DTNs......................................19
    1.4.1 Routing Considerations.................................20
    1.4.2 Classification of Routing Schemes.......................20

1

          1.4.3   Replication-Based Routing..............................21
                  1.4.3.1   Epidemic Routing............................21
                  1.4.3.2   PRoPHET Routing Protocol....................22
                  1.4.3.3   MaxProp Routing Protocol 22
                  1.4.3.4   RAPID Routing Protocol 23
                  1.4.3.5   Spray and Wait Routing Protocol.............23
    1.5   Open Issues in Delay-Tolerant Networking........................24
          1.5.1   Routing...24
          1.5.2   Custody and Congestion 25
          1.5.3   Security..26

1.1   Introduction

1.1.1   History of Delay Tolerant Networking

Inspired by the popularity of computing, in the 1970s, researchers began developing routing technology for non-fixed locations of computers. The field of ad-hoc routing was inactive throughout the 1980s, but the widespread use of wireless protocols reinvigorated the field in the 1990s as mobile ad-hoc networking (MANET). Vehicular ad-hoc networking also became a researcher's area of increasing interest.

Concurrently with but separated from the MANET activities, a proposal on Interplanetary Internet (IPN) had been funded to develop the novel technologies of IPN. The Internet pioneer Vint Cerf and others have developed the initial IPN architecture relevant to the necessity of networking technologies that can cope with the significant delays and packet corruption in deep-space communications.

In 2002, Kevin Fall started to adapt some of the ideas in the IPN to design terrestrial networks. He coined the term delay tolerant networking with DTN as the acronym. The first conference paper presented in 2003 has shown the motivation for DTNs [1]. In the following years, more and more attention has been drawn from researchers including a growing number of academic conferences on delay and disruption-tolerant networking, and growing interests in combining the work from sensor networks and MANETs with the work on DTN. The research work on the topic started from optimizations on classic ad-hoc and delay-tolerant networking algorithms and began to examine issues such as security, reliability, verifiability, and other issues that are well understood in traditional computer networking.

1.1.2   A Delay Tolerant Network

DTN is the area of networking which addresses challenges in disconnected, disrupted networks without end-to-end connection. DTN is designed to operate effectively over extreme distances such as those encountered in space communications or on an interplanetary scale. In such an environment, long latency which is measured in hours or days is inevitable. And the latency is even as long as a year such as an instance in the paper on DTLSR routing protocol [2].

The existing TCP/IP-based Internet protocols operate on a principle of providing end-to-end interpose communication using a concatenation of potentially dissimilar link-layer technologies. These Internet protocols do not work well for some environments tolerating long delays and predictably interrupted communications over long distances due to the fundamental assumptions built into the Internet architecture [3] as follows:

- An end-to-end path exists between source and destination during a communication session.

- For realistic communication, retransmission based on timely and stable feedback from data receivers must be composed for repairing errors.

- The end-to-end packet drop probability is small.

- All routers and end stations support the TCP/IP protocols.

- Applications need not worry about communication performance.

In a delay tolerant network, most of the above assumptions of the Internet are flexible. And the design principles of DTN architecture can be summarized as follows:

Variable-length messages will exist as the communication abstraction to facilitate the ability of the network for scheduling or path selection decisions.

A naming syntax which supports a wide range of naming and addressing conventions is applied to enhance interoperability.

Storage within the network is taken to support store-and-forward operation over multiple paths and potentially long timescales.

Security mechanisms are provided to protect the infrastructure from unauthorized users by discarding traffic as quickly as possible.

1.1.3 Requirements on DTNs

There are some networking scenarios where current Internet protocols do not work well, such as space missions to Mars and testing lake water quality in rural areas. Both these scenarios have something in common with more and more devices incorporating computing and networking technology in less traditional networking environments. Computer networking in such environments faces new challenges and new techniques and protocols are required. And all these challenges can be characterized as follows.

- *Intermittent Connectivity*: If there is no end-to-end path between source and destination, the end-to-end communication using the TCP/IP protocols does not work. New protocols to support the communications without an end-to-end path are required.

- *Long or Variable Delay*: In addition to intermittent connectivity, long propagation delays among nodes and variable queuing delays at each node contribute to end-to-end path delays that can defeat Internet protocols and applications that rely on quick return of acknowledgements or data.

- *Asymmetric Data Rates*: The Internet supports moderate asymmetries of bi-directional data rate for users with cable TV or asymmetric DSL access. But if asymmetries are large, conversational protocols will not work.

- *High Error Rates*: Bit errors over transmission links require correction or retransmission of the entire packet, which can result in more network traffic. For a given link-error rate, fewer retransmissions are needed for hop-by-hop than for end-to-end retransmission.

1.2 The Architecture

The DTN architecture is designed to accommodate network connection disruption with a framework for dealing with heterogeneity.

A multitude of different delivery protocols is used in DTN, which include TCP/IP, raw Ethernet, serial lines, or hand-carried storage devices for delivery. As each of these protocols provides somewhat different semantics, a collection of protocol-specific convergence layer adapters (CLAs) provides the functions necessary to carry DTN protocol data units, called bundles, on each of the corresponding protocols.

1.2.1 Overlay Architecture

The architecture embraces the concepts of occasionally-connected networks that may suffer from frequent partitions and that may be comprised of more than one divergent set of protocols or protocol families.

The end-to-end message overlay is defined as a "bundle layer" that exists at a layer above the transport layers of the networks on which it is hosted and below applications. Devices implementing the bundle layer are called DTN nodes. The bundle layer forms an overlay that employs persistent storage to help combat network interruption. And it includes a hop-by-hop transfer of reliable delivery responsibility and optional end-to-end acknowledgement. A number of diagnostic and management features are also included. For interoperability, it uses a flexible naming scheme capable of overall naming syntax. And also security nodes are designed as options aimed at protecting infrastructure from unauthorized use.

1.2.2 Store and Forward Message Switching

The messages sent in a DTN are with arbitrary length as Application Data Units (ADUs), which are subject to any implementation limitations.

ADUs are transformed by the bundle layer into one or more protocol data units called "bundles," which are forwarded by DTN nodes. Bundles have a defined format containing two or more "blocks" of data. Each block may contain either application data or control information to deliver the bundle to its destination. Blocks serve the purpose of holding information typically found in the header or payload portion of the bundles. Bundles may be fragmented into multiple constituent bundles as fragments or bundle fragments during the transmission. Fragments are themselves bundles, and may be further fragmented. Two or more fragments can be reassembled anywhere in the network to generate a new bundle.

The sources and destinations of a bundle can be identified by Endpoint Identifiers (EIDs), which identify the original sender and final destination(s) of the bundle, respectively. A bundle contains a "report-to" EID used when special operations are requested to direct diagnostic output to an arbitrary entity. An EID may refer to one or more DTN nodes.

While IP networks are based on "store-and-forward" operation, a DTN network does not expect that network links are always available or reliable. It expects that nodes can choose to store bundles for some time. It is anticipated that most DTN nodes will take persistent storages such as disk, flash memory, etc. So the stored bundles can survive when the system restarts.

Each bundle contains four components such as an originating timestamp, a life indicator, a class of service designator, and a length. The information serves bundle-layer routing with a priori knowledge of the size and performance requirements of the requested data transfers. When there is a significant amount of queue occurring in the network, the advantage provided by the information may be significant for making scheduling and path selection decisions [4]. An alternative abstraction would make such scheduling much more difficult. Although simple packets have the same benefits as bundles, larger aggregates provide a way for the network to apply scheduling and buffer management to the bundles that are more useful to applications. An essential element of the bundle-based forwarding is that bundles have a place to wait in a queue until a communication opportunity is available. It highlights the following assumptions:

- Storage is available and well-distributed over the network.

- Storage is sufficiently persistent and robust to store bundles until forwarding.

The "store-and-forward" model is a better choice than attempting to effect continuous connectivity or other alternatives.

For a network to effectively support the DTN architecture, these assumptions must be held as necessary. Even so, the inclusion of long-term storage as a fundamental aspect of the DTN architecture poses new problems, especially with respect to congestion management and denial-of-service mitigation. Node storage in essence represents a new resource that must be managed and protected. Much of the research in DTN revolves around exploring these issues. Security mechanisms, including methods for DTN nodes to protect themselves from handling unauthorized traffic from other nodes, have also been discussed in [5] and [6].

1.2.3   Routing and Forwarding

The DTN architecture provides a framework for routing and forwarding at the bundle layer for unicast, anycast, and multicast information. Because nodes in a DTN might be interconnected using more than one type of underlying network technology, a DTN is best described abstractly using a multigraph, which is a graph where vertices may be interconnected with more than one edge. Edges in this graph are, in general, time-varying with respect to their delay and capacity and transmission direction. When an edge has zero capacity, it is considered to not be connected. Because edges may vary between positive and zero capacity, it is possible to describe a period of time during which the capacity is strictly positive, and the delay and capacity can be considered to be constants [7]. This period of time is called contact. In addition, the product of the capacity and the interval is known as the volume of a contact. If contacts and their volumes are known ahead of time, intelligent routing and forwarding decisions can be made [4]. Optimally using a contact's volume, however, requires the ability to divide large ADUs and bundles into smaller routable units, which can be provided by DTN fragmentation. When delivery paths through a DTN graph are lossy or contact intervals and volumes are not known precisely ahead of time, routing computation becomes especially a challenge, which is an active area of research in the area of delay tolerant networking.

1.2.4   Fragmentation and Reassembly

DTN fragmentation and reassembly are designed to improve the efficiency of bundle transfers by ensuring that contact volumes are fully utilized and avoiding retransmission of partially-forwarded bundles. There are two forms of DTN fragmentation/reassembly.

A block of application data may be divided into multiple smaller blocks by a DTN node and each block is transmitted as an independent bundle. In this case, the final destination(s) are responsible for extracting the smaller blocks from incoming bundles and reassembling them into the original larger bundle and, ultimately, ADU. This approach is called proactive fragmentation because it is used primarily when contact volumes can be predicted in advance.

DTN nodes sharing an edge in the DTN graph may fragment a bundle cooperatively when a bundle is only partially transferred. In this case, the bundle layer at the receiver modifies the incoming bundle to indicate it as a fragment and forwards it normally. The previous-hop sender may learn that only a portion of the bundle has been delivered to the next hop and send the remaining portion(s) when subsequent contacts become available. This is called reactive fragmentation because the fragmentation process occurs after an attempted transmission has taken place. For example, consider a node T and two store-and-forward intervening nodes A and B, moved in opposite directions. While T is transmitting a large bundle to A, a reliable transport layer protocol below the bundle layer at each indicates that the transmission has terminated. But half of the transfer has completed successfully. In this case, T can form a smaller bundle fragment consisting of the second half of the original bundle and forward it to B when available. In addition, A, which has moved out of range of T, can form a new bundle consisting of the first half of the original bundle and forward it to whatever next hop(s) it deems appropriate.

The reactive fragmentation capability is not required to be available in every DTN implementation, as it requires a certain level of support from underlying protocols that may not be present, or present significant challenges with respect to handling digital signatures and authentication codes on messages. When an encrypted message is only partially received, most message authentication codes will fail. When a DTN security is present and enabled, it may therefore be necessary to proactively fragment large bundles into smaller units that are more convenient for digital signatures. Even if reactive fragmentation is not present in an implementation, the ability to reassemble fragments at a destination is required in order to support DTN fragmentation. Furthermore, for contacts with volumes that are small compared to typical bundle sizes, some incremental delivery approach must be used to prevent data delivery live lock.

1.2.5 Custody Transfer

The most basic service provided by the bundle layer is unacknowledged, prioritized unicast message delivery, while it is not guaranteed. It also provides two options for enhancing the delivery reliability as end-to-end acknowledgments and custody transfer. Applications are free to utilize the acknowledgment to implement their own end-to-end information reliability mechanisms. The custody transfer of the DTN architecture has specified only a coarse-grained retransmission capability.

Transmission of bundles with the Custody Transfer Requested option specified generally involves moving the responsibility for reliable delivery of an ADU's bundles among different DTN nodes in the network. For unicast delivery, it will typically involve moving bundles closer by using some routing metrics to their ultimate destination(s) and retransmitting when necessary. The nodes receiving these bundles along the way which agree to accept the reliable delivery responsibility are called custodians. The movement of a bundle and its delivery responsibility from one node to another is called a custody transfer, which is analogous to a database commit transaction [8]. Custody transfer allows a source to delegate retransmission responsibility and recover its retransmission-related resources relatively soon after sending a bundle in almost the minimum round-trip time to the first bundle hop(s). Not all nodes in a DTN are required by the system to accept custody transfers, so it is not a true hop-by-hop mechanism.

The way of the DTN routing performed may be altered for the existence of custodians. In some circumstances, it may be beneficial to move a bundle to a custodian as quickly as possible even if the custodian is much farther away from the bundle's final destination(s) than some other reachable nodes. Designing a system with this capability involves constructing more than one routing graph, which is an area of the research in DTNs.

Custody transfer in DTNs not only provides a method for tracking bundles that require special handling and identifying DTN nodes that participate in custody transfer, but also provides a mechanism for enhancing the reliability of message delivery. Generally speaking, custody transfer relies on underlying reliable delivery protocols of the networks that operate to provide the primary means of reliable transfer from one bundle node to the next. However, when custody transfer is requested, the bundle layer provides an additional coarse-grained timeout and retransmission mechanism and an accompanying (bundle-layer) custodian-to-custodian acknowledgment signaling mechanism. When an application will not request a custody transfer, the bundle layer timeout and retransmission mechanism will typically not be employed. The successful bundle layer delivery depends solely on the reliability mechanisms of the underlying protocols.

When a node accepts custody for a bundle that contains the Custody Transfer Requested option, a Custody Transfer Accepted Signal will be sent by the bundle layer to the Current Custodian EID contained in the primary bundle block. In addition, the Current Custodian EID will be updated to contain one of the forwarding unicast node's EIDs before the bundle is forwarded.

When an application requests an ADU to be delivered with custody transfer, the request is advisory. In some circumstances, a source of a bundle for which custody transfer has been requested may not be able to provide this service. In such circumstances, the subject bundle may traverse multiple DTN nodes before it obtains a custodian. Bundles in this condition are specially marked with their Current Custodian EID field set to a null endpoint. In cases where applications wish to require the source to take custody of the bundle, they may supply the Source Node Custody Acceptance Required delivery option. This is useful to the applications that desire a continuous chain of custody or that wish to exit after being ensured that their data is safely held in a custodian.

In a DTN network where one or more custodian-to-custodian hops are strictly one directional, the DTN custody transfer mechanism will be affected over such hops due to the lack of any way to receive a custody signal back across the path, resulting in the expiration of the bundle at the ingress to the one-way hop. This situation does not necessarily mean that the bundle has been lost. Nodes on the other side of the hop may continue to transfer custody, and the bundle can be delivered successfully to its destination(s). However, in this circumstance that a source has requested to receive expiration BSRs for this bundle, it will receive an expiration report for the bundle, and possibly conclude (incorrectly) that the bundle has been discarded and not delivered. Although this problem cannot be fully solved, a mechanism is provided to help ameliorate the incorrect information that may be reported when the bundle expires after having been transferred over a one-way hop. This is accomplished by the node at the ingress to the one-way hop reporting the existence of a known one-way path using a variant of a bundle status report. These types of reports are provided if the subject bundle requests the report using the "Report When Bundle Forwarded" delivery option.

1.3 The Bundle Protocol

Bundle Protocol defines a series of contiguous data blocks as a bundle—where each bundle contains enough semantic information to allow the application to make progress where an individual block may not. Bundles are routed in a store and forward manner between participating nodes over varied network transport technologies including both IP and non-IP-based transports. The transport layers carrying the bundles across their local networks are called bundle convergence layers. The bundle architecture operates as an overlay network providing a new naming architecture based on Endpoint Identifiers and coarse-grained class of service offerings.

1.3.1 Bundle Service

The bundle layer provides six classes of services for a bundle.

Custody Transfer: Delegation of retransmission responsibility to an accepting node, so that the sending node can recover its retransmission resources. The accepting node returns a custodial-acceptance acknowledgement to the previous custodian.

Return Receipt: Confirmation to the source, or its reply-to entity, that the bundle has been received by the destination application.

Custody-Transfer Notification: Notification to the source, or its reply-to entity, when a node accepts a custody transfer of the bundle.

Bundle-Forwarding Notification: Notification to the source, or its reply-to entity, whenever the bundle is forwarded to another node.

Priority of Delivery: Bulk, Normal, or Expedited.

Authentication: The method of digital signature used to verify the sender's identity and the integrity of the message.

To provide services in the networking environments with intermittent connectivity, large variable delays, and high bit error rates, bundle protocol sits at the application layer of some number of constituent internets, forming a store and forward overlay network. Key capabilities of the bundle protocol can be summarized as follows: [9]

Custody-based retransmission.

Ability to cope with intermittent connectivity.

Ability to take advantage of scheduled, predicted, and opportunistic connectivity in addition to continuous connectivity.

Late binding of overlay network endpoint identifiers to constituent internet addressed.

The bundle protocol uses the native Internet protocols for communications whose location within the standard protocol stack is shown in Figure 1.1. It should be noted that the Internet used here does not only mean TCP/IP and it can also contain some other network protocols. The interface between protocol suites is termed as a convergence layer adapter. Figure 1.1 shows the distinct transport and network protocols (denoted T1/N1, T2/N2, T3/N3).

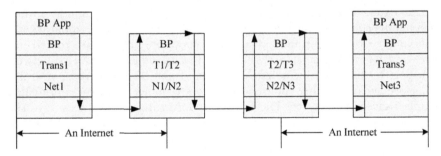

Figure 1.1: The bundle protocol sits at the application layer of the Internet model.

1.3.1.1 Terms

In consequence, there are some basic definitions that should be known as found in [9].

Bundle: A bundle is a protocol data unit of the DTN bundle protocol. Each bundle comprises a sequence of two or more blocks of protocol data, which serve various purposes.

Bundle Node: A bundle node is any entity that can send or receive bundles. In the most familiar case, a bundle node is instantiated as a single process running on a general-purpose computer. But, in general, a bundle node might alternatively be a thread, an object in an object-oriented operating system, a special-purpose hardware device, etc. Each bundle node has the conceptual components defined as a bundle protocol agent, a set of zero or more convergence layer adapters, and an application agent.

Bundle Protocol Agent (BPA): A bundle protocol agent of a node is the node component that offers the bundle protocol services and executes the procedures of the bundle protocol.

Convergence layer adapter: A convergence layer adapter sends and receives bundles on behalf of the BPA utilizing the services of a native Internet protocol that is supported in one of the Internets within which the node is functionally located.

Application Agent: An application agent of a node is the node component that utilizes the BP services to provide communication. The application agent has two elements, an administrative element and an application specific element.

Bundle Endpoint: A bundle endpoint is a set of zero or more bundle nodes whose identities are text strings called "bundle endpoint ID."

1.3.1.2 Service Offered by Bundle Protocol Agent

The bundle protocol agent of each node is expected to provide the following Services to the node's application agent:
Registering a node in an endpoint
Terminating a registration
Switching a registration between Active and Passive states
Transmitting a bundle to an identified bundle endpoint
Canceling a transmission
Polling a registration that is in the passive state
Delivering a received bundle

1.3.2 Bundle Format

Each bundle shall be a concatenated sequence of, at least, two block structures. The first block in the sequence must be a primary bundle block, and each bundle has only one primary bundle block, additional bundle protocol blocks of other types may follow the primary block to support extensions to the bundle protocol, such as the Bundle Security Protocol [10]. At most, one of the blocks in the sequence may be a payload block.

1.3.2.1 Self-Delimiting Numeric Values

The design of the bundle protocol attempts to reconcile minimal consumption of transmission bandwidth with the following two characteristics:
Extensibility to address requirements not yet identified
Scalability across a wide range of network scales and payload sizes

A key strategic element in the design is the use of self-delimiting numeric values (SDNVs). The SDNV encoding scheme is closely adapted from the Abstract Syntax Notation One Basic Encoding Rules for sub-identifiers within an object identifier value [11]. An SDNV is a numeric value encoded in N

octets, the last of which has its most significant bit (MSB) set to zero. The MSB of every other octet in the SDNV must be set to 1. The value encoded in an SDNV is the unsigned binary number obtained by concatenating into a single bit string the 7 least significant bits of each octet of the SDNV.

1.3.2.2 Endpoint IDs in Detail

The destinations of bundles are bundle endpoints, identified by text strings termed endpoint IDs. Each endpoint ID conveyed in any bundle block takes the form of a Uniform Resource Identifier (URI). Each endpoint ID can be characterized in this general structure:

< scheme name > : < scheme-specific part, or "SSP" >

As used for the bundle protocol, neither the length of a scheme name nor the length of an SSP may exceed 1023 bytes. Bundle blocks cite a number of endpoint IDs for various purposes of the bundle protocol. Many, though not necessarily all, of the endpoint IDs referred to in the blocks of a given bundle are conveyed in the dictionary byte array in the bundle's primary block. The array is simply the concatenation of any number of null-terminated scheme names and SSPs.

Endpoint ID references are used to cite endpoint IDs that are contained in the dictionary. All endpoint ID citations in the primary bundle block are endpoint ID references, while other bundle blocks may contain endpoint ID references as well. Each endpoint ID reference is an ordered pair of SDNVs.

The first SDNV contains the offset within the dictionary of the first character of the referenced endpoint ID's scheme name.

The second SDNV contains the offset within the dictionary of the first character of the referenced endpoint ID's SSP.

The encoding enables a degree of block compression. When the source and report-to of a bundle are the same endpoint, for example, the text of that endpoint's ID may be cited twice but appears only once in the dictionary.

The scheme identified by the ¡scheme name¿ in an endpoint ID is a set of syntactic and semantic rules that fully explain the way to parse and interpret the SSP. The set of allowable schemes is effectively unlimited. Any scheme conforming to [12] may be used in a bundle protocol endpoint ID.

1.3.2.3 Formats of Bundle Blocks

The format of the two basic BP blocks has been shown in Figure 1.2. The bundle processing control ("Proc.") flags field in the Primary Bundle Block is an SDNV with variable length. The block length field of the Primary Bundle Block is an SDNV with variable length. More details can be found in Bundle Protocol Specification.

Primary Bundle Block

Version	Proc. Flags
Block length	
Destination scheme offset	Destination SSP offset
Source scheme offset	Source SSP offset
Report-to scheme offset	Report-to SSP offset
Custodian scheme offset	Custodian SSP offset
Creation Timestamp time	
Creation Timestamp sequence number	
Lifetime	
Dictionary length	
Dictionary byte array (variable)	
[Fragment offset]	
[Total application data unit length]	

Bundle Payload Block

Block Type	Proc. Flags	Block length
Bundle Payload (variable)		

Figure 1.2: Bundle block formats.

1.3.3   Bundle Processing

Bundle processing is composed of many sub-processes. In this section, an Interplanetary (IPN) Internet will be used as an example to show how the bundle protocol is used to compose actions [13].

The IPN Special Interest Group's InterPlaNetary Internet by the Internet Society, described at http://www.ipnsig.org, is a DTN that is shown in Figure 1.3. A message transmission from Earth to Mars in the IPN has been explained. The example uses three regions connected by two gateways, with a Domain Name System (DNS) for each region.

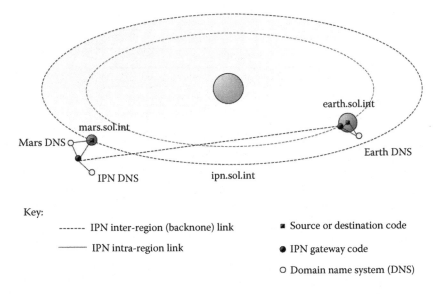

Figure 1.3: IPN Internet

Table 1.1 below shows the names of the nodes accessed in the example. For simplicity, all bundle-layer applications in the Earth and Mars regions use the TCP transport protocol and reside at TCP port 6769.

Table 1.1: Nodes in IPN

Node	IPN Regions	Node Names
Source	earth.sol.int	{earth.sol.int, src.jpl.nasa.gov:6769 }
Earth Gateway	earth.sol.int ipn.sol.int	{eart.sol.int, ipngw1.jpl.nasa.gov:6769}
		{ipn.sol.int, ipngw1.jpl.nasa.gov}
Mars Gateway	ipn.sol.int mars.sol.int	{ipn.sol.int, ipngw2.nasa.mars.org}
		{mars.sol.int, ipngw2.nasa.mars.org:6769}
Destination	mars.sol.int	{mars.sol.int, dst.jpl.nasa.gov:6769}

Before the transmission, the bundle layers of all network nodes synchronize time among themselves. This is needed for the consistent calculation of contact schedules and bundle time-to-live throughout the DTN.

1.3.3.1 Bundle Creation at Source

A source application invokes its bundle layer requesting the transfer of a bundle with a header as shown in Table 1.2. The user data includes instructions to the destination application for processing, storage, disposal, and error-handling of the data. This user data is not visible to the bundle layers handling the transfer (see Figure 1.4).

The source bundle layer verifies the source's signature, creates a bundle, appends its own signature after the bundle header, and stores the result in a persistent storage. The storage is required even if an immediate forwarding opportunity exists because the bundle layer has accepted a custody transfer and must therefore be prepared to retransmit the bundle if it does not receive acknowledgment within the time-to-acknowledge of the bundle that the subsequent custodian has received and accepted the bundle.

Table 1.2: Bundle Header

Item	Value
Source	{earth.sol.int, src.jpl.nasa.gov:6769}
Destination	{mars.sol.int, dst.jpl.nasa.gov:6769}
Class of Service	Custody transfer Normal priority Time-to-live = 36 hours
Signature	<bundle-specific encrypted signature using source's private key>
User Data	Application-specific data, including instructions to the destination application for processing, storage, disposal, and error-handling. (User data is not visible to bundle layers.)

1.3.3.2 Transmission by Source

The source bundle layer consults its routing table and finds that the Earth gateway {earth.sol.int, ipngw1.jpl.nasa.gov:6769} is the next hop capable of accepting custody transfers on a path toward the destination and that TCP is the proper transport protocol. The source bundle layer also determines that it has a continuous connection to the Earth gateway (see Figure 1.5).

The bundle layer transmits a copy of the bundle to the Earth gateway via TCP, starts a time-to-acknowledge retransmission timer, and waits for a custody-transfer acknowledgment from the gateway.

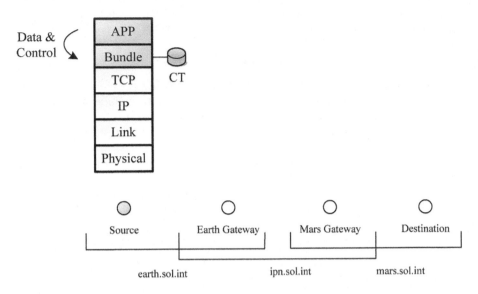

Figure 1.4: Bundle creation at source.

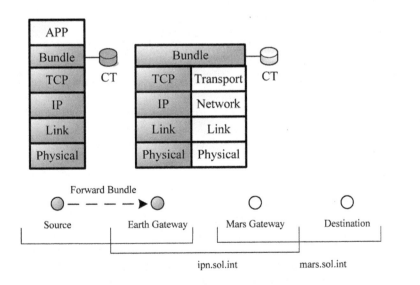

Figure 1.5: Transmission by source.

1.3.3.3   First-Hop Processing and Forwarding

When the Earth-gateway bundle layer receives the bundle via TCP, it terminates the TCP session. Since this is a security boundary for the Interplanetary Internet, the Earth-gateway bundle layer needs to verify the source application's signature and class-of-service (CoS) rights using its stored copies of adjacent-user certificates and certificate-authority (CA) public keys or obtaining such certificates and keys as needed. It compares the signature to its access-control list. After confirming the appropriateness of the transfer, the Earth-gateway bundle layer replaces the signature made by the source bundle layer with its own leaving the source-application's signature intact. Then it stores the received bundle in the persistent storage (see Figure 1.6).

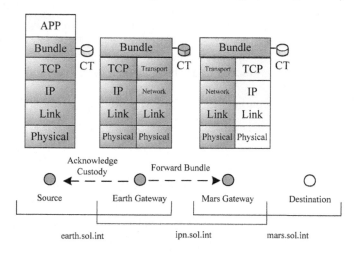

Figure 1.6: First-hop processing and forwarding.

The Earth-gateway bundle layer consults its routing table and finds that the Mars gateway {mars.sol.int, ipngw2.jpl.nasa.mars.org:6769} is the next hop capable of accepting custody transfers on the path toward the destination. It determines that the Mars gateway will be accessible after the following 1100 days, confirms that the bundle's time-to-live is suitable for this hop's delay, and adds the bundle to its contact list for forwarding to that hop.

The Earth-gateway bundle layer then accepts custody of the bundle, updates this information in the bundle header, and confirms this by acknowledgment to the source bundle layer, which deletes its custodial copy of the bundle.

At the next-hop contact time, the Earth-gateway bundle layer establishes contact via the appropriate long-haul transport protocol and forwards the bundle.

1.3.3.4 Second-Hop Processing and Forwarding

When the Mars-gateway bundle layer receives the bundle, it terminates the long-haul transport session, and checks the signature of the Earth-gateway bundle layer using its stored copies of adjacent-router certificates and certificate-authority (CA) public keys. It determines that the bundle has been forwarded by a legitimate source and replaces the signature of the Earth-gateway bundle layer with its own, leaving the source-application's signature intact. Then, it stores the received bundle in its persistent storage (see Figure 1.7).

The Mars-gateway bundle layer consults its routing table and finds that the destination itself is the next hop. It determines that the destination is accessible immediately and the proper transport protocol is TCP. It confirms that the bundle's time-to-live is suitable for this hop's delay.

The Mars-gateway bundle layer then accepts custody of the bundle, updates this information in the bundle header, and confirms this by acknowledgment to the Earth-gateway bundle layer, which deletes its custodial copy of the bundle.

The Mars-gateway bundle layer then establishes contact with the destination bundle layer via TCP and forwards the bundle.

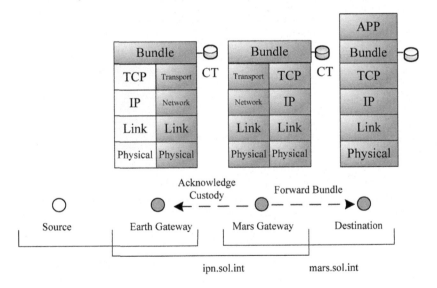

Figure 1.7: Second-hop processing and forwarding.

1.3.3.5 Bundle Reception by Destination

When the destination bundle layer receives the bundle via TCP, it terminates the TCP session and checks the signature of the Mars-gateway bundle layer using its stored copies of adjacent-router certificates and certificate-authority (CA) public keys. It determines that the bundle has been forwarded by a legitimate source. Then it stores the received bundle in its persistent storage, accepts custody of the bundle, and confirms it by acknowledgment to the Mars-gateway bundle layer, which deletes its custodial copy of the bundle (see Figure 1.8).

The destination bundle layer awakens the destination application identified by the entity ID. Depending on the control part of the user data sent by the source, the destination application may generate an application-layer acknowledgment in a new bundle and send it to the source.

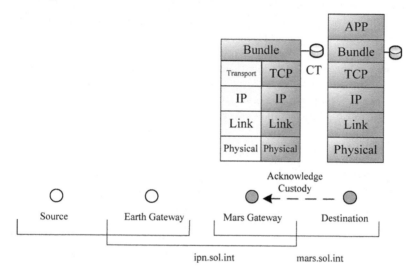

Figure 1.8: Bundle reception by destination.

1.4   Routing Schemes in DTNs

Routing in delay-tolerant networking addresses the ability to transport or route bundles from a source to a destination, which is a fundamental ability all communication networks must have. However, DTNs are characterized by their lack of connectivity resulting in a lack of instantaneous end-to-end paths. In these challenging environments, popular ad hoc routing protocols such as AODV [14] and DSR [15] fail to establish routes. This is due to the fact that these protocols try to establish a complete route first and then, after the route has been established, forward the user data. However, when instantaneous end-to-end paths are difficult or impossible to establish, routing protocols must take a store and forward approach, by which data is incrementally moved and stored throughout the network. A common technique used to maximize the probability of a message to be successfully transferred is to replicate many copies of the message in hope that one will succeed in reaching its destination [16].

1.4.1 Routing Considerations

For all the DTN protocols, there are always many characteristics that must be taken into consideration. The first consideration is whether the information about future contacts is readily available. For example, in the interplanetary communications, the movement of a planet or the moon is the cause of contact disruption, and the long distance between the earth and other planets is the cause of communication delay. However, due to the laws of physics, it is possible to predict the future in terms of the opportunities, when contacts will be available and the duration they will last. The types of contacts are known as scheduled or predictable contacts [4]. On the contrary, in a disaster recovery network, the future location of communicating entities such as emergency responders may not be known. These types of contacts are known as intermittent or opportunistic contacts.

The second consideration is whether mobility can be exploited, and, if so, which nodes are mobile. There are three major cases classifying the level of the mobility in the network. First, it is possible that there are no mobile entities. In this case, contacts appear and disappear based solely on the quality of the communication channel between them. Second, it is possible that some, but not all, nodes in the network are mobile. These nodes, sometimes referred to as data mules [17], are exploited for their mobility. Since they are the primary source of transitive communication between two non-neighboring nodes in the network, an important routing question is how to properly distribute data among these nodes. Third, it is possible that the vast majority, if not all, nodes in the network are mobile. In this case, a routing protocol will most likely have more options available during contact opportunities, and may not have to utilize each one. An example of this type of network is a disaster recovery network where all nodes, generally people and vehicles, are mobile. A second example is a vehicular network where mobile cars, trucks, and buses act as communicating entities.

The third consideration is the availability of network resources. Many nodes, such as mobile phones, are limited in terms of their storage space, transmission rate, and battery life. Others, such as buses on the road, may not be limited. Routing protocols can fully utilize the information to determine how messages should be transmitted and stored to not over-burden limited resources. The efficient resource management is an active research area.

1.4.2 Classification of Routing Schemes

While there are many characteristics of routing protocols, one of the most immediate ways to create a taxonomy is based on whether or not a protocol creates replicas of messages. The routing protocols that never replicate a message are considered as forwarding-based routing whereas the protocols that do replicate messages are considered as replication-based routing [18].

There are both advantages and disadvantages of each type of the routing protocols. Forwarding-based schemes cost generally much less network resources as only a single copy of a message exists in the storage in the network at any given time [19]. Furthermore, when the destination has received the message, no other node can have a copy of it. It can eliminate the need for the destination to provide feedback to the network for indicating the outstanding copies can be deleted. Unfortunately, forwarding-based approaches cannot ensure sufficient message delivery rates in many DTNs [20]. Replication-based schemes, on the other hand, are able to have greater message delivery rates [21], because multiple copies exist in the network while only one copy will reach the destination. However, the tradeoff is that these protocols will consume large valuable network resources. Furthermore, many flooding-based protocols are inherently not scalable. Some protocols, such as Spray and Wait [20], attempt to compromise by limiting the number of possible replicas of a transmitted message. It is important to note that the vast majority of DTN routing protocols are heuristic based without optimal solutions. This is due to optimality being, in the general DTN case, NP-hard [18].

1.4.3   Replication-Based Routing

Replication-based protocols have recently obtained much attention in the scientific community, as they can sustain substantially better message delivery ratios than that by forwarding-based protocols. This type of routing protocol allows for a message to be replicated. Possible research issues on the replication-based routing schemes include:
Network congestion in clustered areas
Consumption of network resources including bandwidth, storage, and energy
Network scalability.

Since network resources may quickly become constrained, deciding which messages to transmit first and which messages to drop first play critical roles in many routing protocols.

1.4.3.1   Epidemic Routing

Epidemic routing[16] is flooding based in nature, as nodes continuously replicate and transmit messages to newly discovered contacts that do not already possess a copy of the message. In the simplest case, epidemic routing is flooding. However, more sophisticated techniques can be used to limit the number of message transfers. Epidemic routing has its roots in ensuring distributed databases remain synchronized. And other techniques such as rumor mongering can be directly applied to routing.

1.4.3.2   PRoPHET Routing Protocol

Epidemic routing is particularly resource hungry because it deliberately makes no attempt to eliminate replications that would be unlikely to improve the delivery probability of messages. This strategy is effective if the opportunistic encounters between nodes are purely random. But in realistic situations, encounters are rarely totally random. Data mules move in a society and accordingly tend to have greater probabilities of meeting certain mules than others. The Probabilistic Routing Protocol using History of Encounters and Transitivity (PRoPHET) protocol uses an algorithm that attempts to exploit the non-randomness of real-world encounters by maintaining a set of probabilities for successful delivery to known destinations in the DTN (*delivery predictabilities*) and replicating messages during opportunistic encounters only if the mule that does not have the message appears to have a better chance of delivering it. This strategy has been first documented in [22].

An adaptive algorithm is used to determine the delivery predictabilities at each mule. The mule M stores the delivery predictabilities $P(M, D)$ for each known destination D. If the mule has not stored a predictability value for a destination, $P(M, D)$ is assumed to be zero. The delivery predictabilities used by each mule are recalculated at each opportunistic encounter according to three rules.

1.4.3.3   MaxProp Routing Protocol

MaxProp [21] has been developed at the University of Massachusetts. MaxProp is flooding based in nature. If a contact is discovered, all the messages that the contact has not held will be attempted to be replicated and transferred. The intelligence of MaxProp comes in determining which messages should be transmitted first and which messages should be dropped first. In essence, MaxProp maintains an ordered-queue based on the destination of each message, ordered by the estimated likelihood of a future transitive path to that destination.

To obtain these estimated path likelihoods, each node maintains a vector with size $n - 1$ (where n is the number of nodes in the network) consisting of the likelihoods that the node has encountered each of the other nodes in the network. Each of the $n - 1$ elements in the vector is initially set to $1/(|n \quad | - 1)$, meaning the node is equally likely to meet any other node next. When the node meets one node j, the jth element of its vector is incremented by 1. And then the entire vector is normalized such that the sum of all entries is 1. When two nodes meet, they first exchange their estimated node-meeting likelihood vectors. Ideally, every node will have an up-to-date vector from each other node. With these n vectors at hand, the node can then compute a shortest path via a depth-first search where path weights indicate the probability that the link does not occur. These path weights are summed to determine the total path cost and computed over all possible paths to the destinations desired (destinations for all messages currently being held). The path with the least total weight is chosen as the cost for that particular destination. The messages are then ordered by destination costs, and transmitted or dropped in that order.

In conjunction with the core routing described above, MaxProp allows for many complementary mechanisms to enhance the message delivery ratio in general. First, acknowledgment messages can be injected into the network by the nodes that have successfully received a message. These acknowledgements are 128-bit hashes of the message that are flooded into the network. The nodes can be instructed to delete the extra copies of the message from their buffers. It will help to free space so outstanding messages will not be dropped too often. Second, messages with lower hop-counts will be given higher priority. It helps promote initial rapid message replication to give new messages a head start. Without the head start, newer messages can be quickly starved by older messages since there are generally fewer copies of new messages in the network. Third, each message maintains a hop list indicating the nodes that have been previously visited, to prevent the visit of a node a second time.

1.4.3.4 RAPID Routing Protocol

RAPID [18], which is an acronym for Resource Allocation Protocol for Intentional DTN routing, has been developed at the University of Massachusetts, Amherst. The goal of RAPID is to intentionally affect a signal routing metric. RAPID has been instrumented to intentionally minimize one of three metrics: average delay, missed deadlines, and maximum delay.

The core of the RAPID protocol is based on the concept of a utility function. A utility function assigns a utility value, U_i, to every packet i, which is the metric to be optimized. U_i is defined as the expected contribution of packet i to this metric. RAPID replicates packets, which locally results in the higher increase of utility. For example, assume that the metric to optimize is the average delay. The utility function defined for average delay is $U_i = -D(i)$, which is the negative value of the average delay. Hence, the protocol replicates the packet that results in the greatest decrease in delay. RAPID, like MaxProp, is flooding based, and will therefore attempt to replicate all packets if network resources are available.

1.4.3.5 Spray and Wait Routing Protocol

Spray and Wait [25] is a routing protocol that attempts to gain the delivery ratio benefits of replication-based routing as well as the low resource utilization benefits of forwarding-based routing. It achieves resource efficiency by setting a strict upper bound on the number of copies per message allowed in the network.

The operation of the Spray and Wait protocol is composed of two phases, the spray phase and the wait phase. When a new message is created in the system, a number L is attached to that message indicating the maximum allowable copies of the message in the network. During the spray phase, the source of the message is responsible for spraying or delivering one copy to L distinct relays. When a relay receives the copy, it enters the wait phase, where the relay simply holds that particular message until the destination is encountered directly.

There are two main versions of Spray and Wait, vanilla and binary. The two versions are identical except for the ways for L copies to reach L distinct nodes during the spray phase. The simplest way to achieve this, known as the vanilla version, is for the source to transmit a single copy of the message to the first L distinct nodes it has encountered after the message has been created. The second version is referred to as Binary Spray and Wait, by which the source starts, as before, with L copies. It then transfers the floor of half of the L copies to the first node it has encountered. The first node transmits half of the total number of copies it has. Each of the nodes, which have accepted the copies, then transfers half of the total number of copies it has to the future nodes it meets, which have no copies of the message. When a node eventually gives away all of its copies, except for one, it switches into the wait phase where it waits for a direct transmission opportunity to the destination. The benefit of Binary Spray and Wait is that messages are disseminated faster than in the vanilla version.

1.5 Open Issues in Delay-Tolerant Networking

1.5.1 Routing

The DTN architecture claims its applicability to a wide range of operating environments, and is therefore intended to support pluralism among the naming formats, routing algorithms, and network technology. The routing problem can be coarsely divided into whether the routing graph is assumed to be connected or not. In addition, schemes for routing bundles may involve creation and deletion of single or multiple copies of a bundle, various degrees of knowledge about the topology and traffic pattern including past, current, and future contact, traffic load, and buffer occupancy, fragmentation, various levels of granularity in the decision making, resource reservations, different routing for different classes of service or custody bundles, and different options for the loci of the routing computation.

DTN concepts are beginning to find their way into the MANET literature, which has focused largely on routing in relatively dense mobile ad-hoc networks where end-to-end connectivity between any pair of nodes is possible. Combining DTN and MANET concepts together suggests that nodes may operate using some combination of simple forwarding and more delay-tolerant store-carry-forward operation. This coupling supports the ability to beneficially place nodes to act as routers or data ferries to enable communication even in otherwise sparse and partitioned networks.

Future DTN nodes will likely have to support a number of different routing strategies and protocols in order to operate efficiently in the vast diversity of the environments in which the node may find itself. For example, a node may be well connected locally and can use the standard Internet routing scheme globally, and may then be subject to significant disruption, and so may have to switch over to a more complex routing scheme.

DTN routing may eventually involve not only path or next hop selection toward a destination using a single metric of goodness, but also possibly multiple routing solutions depending on the types of bundles being moved. For example, a long path including a reliable custodian may be preferable to a shorter path lacking such a custodian. In addition, DTN routing selects not only next hops at each forwarding node but also next protocol. Thus, a routing solution may involve not only a set of paths, but also a set of appropriate encapsulating protocols used to facilitate delivery using a heterogeneous set of transports.

1.5.2 Custody and Congestion

DTN custody transfer is a service which is optionally provided to a bundle as it is delivered through a DTN. When it is used, custody transfer keeps the track of a current responsible entity or custodian for each bundle. The custodian is required to keep the bundle safe in the persistent memory until another custodian has received it successfully. Bundles may be moved from one custodian to another toward the bundle's destination with an acknowledged transfer accomplished for each. There are circumstances where this acknowledgment procedure can fail when the connection breaks during a transfer operation, or the network does not support bi-directional data transfer. The custody transfer model and use of the persistent storage at intermediate nodes provides the ability to delegate the responsibility for reliable data transfers to portions of the network other than the original sender without violating the guiding end-to-end principal in IP networks. It is possible and necessary in the DTN context because it is assumed that the original source of data may become unreachable or inoperable before the transmitted data reaches its ultimate destinations.

Not every node in a DTN needs to offer custody transfer. A node may refuse to accept custody for messages for implementation or policy reasons because not enough free storage space is currently available or some other reasons. If every node and every bundle operates using custody transfer or some equivalent capability, it is adequate for a stable network with sufficient storage resources. But it is not suitable when the source rate exceeds the network delivery rate beyond the network's buffering capability resulting in the main problem of DTN congestion.

Congestion control has been explored much less extensively in DTNs [23]. The DTN architecture specification indicates congestion is still an open topic.

DTN congestion occurs when storage resources become scarce due to the presence of too much bundle data or too many bundle fragments. A node has several options to mitigate the situation facing these situations in the following order of preference: drop expired bundles, move bundles somewhere else, cease accepting bundles with custody transfer, cease accepting regular bundles, drop unexpired bundles, and drop unexpired bundles for which the node has custody. Given that expired bundles are subject to being discarded prior to the onset of congestion, there may be no such bundles to discard. Moving bundles somewhere else may involve interaction with routing computations, which is a reasonable approach if storage exists near the congestion point [23]. It is also straightforward to cease accepting bundles with custody. It amounts to a form of flow control operating at the DTN hop-by-hop and can result in backlogs of custody transfers as they accumulate upstream of congested nodes. To cease accepting regular bundles can make the node essentially disconnect from its neighbors for some period of time. DTN can tolerate such disconnections but results in, once again, upstream congestion. Some protocols could be developed to propagate the policy-based early expiration times implemented by certain nodes, which has received no attention to date. Discarding bundles of which a node has taken custody defeats much of the delay tolerant aspects of DTN. The value of custody transfer and behavior of DTN congestion is still not fully investigated. It is likely that they will remain poorly solved until the DTN architecture is more widely deployed and carries significant traffic loads.

1.5.3   Security

The DTN bundle security protocol specification [24] defines basic data integrity and confidentiality mechanisms for bundles. Two different data integrity blocks are defined in [24]. One is for end-to-end integrity. And another one is for hop-by-hop integrity between adjacent DTN nodes. The rationale for the separation is to provide for different types of canonicalization and key management that are likely to be used for hop-by-hop vs. end-to-end cryptographic services.

Some DTNs such as wireless sensor networks may have nodes such that they cannot themselves encrypt, decrypt, and sign or verify bundles due to the lack of resources at each node. In addition, there may be some DTNs in which portions of the physical network topology are contained in physically secured facilities. Cryptographic protection at the bundle layer may not be necessary in the network segments. To face these challenges, DTN security allows for intermediate DTN nodes to apply or check the validity of the cryptographic credentials. The relevant nodes in these cases are referred to as the security source and security destination, respectively, which can be different from the bundle source and destination. It still remains to be proved whether these features are useful or not in the future DTN. However, they do represent subtle differences from how cryptographic services are used in most networks today.

There are a number of open issues in DTN security, some of which may be more tractable than others. First of all, the interaction of fragmentations and the application of cryptographic mechanisms can be challenging. The case that supports cryptographic services is optional, where it is possible for a set of fragments to be reassembled among which only one of the fragments contains cipher text. Clearly for such combinations, an additional deployment will be required. Secondly, while the bundle security protocol defines cryptographic services, it has not provided any way to manage the required keys. So the appropriate solutions for management and distribution of the required keys will have to be explored to operate in the DTN environments, where regular low-latency communication may be infrequent. At last, the models for the authorization of traffic in DTNs to target the issues of authentication, authorization, and accounting (AAA) should be developed.

Bibliography

[1] K. Fall, "A delay-tolerant network architecture for challenged internets," SIGCOMM 2003.

[2] M. Demmer and K. Fall, "DTLSR: Delay Tolerant Routing for Developing Regions," SIGCOMM Workshop on Networked Systems for Developing Regions (NSDR), August 2007.

[3] V. Cerf et al., "Delay-Tolerant Network Architecture," IETF RFC 4838, informational, April 2007.

[4] S. Jain, K. Fall, and R. Patra, "Routing in a Delay Tolerant Network," SIGCOMM, 2004.

[5] S. Symington, S. Farrell, and H. Weiss, "Bundle Security Protocol Specification," Work in Progress, October 2006.

[6] S. Farrell, S. Symington, and H. Weiss, "Delay-Tolerant Networking Security Overview," Work in Progress, October 2006.

[7] J. Alonso and K. Fall, "A Linear Programming Formulation of Flows over Time with Piecewise Constant Capacity and Transit Times," Intel Research Technical Report IRB-TR-03-007, June 2003.

[8] K. Fall, W. Hong, and S. Madden, "Custody Transfer for Reliable Delivery in Delay Tolerant Networks," Intel Research Technical Report IRB-TR-03-030, July 2003.

[9] K. Scott and S. Burleigh, "Bundle Protocol Specification," IETF RFC 5050, November 2007.

[10] Symington, S., "Bundle Security Protocol Specification," Work Progress, October 2007.

[11] Abstract Syntax Notation One (ASN.1), "ASN.1 Encoding Rules: Specification of Basic Encoding Rules (BER), Canonical Encoding Rules (CER) and Distinguished Encoding Rules (DER)," ITU-T Rec. X.690 (2002) | ISO/IEC 8825- 1:2002," 2003.

[12] T. Hansen, T. Hardie, and L. Masinter, "Guidelines and Registration Procedures for New URI Schemes," RFC 4395, BCP 115, February 2006.

[13] F. Warthman, "Delay-Tolerant Networks (DTNs): A Tutorial v1.1," Wartham Associates, 2003.

[14] C. E. Perkins and E. M. Royer, "Ad-hoc on-demand distance vector routing," In the Second IEEE Workshop on Mobile Computing Systems and Applications, February 1999.

[15] D. B. Johnson and D. A. Maltz, "Mobile Computing, chapter on Dynamic source routing in ad hoc wireless networks," Kluwer Academic Publishers, pp: 153–181, February 1996.

[16] A. Vahdat and D. Becker, "Epidemic routing for partially connected ad hoc networks," Technical Report CS-2000-06, Department of Computer Science, Duke University, April 2000.

[17] R. C. Shah, S. Roy, S. Jain, and W. Brunette, "Data MULEs: Modeling a Three-tier Architecture for Sparse Sensor Networks," In Proc. IEEE SNPA Workshop, May 2003.

[18] A. Balasubramanian, B. N. Levine and A. Venkataramani, "DTN routing as a resource allocation problem," SIGCOMM, August 2007.

[19] D. Henriksson, T. F. Abdelzaher, and R. K. Ganti, "A caching-based approach to routing in delay-tolerant networks," Proceedings of 16th International Conference on Computer Communications and Networks, 2007. ICCCN 2007, 2007.

[20] T. Spyropoulos, K. Psounis, and C. S. Raghavendra, "Spray and wait: An efficient routing scheme for intermittently connected mobile networks," In WDTN '05: Proceeding of the 2005 ACM SIGCOMM workshop on Delay-tolerant networking, 2005.

[21] J. Burgess, B. Gallagher, D. Jensen, and B. N. Levine, "MaxProp: Routing for vehicle-based disruption-tolerant networks," IEEE INFOCOM, April 2006.

[22] A. Lindgren, A. Doria, and O. Scheln, "Probabilistic routing in intermittently connected networks," The Fourth ACM International Symposium on Mobile Ad Hoc Networking and Computing (MobiHoc 2003), 2003.

[23] M. Seligman, K. Fall, and P. Mundur, "Alternative custodians for congestion control in delay tolerant networks," SIGCOMM, 2006.

[24] S.F. Symington, S. Farrell et al., "Bundle Security Protocol Specification," Internet-Draft, March 2009.

[25] T. Spyropoulos, K. Psounis, and C. S. Raghavendra, "Spray and Wait: An Efficient Routing Scheme for Intermittently Connected Mobile Networks," Proceedings of the 2005 ACM SIGCOMM Workshop on Delay Tolerant Networking, New York: ACM, 2005.

Chapter 2

DTN Routing: Taxonomy and Design

Thrasyvoulos V. Spyropoulos, Rao Naveed Bin Rais, Thierry Turletti, Katia Obraczka, and Athanasios Vasilakos

2.1	Introduction	32
2.2	Routing in Intermittently Connected Networks	34
	2.2.1 Routing as Opportunistic Forwarding	35
	2.2.2 Message Replication	36
	2.2.2.1 Greedy Replication	36
	2.2.2.2 Controlled Replication	37
	2.2.2.3 Utility-Based Replication	38
	2.2.3 Message Forwarding	39
	2.2.4 Message Coding	40
	2.2.5 DTN Routing as Resource Allocation	42
2.3	DTN Routing Utility Functions	42
	2.3.1 Destination Dependent (DD) Utility	42
	2.3.2 Destination Independent (DI) Utility	45
	2.3.3 Additional Considerations	47
	2.3.4 Examples of DTN Routing Protocols	47
2.4	A Taxonomy of DTNs	47
	2.4.1 Connectivity	49
	2.4.2 Mobility	52
	2.4.3 Node Resources	54
	2.4.4 Application Requirements	55
2.5	DTN Routing Design Guidelines	56
2.6	Case Studies	57
	2.6.1 Pocket Switched Networks	58

2.6.2   Metropolitan Networks with Heterogeneous Nodes.........59
2.6.3   Applications with Priorities 61

Communication networks (wired or wireless) have traditionally been assumed to be connected at least most of the time. However, emerging applications such as emergency response, special operations, smart environments, VANETs, etc. coupled with node heterogeneity and volatile links (e.g. due to wireless propagation phenomena and node mobility) will likely change the typical conditions under which networks operate. In fact, in such scenarios, networks may be mostly disconnected, i.e., most of the time, end-to-end paths connecting every node pair do not exist.

To cope with frequent, long-lived disconnections, *opportunistic routing* techniques have been proposed in which, at every hop, a node decides whether it should forward or store-and-carry a message. Despite a growing number of such proposals, there still exists little consensus on the most suitable routing algorithm(s) in this context. One of the reasons is the large diversity of emerging wireless applications and networks exhibiting such "episodic" connectivity. These networks often have very different characteristics and requirements, making it very difficult, if not impossible, to design a routing solution that fits all.

In this chapter, we first break up existing routing strategies into a small number of common and tunable *routing modules* (e.g. message replication, coding, etc.), and then show how and when a given *routing module* should be used, depending on the set of *network characteristics* exhibited by the wireless application. We further attempt to create a taxonomy for intermittently connected networks. We try to identify generic *network characteristics* that are relevant to the routing process (e.g. network density, node heterogeneity, mobility patterns) and dissect different "challenged" wireless networks/applications based on these characteristics. Our goal is to identify a set of useful *design guidelines* that will enable one to choose an appropriate routing protocol for the application/network in hand. Finally, to demonstrate the utility of our approach, we take up some case studies of challenged wireless networks, and validate some of our routing design principles using simulations.

2.1   Introduction

Traditionally, communication networks (wired or wireless) have always been assumed to be connected almost all the time, i.e., there exists at least one end-to-end path between any pair of nodes in the network most of the time. When partitions occur, they are considered transitory failures and core network functions such as routing react to these failures by attempting to find alternate paths. Even in wireless multi-hop ad-hoc networks (e.g., MANETs), where links are more volatile due to wireless channel impairments and mobility, partitions are still seen as exceptions and assumed infrequent and short-lived.

However, for some emerging applications like emergency response, special operations, smart environments, habitat monitoring, and VANETs, which are motivated by advances in wireless communications as well as ubiquity of portable computing devices, the assumption of "universal connectivity" among all participating nodes no longer holds. In fact, for some of those scenarios/applications, the network may be disconnected most of the time; in more "extreme" cases, there may never be an end-to-end path available between a source and a destination. Besides the application scenarios themselves, other factors contributing to frequent, arbitrarily long-lived connectivity interruptions include node heterogeneity (e.g., nodes with different radio ranges, resources, battery life, etc.), volatile links (e.g. due to wireless propagation phenomena, node mobility, etc.), and energy efficient node operation (e.g., duty cycling).

Networked environments which operate under such intermittent connectivity are also referred to as episodically connected, delay tolerant, or disruption tolerant networks (or DTNs). Clearly, traditional routing, including MANET routing protocols like OLSR [1], AODV [2], and DSDV [2] cannot deliver adequate performance in DTNs. Consequently, a number of new routing approaches have been proposed to cope with frequent, arbitrarily long-lived connectivity disruptions. They can be classified into three categories: *deterministic* or *scheduled, enforced,* and *opportunistic* routing. Deterministic routing solutions are used when contact information is known a priori. Jain et al. [3] showed how little or full information about contacts, queues, and traffic can be utilized to route messages from a source to a destination in the case of disruptions. They have presented a modified Dijkestra algorithm based upon information on scheduled contacts and compare the proposed approach against an optimal LP formulation. In order to deliver messages to otherwise disconnected parts of network (islands), enforced routing solutions like message ferries [4] and data mules [5] can be employed, where special-purpose mobile devices move over predefined paths in order to provide connectivity. Epidemic dissemination [6] is the basic form of opportunistic routing and works as follows. When node A encounters node B, it passes to B replicas of messages A is carrying which B does not have. In other words, epidemic routing is to episodically connected environments what flooding is to "traditional," well-connected networks. While on one hand epidemic routing offers minimum delivery delay, it may be prohibitively expensive since it consumes considerable network resources due to the excessive amount of message duplicates generated.

Our focus here is on opportunistic approaches to DTN routing, i.e., where no contact information is known a priori and no network infrastructure (e.g., special-purpose nodes with controlled trajectories) exists to provide connectivity. Besides the question of when contact opportunities happen between nodes, a number of other factors also affect data forwarding, including available storage at peering nodes, contact duration, available bandwidth, message priority and/or expiration time, etc.

An ever-growing number of protocols addressing these "opportunistic" DTN scenarios have been proposed. However, it is not at all clear how existing solutions can be applied to a variety of DTN applications given their requirements and underlying network characteristics (e.g., connectivity, node mobility and capability, etc).

In this chapter, we address this question and thus help map the design space of opportunistic DTN routing. We can summarize the contributions of this work as follows:

- First, we dissect opportunistic routing solutions identifying their basic *building blocks* in terms of the forwarding scheme employed, including *message replication, forwarding*, and *(source and network) coding.*

- We also identify a number of features that can be used to classify DTNs. Classifying DTNs according to their connectivity, mobility, and capability (e.g., storage, battery life, processing) of the participating nodes will be key in deciding what routing mechanism(s) to use in order to achieve adequate application-level performance.

- We then proceed to map the opportunistic routing design space by drawing the correspondence between the proposed DTN taxonomy and the basic opportunistic routing building blocks.

- Finally, through simulations, we conduct case studies of a number of challenged wireless network scenarios in order to validate some of our DTN opportunistic routing design principles and recommendations.

The remainder of this chapter is organized as follows. Section 8.4 discusses the routing strategies in intermittently connected network by dissecting the existing solutions into a small number of common and tunable routing modules. Important utility functions for routing decisions are described in Section 2.3. Section 2.4 presents a DTN taxonomy by detailing the network characteristics that are important in designing a routing protocol. DTN routing design guidelines and a discussion are presented in Section 2.5, and in the end, we provide some case studies of challenged wireless networks in Section 2.6.

2.2 Routing in Intermittently Connected Networks

In this chapter, we focus on opportunistic routing approaches, i.e., where no information about connectivity or mobility is assumed to be known a priori and no special-purpose nodes (e.g., data mules or ferries) are used. The basic principle governing opportunistic routing is that when two nodes meet one another, they must decide whether to forward a message, and/or to carry it further. It represents a shift from basic "store-and-forward" to the so-called "store-carry-and forward."

Due to its inherent characteristic of running without a priori knowledge, opportunistic routing is quite general and is also applicable to both scheduled and enforced connectivity scenarios since they may suffer from some non-determinism and uncertainty. For example, a bus that is scheduled to reach a bus stop at a certain instant may get stuck in a traffic jam, causing a deviation in its schedule, which ultimately may affect deterministic routing. Also, there can be other factors affecting scheduled behavior like weather, radio interference, and system failure.

Even though our focus in this chapter is on networks/applications exhibiting frequent and long-lasting disruptions in connectivity, we should point out that node mobility has been shown to increase capacity of *connected* wireless networks [26]. Thus, DTN routing approaches can be employed in *connected* networks to harness node mobility for capacity reasons.

Additionally, it is important to note that we are only targeting applications which disseminate data in a point-to-point manner. Multicast or broadcast applications require different routing strategies; however, we argue that insight from this work is also relevant for multipoint data dissemination services.

2.2.1 Routing as Opportunistic Forwarding

As previously pointed out, traditional routing protocols (including MANET routing) do not work well in environments prone to frequent and long-lived disruptions; these routing protocols assume almost always connected network and require an end-to-end path to exist in order for a source to send data to a destination. Paths are discovered either in a proactive (i.e., table-driven routing) or reactive (on-demand routing) manner. This is not the case in a DTN-like environment, as it is possible that a path may never be available between source-destination pairs. Hence, the store-carry-and-forward routing paradigm is utilized in such scenarios; this means that a set of *independent*, *opportunistic* (i.e., no certainty about whether there will ever be a path to the destination) *forwarding decisions* will attempt to *eventually* deliver messages to destinations.

In the remainder of this section, we define opportunistic routing based on the evolution of the message vectors at nodes as they encounter other nodes. It is important to note that as energy is a precious resource in mobile nodes, any node can turn to *sleep* mode to conserve battery lifetime. Thus, it is possible that two nodes are within communication range of each other but are unable to exchange any information, if any of them is in *sleep* mode. For clarity, we define "encounter of two nodes" for the case when two nodes are within communication range of each other and are in *power on* mode.

Opportunistic Routing: If node A with a set of messages $S_{msg}^{(A)}(t)$ and a set of context information, $S_{ctxt}^{(A)}(t)$ at time t, encounters nodes

$B_1, \ldots, B_n$, each with message vectors $S_{msg}^{(i)}(t), i \in [1, n]$ and context information $S_{ctxt}^{(i)}(t), i \in [1, n]$. Then opportunistic routing does the following:

- $S_{msg}^{(i)}(t + \Delta t) = f(S_{msg}^{(A)}(t), S_{msg}^{(1)}(t), \ldots, S_{msg}^{(n)}(t), S_{ctxt}^{(1)}(t), \ldots, S_{ctxt}^{(n)}(t)),$
  $\forall i \in \{A, 1, \ldots, n\},$

- $S_{ctxt}^{(i)}(t + \Delta t) = f(S_{ctxt}^{(A)}(t), S_{ctxt}^{(1)}(t), \ldots, S_{ctxt}^{(n)}(t)), \forall i \in \{A, 1, \ldots, n\},$

where Δt is a random variable and is the time it takes to forward a message (medium access, transmission and propagation delay, etc.).

We use the same notation to define opportunistic routing's three basic primitives, namely *replication, forwarding,* and *coding*. These primitives are the basic building blocks of mobility-assisted routing, based upon which, every opportunistic routing protocol can be constructed.

Next, we look into these three primitives in more detail, also providing specific examples. Let us assume that a node A which has a set of neighbors B_j encounters node $B_i, j \neq i$. A has then to decide whether to forward message m to B_i.

2.2.2   Message Replication

A relay A carrying a copy of m can decide to spawn a new copy of m and forward it to the newly encountered node, (B). This decision will depend on the message vectors of the two nodes (e.g., if the new neighbor does not have a copy of the message in question) as well as on the "context" of the two nodes (e.g., the new neighbor tends to see the message destination often). In other words, if nodes have infinite buffer space and if $m \notin S_{msg}^{(B)}(t)$, then

$$
\begin{aligned}
S_{msg}^{(B)}(t + \Delta t) &= S_{msg}^{(B)}(t) \cup f_{rep}(S_{ctxt}^{(A)}(t), S_{ctxt}^{(B)}(t)), \\
S_{msg}^{(A)}(t + \Delta t) &= S_{msg}^{(A)}(t),
\end{aligned}
$$

where $f_{rep}(\cdot)$ is either $\{m\}$ or $\{\emptyset\}$ (the empty set). Several studies such as [27, 28, 23] have reported the benefits of replication for DTN routing. Note that in cases where more than two nodes encounter each other at the same time, $f_{rep}(\cdot)$ would contain context information of all the nodes that meet each other at that time.

2.2.2.1   Greedy Replication

The simplest version of copy replication is performed in a "greedy" manner. When node A encounters any node, say B, and B does not have a copy of m, A will spawn and forward a copy of m to B; that is, $f_{rep}(S_{ctxt}^{(A)}(t), S_{ctxt}^{(B)}(t)) = \{m\}$:

If nodes have infinite buffer space and if $m \notin S_{msg}^{(B)}(t)$ then

$$
\begin{aligned}
S_{msg}^{(B)}(t + \Delta t) &= S_{msg}^{(B)}(t) \cup \{m\}, \\
S_{msg}^{(A)}(t + \Delta t) &= S_{msg}^{(A)}(t).
\end{aligned}
$$

This is a fast and robust method to distribute copies, creating a number of "copy custodians" that will look for the destination concurrently. Greedy replication is the basic primitive used by epidemic routing [6]. Epidemic routing has many variants and has been used by researchers as a baseline to evaluate DTN routing protocols, as it offers minimum average message delay at the cost of consuming maximum network resources. Prioritized Epidemic Routing (PREP) [54] is a recent greedy replication-based protocol, where the stored bundles are prioritized based upon their expiry time and distance to destination in order to better utilize resources.

Generating and passing a new copy to *every* node encountered may produce considerably high overhead in terms of buffer space for storage and energy spent on transmission/reception. Variants of replication that control the number of copies/custodians of a message circulating in the network at any given point are quite effective in reducing overhead and still achieving adequate performance. They are described below.

2.2.2.2 Controlled Replication

Here, there is some "context" associated with each given message m. This context keeps track of the number of copies that have been created for m. If the perceived number of generated copies is smaller than some desired value L, then $f_{rep}(m, S_{ctxt}^{(A)}(t)) = \{m\}$. Otherwise, $f_{rep}(m, S_{ctxt}^{(A)}(t)) = \{\emptyset\}$. Below are some examples of controlled replication strategies:

- In *copy-limited replication*, each message copy generated is accompanied by a number of forwarding tokens $(fwd(m) \geq 1)$. This number indicates how many extra copies of the message the new node can further create itself and replicate.

$$
\begin{aligned}
fwd(m) > 1 &\Rightarrow S_{msg}^{(B)}(t + \Delta t) = S_{msg}^{(B)}(t) \cup \{m\}, \\
fwd(m) = 1 &\Rightarrow S_{msg}^{(B)}(t + \Delta t) = S_{msg}^{(B)}(t).
\end{aligned}
$$

- In *time-limited replication*, each new message generated (say at time T_s) may be further replicated to nodes other than the destination, only for an amount of time T_{rep}. If t is the time a node B is encountered and B is not the message destination, then

$$
\begin{aligned}
t \leq T_s + T_{rep} &\Rightarrow S_{msg}^{(B)}(t + \Delta t) = S_{msg}^{(B)}(t) \cup \{m\}, \\
t > T_s + T_{rep} &\Rightarrow S_{msg}^{(B)}(t + \Delta t) = S_{msg}^{(B)}(t).
\end{aligned}
$$

- In *probability-limited replication* [24], a node decides to forward a copy of a message to any node it encounters with a specific probability p_i, where i indicates the service class to which the message belongs.

Controlled replication has been shown to achieve competitive delays with only a small fraction of the copies used by uncontrolled replication policies such as epidemic routing [6]. It is the strategy used in protocols like Spray and Wait [27, 23], more specifically the copy-limited version.

Controlled replication performs especially well when nodes are homogeneous and move frequently around the network. However, if candidate relays have very different capabilities, greedy and even controlled replication may waste valuable message copies by forwarding them to nodes that are of little use in the delivery process. In heterogeneous scenarios, one may want to consider the capabilities, characteristics, or context of candidate relays and hand over a copy of a message only if the perceived "utility" of a node as a copy custodian is high enough.

2.2.2.3 Utility-Based Replication

Here, the forwarding decision depends on the context of the current custodian and that of the candidate relay. Specifically, we assume that a set of parameters related to the nodes in question are evaluated to estimate the nodes' "utility" or "fitness" as a relay for a given message bound to a certain destination. This utility may correspond, for example, to the probability of the new node encountering the destination in the future. This and other utility functions will be discussed in detail in Section 2.3).

There are basically two variants of utility-based replication, namely *uncontrolled* and *controlled* utility-based replication, both of which are described below using our message vector notation:

- *Uncontrolled utility-based replication:* If $m \notin S_{msg}^{(B)}(t)$ AND $f_{rep}(S_{ctxt}^{(A)}(t), S_{ctxt}^{(B)}(t)) = \{m\} \Rightarrow S_{msg}^{(B)}(t + \Delta t) = S_{msg}^{(B)}(t) \cup \{m\}$.

- *Controlled utility-based replication:* If $m \notin S_{msg}^{(B)}(t)$ AND $f_{rep}(S_{ctxt}^{(A)}(t), S_{ctxt}^{(B)}(t)) = \{m\}$ AND $fwd(m) > 1 \Rightarrow S_{msg}^{(B)}(t + \Delta t) = S_{msg}^{(B)}(t) \cup \{m\}$.

Uncontrolled utility-based replication has been used to reduce the overhead of epidemic routing [20, 14]. As an example, rather than handing over a copy to every new node encountered, each node maintains a probability measure of future encounters using the history of past encounters; based on this probability, a node forwards a new copy to a new neighbor only if the neighbor has a high enough (or higher than the current relay's) probability of a future encounter with the destination.

Controlled, utility-based replication, on the other hand, has been proposed in [10] to improve the quality of forwarding decisions made by Spray and

Wait in heterogeneous environments. Encounter-Based Routing (EBR) [53] is another example of controlled, utility-based replication, in which future rate of node encounters is predicted using number of past encounters with nodes, and encounter metric is computed locally at each node. Number of replicas of a message delivered to a relay node depends upon the ratio of encounter value that the relay advertises.

2.2.3 Message Forwarding

Unlike replication, under a copy forwarding, a relay A carrying a message m may decide to hand that message over to a node B it encounters; by doing so, A relinquishes its copy of m and ceases to be one of its custodians. Clearly, forwarding incurs minimal message duplication overhead. It is beneficial when the initial relay(s) chosen is(are) not the best one(s). Using our message vector evolution notation, we can define forwarding as follows. If $m \notin S_{msg}^{(B)}(t)$ then

$$
\begin{aligned}
S_{msg}^{(B)}(t + \Delta t) &= S_{msg}^{(B)}(t) \cup f_{rep}(S_{ctxt}^{(A)}(t), S_{ctxt}^{(B)}(t)), \\
S_{msg}^{(A)}(t + \Delta t) &= S_{msg}^{(A)}(t) - f_{fwd}(S_{ctxt}^{(A)}(t), S_{ctxt}^{(B)}(t)),
\end{aligned}
$$

where $f_{rep}(\cdot)$ and $f_{fwd}(\cdot)$ take values either $\{m\}$ or $\{\emptyset\}$ (the empty set).

Forwarding a message can be performed either using a utility function or in a probabilistic manner (e.g., tossing a coin to decide, at each contact, if a message should be forwarded or not). If a utility function approach is used, each node i maintains a value for the utility function $U_i(j)$ for every other node j in the network. $U_i(j)$ which can be interpreted as the probability that node i will forward a message to node j, may be based on a number of different parameters (e.g., encounter history, mobility, friendship index with j, etc.). In general, $U_i(d)$ is a function of the context $S_{ctxt}^{(i)}(t)$ of node i, and possibly of that of node d, the destination, $S_{ctxt}^{(d)}(t)$. That is, $U_i(d) = g(S_{ctxt}^{(i)}(t), S_{ctxt}^{(d)}(t))$. If a node i carrying a message copy for a destination d encounters a node j with no copy of the message, then

- **Rule 1: Absolute utility criterion** If $U_j(d) > U_{th}$ for some U_{th} threshold value OR

- **Rule 2: Relative utility criterion** If $U_j(d) > U_i(d)$ (*relative utility criterion*), then

$$
\begin{aligned}
S_{msg}^{(B)}(t + \Delta t) &= S_{msg}^{(B)}(t) \cup \{m\} \\
S_{msg}^{(A)}(t + \Delta t) &= S_{msg}^{(A)}(t) - \{m\}
\end{aligned}
$$

Scale Free Routing (SFR) [52] is an example of routing protocol that is based on message forwarding, where single copy per message is used, and there is no replication. Forwarding is based upon some utility function, but if utility function is lower than a certain threshold, nodes with highest mobility to move farthest in the network are chosen as relays and message is forwarded to these relays, which are called Ballistic Nodes. This protocol is based upon the concept of Levy Walks.

2.2.4 Message Coding

Messages may be coded and processed at the source, i.e., *source coding* or as they traverse the network, i.e., *network coding*. In the following subsections, both of these coding variants are presented.

Source Coding: Source coding aims at increasing delivery reliability and reducing worst-case delay. A notable example is *erasure coding* [29], in which the coding is performed by the source, a coded part of a message is further treated as any other message in the network, and there are no specific implications on routing/forwarding.

A variation of source coding known as *distributed source coding* tries to minimize propagating redundant information in the network, and thus reduce overhead. Sensor networks, which aimed at a variety of monitoring applications (e.g., environmental and habitat monitoring), are typical target scenarios for distributed source coding [30]. The basic idea behind distributed source coding is to take advantage of the data's inherent spatial and temporal locality to suppress propagation of unnecessary information. For example, in a sensor network tasked to measure the temperature field of a given region, nodes that are in close proximity to one another are expected to report similar temperature values. Through DSC strategies, nodes can identify such redundancies and perform *in-network aggregation* to reduce the volume of data transmitted in the network [31]. Another example of DSC is the growth codes [32], which use coding redundancy at neighbors to avoid the impact of loss.

Network Coding: Network coding has been proposed as a way to increase the capacity of wireless network [33, 11]. The main idea behind network coding is to allow mixing of messages at intermediate nodes in the network. In this way, a receiver reconstructs the original message, once it receives enough encoded messages. The network coding is shown to achieve maximum information flow in a network, which is not attainable with traditional routing schemes.

Linear network coding has been shown to achieve the capacity of information networks [34]. This coding scheme permits a node to apply a linear transformation to a vector (a block of messages over a certain base field) before passing it further in the network. It can be used to reduce the time to deliver a given flow, maximize the throughput, reduce the number of transmissions (and thus energy expended), etc.

Random network coding, where coding coefficients are chosen by each node randomly from a large enough field (often Z^8), and in a distributed manner, is an efficient method to implement network coding in practice (coding coefficients are sent as part of the packet, with only a small overhead) [35]. To take advantage of the benefits of network coding in a wireless, often "challenged," environment, the following modifications of greedy replication have been proposed [33]: instead of transmitting single packets, linear combinations of packets are generated and transmitted; assume a node A has a set of linear combinations of N packets $S_{msg}^{(A)} = \{\hat{m}_1, \hat{m}_2, \dots, \hat{m}_m\}$ and encounters another node B. Then, it creates a linear combination of all its messages in the queue

$$\hat{m}_{new} = \sum_{i=1}^{m} c_i \hat{m}_i \ (\text{mod}) \ ,$$

Finally, depending on the context of nodes A and B, $f_{code}(S_{ctxt}^{(A)}(t), S_{ctxt}^{(B)}(t)) = \{\hat{m}_{new}\}$ or $\{\emptyset\}$, and

$$S_{msg}^{(B)}(t + \Delta t) = S_{msg}^{(B)}(t) \cup f_{code}(S_{ctxt}^{(A)}(t), S_{ctxt}^{(B)}(t)).$$

When enough independent combinations ($\geq N$) of the N messages, belonging to a given coding generation, have been received, a node can *decode* them to get the original N messages. Finally, the forwarding function $f_{code}(\cdot)$ might be, for example:

- a random coin toss, i.e. $f_{code}(S_{ctxt}^{(A)}(t), S_{ctxt}^{(B)}(t)) = \{\hat{m}_{new}\}$ with some probability $p \leq 1$ [33].

- based on a utility function as described in Section 2.3.

One key problem with the network coding approach described above is that coding *every* single message together may result in never collecting enough independent combinations of messages to successfully decode. Some control is needed on how many and which messages will be coded together. This is known as generation control. Coding messages from many different sessions and from large time or sequence number windows (large generations) might result in high delivery delays. On the other hand, using small generations limits the amount of gains achievable by network coding. Finally, even controlling the generations in a distributed manner might pose significant challenges.

For these reasons, it has been suggested to implement network coding hop-by-hop, in an *opportunistic* fashion [11]. Assume that a node A with message vector $S_{msg}^{(A)}$ encounters a set of nodes $B_i, \dots, B_n$ with message vectors $S_{msg}^{(B_1)}, \dots, S_{msg}^{(B_n)}$. Let us further define the n sets $S_i^{(A)}$, such that

$$S_i^{(A)} = S_{msg}^{(A)} \cap \overline{S_{msg}^{(B_i)}}, i \in [1, n].$$

In other words, $S_i^{(A)}$ is the subset of A's messages that neighbor B_i does not have. Then, opportunistic network coding looks for a combination of messages in $\cup_i S_i^{(A)}$ that will result in maximizing the number of neighbor nodes, B_1 to B_n, that will be able to decode a new packet. A then *broadcasts* this message combination. Opportunistic network coding simply takes advantage of favorable traffic patterns to locally save some transmissions, without requiring any generation control or imposing additional delays.

2.2.5   DTN Routing as Resource Allocation

In traditional DTN routing (replication or forwarding), routing is mostly performed based upon some utility function(s). The main aim is always to find a path to a destination with the available information. All the routing strategies mentioned in the previous subsections are no exception to this, and thus they have an incidental effect on routing metrics (like average delay, delivery ratio). Another angle to look at DTN routing is to treat it as a resource allocation problem. The purpose is to have an intentional effect on the DTN routing, rather than an incidental one, in order to maximize the performance of specific routing metrics. RAPID [46], [47] is the first protocol in this respect. In RAPID, messages are ordered with respect to their utilities, keeping in view the goal of maximizing specific metrics (e.g. delay). The protocol translates a routing metric to per-packet utilities, and at every transfer opportunity, it is verified if the marginal utility of replication justifies the resources used. In a way, it is a replication-based protocol, but what differentiates it from the traditional replication scheme is resource allocation.

Erramilli et al. [55] have done a study that is based upon prioritizing messages to better manage network resources in a resource-constrained environment, where they use delegation forwarding [58] as their forwarding algorithm. Another protocol that is based upon the resource allocation concept is OR-WAR (Opportunistic Routing with Window-Aware Replication) [56] that uses a message utility-based differentiation mechanism, which allows allocation of more resources for messages with high utilities. Thus, it replicates messages in order of high utilities first, and it removes messages in the reverse order, if needed. Again, this is a replication routing scheme, but the delivery of number of copies depends upon evaluation of the contact window.

2.3   DTN Routing Utility Functions

We now turn our attention to utility functions that can be used in message replication (or forwarding) by the routing primitives previously discussed. Candidate utility functions could be broadly categorized into *destination-dependent* ("DD") and *destination-independent* ("DI") functions. These utility functions are very useful especially when the network as well as the participating nodes are heterogeneous. Many utility functions have been presented in [37], and are applied to heterogeneous environments in [36].

2.3.1   Destination Dependent (DD) Utility

One node may be the best relay for one destination (d_1), and another node may be the best relay for a different destination (d_2). In other words, for DD utility functions, it is possible that the following is true:

$$U_i(d_1) > U_j(d_1) \text{ but } U_i(d_2) < U_j(d_2), d_1 \neq d_2. \tag{2.1}$$

Below we describe a number of parameters that can be used to build destination-dependent utility functions.

- *Age of Last Encounter*: It has been suggested that keeping track of past encounters with a given node can be helpful in successfully predicting future encounters. For example, each node could maintain a timer for every other node in the network that records the time elapsed since the two nodes last "saw" each other [38]. These timers could then act as indirect location information. Additionally, a node can keep a record of its encounters with another node by noting the last encounter time and the node's position at the time of encounter [39]. Although keeping the last encounter time for nodes does not provide any guarantee that a node would meet a destination in the future, it can be useful in predicting the current location of a destination.

   Because, nodes tend to move in a continuous manner (i.e., they don't ordinarily perform jumps in space), often, a smaller timer value implies a smaller distance to the destination, if we assume that the average speed of nodes does not vary too much. In case nodes are heterogeneous in terms of their characteristics and capabilities, some other parameters should be used in combination with age of last encounter in order to choose a "suitable" relay node. Note that the age of last encounter with a destination is related to the *instantaneous* fitness of a node as a candidate relay for that destination. Besides, choosing relays on the basis of age of last encounter may not be useful, if the node movement is not periodic. Obviously, nodes cannot keep an "age-of-last-encounter" entry for a destination forever. Depending upon the dynamics of the network, the mobility pattern of nodes, etc., the time-to-live value of this parameter should be set accordingly.

- *History of Past Encounters*: The age of last encounter is only a single "snapshot" of the history of past encounters and may not necessarily predict future encounters successfully. Instead, a node could maintain a "richer" set of information about past encounters with another node, like *frequency of encounters, average inter-encounter time, higher moments of inter-encounter time, average encounter duration*, etc. Such information could help identify more accurately good candidate next hops; on the other hand, keeping more information about encounters increases the overhead in terms of context data that needs to be stored. Also, depending upon the application requirements, a combination of past encounter parameters can be used to choose the best possible relay for a destination. Another consideration is how long to keep this history about a certain destination at a node as it may not be useful, or even misleading after a certain threshold of time depending upon the dynamics and mobility pattern of participating nodes. An example of this kind of utility function is Encounter-Based Routing (EBR) [53], in which future rate of node encounter is predicted using information about past encounters with a node.

- **Pattern of Locations Visited:** In the real world, mobile users move with certain purposes in mind (e.g., going to work, going to a class, going from work to lunch, etc.). Additionally, they may follow specific paths in between these locations due to geographical constraints. As a result, people tend to follow a *movement pattern* in their daily activities. These patterns are a function of a variety of parameters including professional activity, work and home location, etc. More importantly, most people also tend to spend the majority of their time in a small subset of *preferred* locations, as opposed to indiscriminately roaming everywhere (unless this is part of their job, e.g., taxi driver, salesman, etc.). "Location preference" as well as the periodic nature of human mobility (diurnal and weekly patterns) have been consistently demonstrated in a variety of real mobility traces [13]. Mobility patterns (known a priori or "learned" online by collecting appropriate statistics) could help identify a *profile* for a given node; nodes with a matching or similar mobility profile as the destination could be considered good candidate relays for messages to that destination [19, 42, 15].

- **Social Networks:** Humans are involved in complex social relationships (networks), and people who are socially related to each other (e.g., friends, students in the same class, and colleagues in the same department) are expected to interact more often with each other. These social features can have important implications for networks formed by communication devices operated or carried by humans (e.g., vehicles, PDAs, laptops). Knowledge about existing social links could allow one to choose a "data relay" that has a much better chance of encountering the destination soon. Note that one way to gather information about social networks is by keeping a history of past encounters. However, there is additional data that is relevant in the context of social networks. For example, suppose that it is known a priori that A is a good friend of D, but B hardly knows D; then, even with no past encounter information of D at A or B, A can be considered a *better* relay for D than B. Another way to do this is by labeling the nodes with community names and by making nodes advertise the communities they belong to as they move and meet other nodes. The social network information about nodes can also be gathered by observing and estimating their mobility pattern.

Bubble [57] is one of the recent social-based forwarding protocols, in which forwarding is based upon identifying "hubs" and "centrality points" in the network. Having no information about a destination, a message is forwarded towards a more "popular" area or node, and then the forwarding mechanism tries to find the destination itself, or a node having the same "community" as the destination node. The logic behind finding a popular node first is that in a social network, some nodes tend to see other nodes more often than others.

- ***Traditional Routing Table Entry***: In a network that is often disconnected, it is possible to have network connectivity in parts of the network (connectivity islands). So, in such cases, each node could maintain a limited-range (e.g., n-hop) view of the topology in a proactive manner (link-state, distance vector) to improve performance. In many scenarios, complementing traditional routing mechanisms with "mobility-assisted" primitives to overcome partitions or other route failures may be a more suitable solution than replacing traditional routing altogether.

2.3.2 Destination Independent (DI) Utility

In this case, the "utility" of a given node is independent of any destination; rather, it depends on some characteristic(s) exhibited by a node. This implies that one node may be the best relay for most or all destinations. In other words, for DI functions it holds in general that:

$$U_i(d_1) \geq U_j(d_1) \Rightarrow U_i(d) \geq U_j(d), \text{for most or all } j, d. \qquad (2.2)$$

Examples of nodes which are highly preferable as relays for any destination could be nodes with high and frequent mobility (e.g., vehicles), nodes with many "friends" (e.g., *hubs* in scale-free networks), nodes with more resources (e.g., buses [25]), or nodes with high cooperative behavior (e.g., APs, routers/gateways, ferries). Below, we describe in more detail some destination-independent parameters that should be considered when making forwarding decisions.

- ***Amount of Mobility***: In some wireless network deployments, nodes may vary in different ways, e.g., some might be more mobile than others. In the case of a campus environment, nodes carried by humans may tend to be more static, while nodes attached to campus transportation vehicles (e.g., [25]) move around the campus periodically, some of which are following regular trajectories. These more mobile nodes tend to traverse a wider portion of the network in the same amount of time than the more static nodes, and thus encounter a larger subset of other wireless nodes. As a result, they represent highly desirable relays, if a DTN-like routing strategy is employed. One way to identify such relays could be, for example, to use *labels* that represent the type of mobility exhibited by nodes, e.g., "BUS," "TAXI," "PEDESTRIAN," "BASE STATION," etc. In some scenarios, it would not be too burdensome to manually configure a label (e.g., by setting some software parameter when installing a radio, say, on the top of a bus). Nevertheless, algorithms that estimate the "degree of mobility" *online* could also be deployed in self-organized, more dynamic environments [10].

- **Node Resources**: When forwarding a message to a node, the resources and capabilities of that node should be considered. Even if a certain node has some ties to the destination (e.g., close friendship), giving a message copy to that node might be a waste of resources, if it is almost out of battery. Chances are it will either turn itself off or run out of battery before it gets a chance of delivering the message. Similarly, if a candidate relay has its buffer almost full, it might be more prudent to prefer another node instead. This may not only result in smaller queuing delays, but may also reduce the probability of the message getting dropped later. Consequently, nodes may maintain the current status of their resources, which can be used to identify nodes that are "good" (or "bad") relays independent of the destination.

- **Cooperative Behavior**: Message forwarding is not free and consumes node resources including battery life and buffer space. So, it is possible that some nodes refuse to forward messages on behalf of others because either they have limited resources, or they are pre-configured with specific forwarding policies, or because they have been either compromised or are owned by an attacker. So, forwarding a message to such nodes would be disadvantageous. Consequently, forwarding decisions should also consider how cooperative nodes are in forwarding messages. Approaches to boosting cooperation among nodes include offering incentives to cooperating nodes, and/or penalizing non-cooperative ones. This also has implications in building trust among participating nodes, which is the topic of the DI parameter discussed below.

- **Trustworthiness**: Although a number of research efforts have been devoted to addressing various problems related to data delivery in wireless networks (e.g., media access, routing, and transport protocols), securing wireless communication is among the biggest challenges. This is due to a number of factors, notably the shared, uncoordinated access to the wireless medium, as well as its inherent unreliability and non-determinism. The peer-to-peer, non-hierarchical nature of many emerging wireless applications requires the collaboration among participating nodes so that data delivery can be accomplished. Malicious peers could exploit this to intervene with the network's normal operation or extract sensitive information, such as passwords, credit card numbers, etc., from packet streams. In other cases, malicious users could pretend to carry and forward other nodes' traffic, while in fact, they don't do so, which may create drastic forwarding problems. More importantly, wireless node resources like bandwidth and battery power will be scarce and valuable in the foreseeable future. Thus, non-malicious yet selfish users might be tempted to refuse carrying others' traffic. For these reasons, the utility of a node as a message relay might also be a function of the trust other nodes have in it, a trust which could be based on signed certificates, PGP-like architectures [43], reputation systems [44], etc.

2.3.3 Additional Considerations

It is certainly possible (and probably desirable) to define utility functions that take into account both the general, destination-independent *fitness* of a node as well as destination-specific information. For example, we can combine history of past encounters (DD utility) with nodes' mobility patterns, and/or their resources (DI utility) in order to define a hybrid utility function that is able to deliver messages to destinations more efficiently.

Most utility functions discussed above are based solely on a snapshot of the past (e.g., the last time node X encountered node Y). However, in real life scenarios node interactions may exhibit rich and intricate structure; it would thus be beneficial to explore learning techniques that try to use history over a window of time and/or feedback (e.g., from the destination) to make better routing decisions.

2.3.4 Examples of DTN Routing Protocols

In Section 8.4, we have described three basic primitives based on which DTN routing can be built. We now proceed to identify the use of these primitives in some existing DTN routing protocols. Table 2.1 summarizes this correspondence between DTN primitives, their variants, and existing DTN solutions. The table shows examples of DTN-routing protocols and categorizes them in terms of the three main primitives (i.e., replication, forwarding and coding). The first column represents the properties based on which the routing protocols are built, and the second column shows the routing protocol examples.

Take for example Epidemic Routing [6]: it is a typical case of "uncontrolled," i.e., with no constraints on the number of copies generated, message replication using a greedy approach; on the other hand, Spray and Wait [27] is an example of "controlled" greedy replication as it limits the number of copies for each message. Replication can also be made "smart" by using some utility functions as in [10]. Spray and Focus [45] is an example of a protocol that combines greedy replication with smart forwarding mechanisms. Performance and efficiency can further be improved if smart forwarding is used with smart replication. On the other hand, smart forwarding mechanisms can be used with source coding schemes such as Erasure Coding [29], and replication can be used with coding schemes [11, 32].

2.4 A Taxonomy of DTNs

In this section, we classify DTNs according to a set of characteristics relevant to routing. For example, a well-connected network whose nodes exhibit little or no mobility would imply that traditional MANET routing algorithms (e.g. OLSR [1], AODV [2], etc.) might be appropriate. Similarly, a network where nodes have little or no energy limitations (e.g., vehicles) would likely render

Table 2.1: Routing Primitives and Their Use by Existing DTN Routing Protocols

	Forwarding	Replication	Coding
Greedy		Epidemic [6] PREP [54]	
Controlled		Spray and Wait [27] SWIM [23]	
Utility Based	FRESH [38] Scale-Free [52] Spray and Focus [45]	History-Based Epidemic [14] Probabilistic Flooding (Prophet) [20] Smart Replication [10] MV Routing [51] Encounter Based [53]	
Resource-Allocation		RAPID [46], [47] ORWAR [56]	
Mobility Characteristics	Mobyspace [19], [42] Solar [15] Scale-Free [52]	Maxprop [25]	
Routing Table Entry	Island Hopping [12]		
Network (end-to-end)			LeBoudec [33]
Opportunistic			COPE [11]
Distributed Source Coding			Growth Codes [32]

routing protocols that focus on minimizing energy consumption inadequate. We start by describing the network features used in our DTN taxonomy.

2.4.1 Connectivity

Connectivity is an important characteristic of wireless networks. Two well-known definitions of network connectivity are (i) the probability that a path exists between two randomly chosen nodes [7], or (ii) the percentage of nodes connected to the largest connected component [7]. Although these two definitions are slightly different, they have similar implications from a macroscopic point of view.

Traditional routing techniques assume the "Internet model" where networks are always connected. Partitions are treated as faults and routing attempts to mend them as soon as they are detected. Typically, alternate routes can be found and disconnections, if they happen, are ephemeral events. In multi-hop wireless ad hoc networks, or MANETs, due to node mobility, wireless channel impairments, limited node capabilities, etc, the assumption that the network is always connected no longer holds and routing had to be re-thought. However, partitions are still considered exceptions to normal operation and routing reacts by trying to find alternate paths. In case it fails and disconnections persist, data queued at nodes waiting to be forwarded starts to get dropped as queues fill up. In fact, it is well-known that the so-called reactive (or on-demand) routing protocols such as DSR [2] and AODV [2] perform poorly when disconnections are frequent and persist for arbitrarily long periods of time.

Recently, it has been recognized that in some classes of wireless networks, namely DTNs, connectivity will be consistently below 1 (or 100%). As a result, the whole spectrum of possible connectivity values all the way from 0 (very sparse networks) to 1 (connected networks) need to be considered when designing routing algorithms.

It is well-known from percolation theory that, in networks consisting of randomly placed (or randomly moving) nodes, connectivity exhibits a *phase transition* behavior [8] as depicted in Figure 2.1. Specifically, if connectivity is scaled by changing the nodes' transmission range, then the following can be observed [9]: (i) for (a large number of) low transmission range values, connectivity values are quite low: no large cluster exists, but rather very small clusters (few with 1 node), whose sizes are exponentially distributed, are found; (ii) when transmission range crosses some threshold value, connectivity starts increasing rapidly and quickly enters a region where a giant component is formed containing a large percentage of nodes, while the rest of the nodes form smaller clusters (again of exponentially distributed size).

This phase transition behavior has some important implications: *random networks*, i.e., those formed by randomly placing nodes (e.g., sensors scattered uniformly in the field) or randomly moving nodes (e.g., random direction), will be either *sparse* or *almost connected*, in most cases. But, if transmission

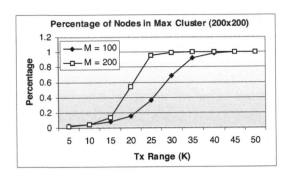

Figure 2.1: Expected percentage of total nodes in largest connected component, as a function of the number of nodes (M) and transmission range (K)
$(200 \times 200$ grid).

range or number of nodes is low, we can have the case where nodes tend
to form clusters (or connectivity islands) due to their mobility patterns. So,
in the following, we focus on three different kinds of networks according to
their connectivity, namely: *almost connected networks*, *sparse networks*, and
connectivity islands.

Almost connected networks: Also known as "flaky nets," these networks more closely resemble the traditional MANET viewpoint of a connected
graph. However, the graph here often exhibits partitions. A good percentage
of end-to-end pairs are connected at any time, even though the paths might
not be long-lasting. Traditional proactive (e.g., link-state) or reactive routing
protocols (e.g., DSR, AODV) could still deliver a part of the traffic successfully (although with a higher overhead for route discovery and maintenance).
Yet, they are unable to deliver any traffic between nodes that lie in different
partitions.

Mobility-assisted routing schemes can be beneficial in bridging disconnected parts of the network and are able to deliver traffic between any two
nodes. Yet, hybrid protocols that can also take advantage of the existence of
large connected clusters are desirable.

Sparse networks: This is a more challenging scenario. In these networks,
transmission range is much lower and no large clusters exist. Most nodes have
only a few neighbors or are isolated most of the time. Every now and then,
two such nodes come into contact, at which time they can exchange data
or other useful information, and soon go back to having no neighbors. It
is evident that traditional or even MANET routing protocols would fail to
satisfy most end-to-end traffic requests, as very few contemporaneous paths
exist. More importantly, the small size or nonexistence of clusters imply that
routing modules that aim at maintaining multi-hop neighborhood information
(2-hop, k-hop, etc.) do not have much value to offer.

Instead, a message has to get routed predominantly by being carried using relays. Occasionally a new candidate relay is encountered and the routing protocol needs to decide whether it should hand over custody, replicate some of its messages, or continue carrying them. Consequently, node mobility is a crucial feature in these sparse networks, both in terms of how mobile nodes are, as well as how structured node mobility is (i.e., whether mobility patterns exist). Similar to network connectivity, mobility is another important feature and will be discussed in detail in Section 2.4.2 below.

It is thus important to *discover nodes that move frequently and quickly around the network* as well as *nodes whose mobility pattern might be correlated with that of the destination.* To do so, nodes may exchange useful information about themselves or other nodes encountered recently. If such information can be collected often enough (before it becomes irrelevant/obsolete), mobility-assisted routing policies can be used to deliver close-to-optimal performance.

Another important implication of sparse networks is that whenever two nodes encounter each other, there is only a small probability that other nodes are also within range. As a result, there is little contention, on average, at the MAC layer for each transmission, and there is also little (in-channel) interference. This suggests that available bandwidth (or buffer space) per contact is the limiting factor as far as performance is concerned. More importantly, it suggests that forwarding or scheduling techniques that aim to choose the right neighbor (e.g., transmit to the "best" neighbor according to some utility function) [10] or combine packets for different neighbors (e.g. opportunistic network coding [11]) offer little gain here.

"Connectivity Islands": It has been observed that in real world deployments, node location does not typically follow a uniform distribution. Similarly, node mobility is usually non-uniform. In fact, it is often the non-uniform mobility process that creates the non-uniform node location distribution. Thus, even though the phase transition phenomenon described earlier might imply that networks are either sparse or almost connected, in the real world different connectivity structures might be observed. For example, in vehicular networks nodes may tend to gather around different concentration points for reasons dependent on the transportation network (e.g., traffic lights, junctions, toll, etc.) or application (e.g., taxi booths at airports, popular locations, etc.) [12]. Other real world examples include *First Mile Solutions* [40] and *VLINK* [41].

This non-uniform placement or mobility of nodes can also be observed in a variety of other scenarios. Consider, for example, a campus with people mostly moving within their own departments [13], or herds of animals mostly moving together in packs [14]. These networks can be seen as a set of separated islands of (full) connectivity, formed around a concentration point, with few or no contemporary paths between concentration points.

Connectivity Islands lie in between *Almost-connected* and *Sparse Networks.* On the one hand, their sizable clusters imply that proactive routing approaches could help collect and maintain useful information about *immediately reachable* nodes. On the other hand, a large number of nodes outside the local cluster are not immediately reachable using traditional techniques. Instead, mobility-assisted routing should be used to move messages between different "islands" where no immediate path is available. Consider, for example, a scenario where some *anchor nodes* are stable over time and can serve as "connectivity points" (e.g., VANET concentration points at traffic lights are expected not to change often, but *attached nodes* change often). In these cases, routing can be done hierarchically where at the macroscopic level, relatively stable paths can be constructed and used to route traffic between "islands" while store-carry-and-forward is used on a microscopic level to forward messages when no routes exist, likely between "islands" [12]. More importantly, if the nodes that are associated with a given concentration point are stable over time (e.g., nodes affiliated with a given department), macroscopic information about the mobility pattern [15] or community structure [16] between nodes could be used to route traffic across disconnected parts.

To summarize, if a routing table entry exists for a given destination on the *microscopic level* (i.e., populated by traditional routing techniques, such as proactive link-state (e.g., OLSR) or on-demand distance-vector (e.g., AODV)), then no special measures are needed. If, however, no paths exist to that node, a routing entry can indicate a possible course of action on the *macroscopic level*, e.g., "send to connectivity island X." This latter action could be performed by, say, finding a node that is affiliated with X [15] or replicated/sprayed to a number of nodes, with the hope that one of them will soon visit X.

2.4.2   Mobility

Node mobility is another important factor to be considered when choosing adequate routing approaches, especially as the network becomes sparser. In particular, we will discuss two aspects related to node mobility, namely its *amount* and *structure*.

Amount of Mobility: The "amount of mobility" of a node can be defined as the percentage of the network traversed or "covered" by the node within a given amount of time. Alternately, it can also be expressed as the number of new nodes (and thus either destinations or candidate relays) a given node encounters within a given time window. The following characteristics are needed to quantify mobility.

- *Node Speed*: Intuitively, the faster a node is moving, the more new area it should cover in a given amount of time, all other parameters unchanged. Additionally, if nodes move fast, they would have more chances to meet more nodes, thus increasing the number of contacts. On the other hand, if node speed is too high, contact duration is reduced, directly affecting routing protocol performance.

- *Pause Time and Frequency*: Depending upon the environment and the application, mobile nodes may tend to stay at a particular position for extended periods of time. We call this duration the pause time. For example, in an exposition hall, nodes may move from one place to another and stay at the other place for some time before moving further. Again depending upon the application, the pause time may be used to deliver messages to destinations as it increase the contact duration when the node is in static position. On the other hand, the nodes that have longer pause times may not be as useful in the delivery process as mobile nodes. In some cases, the nodes that are static are more useful to relay messages because of their placement in the area (e.g., throwboxes [17], bus stops, etc.).

The nodes' periodicity of visiting places or their frequency can also be exploited in the delivery process of messages.

- *Mixing Time*: This is essentially the time it takes a node, starting at a given state of the mobility model, to arrive at its stationary distribution; the higher the mixing time, the more time it takes the average node to reach a randomly chosen destination.

In general, the larger the amount of *average* node mobility, the better the performance of routing protocols that rely on such mobility. Furthermore, in a number of situations it holds that the higher the average node mobility, the less sophisticated the design of a protocol needs to be. This seems to be in contrast with the traditional viewpoint that node mobility has a negative effect on routing protocol performance.

Structure of Mobility Model: The structure of the mobility model is equally important, and becomes significantly more so for sparser and "less mobile" networks. The following information about the structure of a node's mobility pattern is particularly important from a routing protocol's perspective:

- *Homogeneous vs. Heterogeneous Mobility*: Depending on the particular DTN application, participating nodes may all have the same capabilities and behavior. Conversely, in a heterogeneous deployment, nodes may differ from one another. For example, one could reasonably assume that nodes in a sensor network have homogeneous capabilities and behavior (e.g., duty cycle operation). However, people forming a Pocket Switched Network [18] might have largely different mobility patterns from one another.

Node heterogeneity affects protocol design in a number of ways. For example, some nodes will be better relays than others for delivering traffic. Some relays might be preferable for any destination, [1] as in the case of nodes that move fast and frequently around the networks (e.g. vehicles).

---

[1]There are also cases where some nodes are better relays for certain destinations. Destination-dependent and destination-independent choice of relays will be discussed in detail in Section 2.3.

Protocols that are "smart" enough to discover and pick such advantageous relays are expected to perform better the more heterogeneous a network is. Attention is needed though to make sure not to overload a few nodes with relaying responsibilities; this will possibly have detrimental effects due to congestion or battery drainage. Alternatively, if the network is homogeneous, then simple greedy solutions may be adequate to achieve good performance.

- *Spatial or Temporal Correlation*: In addition to differences in the mobility pattern between nodes, individual nodes may exhibit specific mobility patterns which could be leveraged to improve routing performance. For instance, a given node may visit some locations (e.g., a person's home or office) often in a way that exemplifies spatial correlation of movement. Also, a given node may exhibit different mobility behaviors depending on the time of day (temporal correlation). For example, most employees might head to the company's cafeteria between 12 and 1 p.m. Finally, there might also exist correlations between the mobility of different nodes both in space (e.g., nodes that tend to visit the same locations [19]) and time (e.g., nodes that leave their "home" location at around the same times). In such cases, good relays may be *destination-specific*, that is, a given node may be the best relay to deliver a message to destination X but may never do so for another destination Y. In some other cases, good relays may be *time-specific*, which means that a given node can act as the best relay at a specific time for a destination (or during a specific time interval), and another node would serve as relay for another time interval. Protocols that possess the necessary intelligence to distinguish between relays in general, and more specifically, take advantage of mobility patterns they exhibit, are desirable.

- *Other Considerations*: In addition to the previous generic mobility characteristics, a given set of networked nodes may also exhibit mobility attributes that may result in special structures which should be accounted for by routing. This is the case of *disconnected islands* as discussed in Section 2.4.1. In several applications, a set of mobile nodes can create well-connected clusters (e.g., a military platoon, a nomadic community [20], wildlife herd or pack [14]) which may be far enough away from one another that they cannot communicate among themselves. It has been shown that, in these cases, hybrid protocols that take explicit advantage of this structure, using regular routing protocols within a cluster and mobility-assisted techniques to bridge such clusters, can achieve good performance [21, 12].

2.4.3   Node Resources

Although network and node resources are becoming less and less of an issue in wired networks, it is not typically the case for their wireless counterparts.

Depending on the application, node capabilities such as bandwidth, storage, and battery lifetime may vary largely. Resource availability or lack thereof should play an important role in the design and performance of a routing protocol.

- **Bandwidth:** In networks which operate over a common shared wireless medium, the available bandwidth is always a valuable and often scarce resource. If bandwidth is limited, then routing protocols should be efficient, especially in terms of signaling and control information exchange. Furthermore, the more limited the available bandwidth, the more prudent the choice of forwarding opportunities needs to be.

- **Storage:** Sensor networks are the typical case where available memory at nodes might be limited relative to the amount of information that needs to be stored locally. Besides affecting the choice of the routing algorithm to be used, storage limitation also influences relevant routing protocol parameters (e.g., TTL) as well as mechanisms such as buffer replacement policies and garbage collection [22, 23]).

- **Battery Lifetime:** Power awareness is usually an important feature in routing protocols for wireless networks.[2] In the case of DTNs, it becomes even more critical, especially in the case of deployments in remote, hard to access regions where nodes may be left unattended for extended periods of time. There is also a recent work [64] that considers making throwboxes energy efficient in order to increase their lifetime while maintaining high efficiency of the system in terms of delivery ratio and latency. In order to minimize the energy waste in DTN, optimal searching/probing intervals are calculated using statistical information of contact opportunities in [59], [60], and [61] and energy efficient sleep scheduling mechanisms are constructed in [62] and [63].

Heterogeneous Node Capabilities: In addition to different mobility patterns, nodes may also have largely varying capabilities, like battery life, processing power, storage capability, etc. Imagine, for example, a scenario where some of the wireless nodes are vehicles (with little or no energy and storage limitations) while others are small PDAs carried by pedestrians. In such a scenario, it is important for the routing protocol to be able to identify the more capable nodes as they are possibly better candidates for relaying traffic than nodes that have barely enough resources to handle their own traffic.

2.4.4 Application Requirements

The discussion so far focused on network and individual node features and capabilities. In this section, we consider application-specific requirements, which

---

[2]There are of course some notable exceptions, e.g., VANETs.

must be taken into account when choosing and/or designing DTN routing mechanisms.

- *Message Content/Priority*: Despite the inherent delay-tolerance of most DTN driving applications, there can be situations where some messages may be more important than others. For example, in a VANET network it is reasonable to assume that an accident notification message will have higher priority than a chat message, or announcements of nearby shops. In some cases, users might be willing to "pay" more for some of their traffic to get through quickly. Under such heterogeneous traffic requirements, different forwarding policies will be needed to serve the different types of traffic. What is more, not only is it important to ensure that a given protocol can deliver the desired performance (this is not always the case in such a partitioned environment), but the coexistence of the different protocols must be harmonic, as well.

- *Reliability*: In addition to different priority requirements, some messages may need to be sent reliably. Unlike in conventional networks, acknowledging messages end-to-end in partitioned networks is not a trivial task and may often have a significant performance overhead (e.g., flooding an ACK message after successful reception at the destination). Furthermore, if a whole session of messages needs to be sent reliably, the considerably large delays of the *loosely* closed feedback loop may significantly reduce the ability to "pipeline" data through the network. What is more difficult in terms of reliability in a disruption-tolerant kind of network is the ability to reliably deliver data in a certain order.

2.5   DTN Routing Design Guidelines

In the previous two sections, we have discussed different properties of DTNs like connectivity, mobility, node resources etc., and have dissected DTN-based routing solutions with respect to their characteristics (replication, forwarding, coding, etc.). Now, we will try to summarize the discussion by providing a correspondence between DTN-based routing solutions and the characteristics of different networks/applications. Having known, a priori, a given set of application characteristics/requirements, we can choose/build a specific kind of routing solution. For example, where connectivity and mobility are low, but the nodes have enough resources in terms of energy, bandwidth, and buffering, and we need a reliable solution, the epidemic routing (or its variant) can be employed. On the other hand, if the connectivity is low in an environment where nodes are highly mobile and nodes' resources are restricted and expensive (energy, buffering etc.), message replication schemes (or smart replication schemes) are the better choices to be utilized/built. Similarly, with low or high connectivity, low mobility and any kind of mobility model (homogeneous, heterogeneous, or correlated), Spray and Focus [45] seems to fit in, when the nodes' resourced are not limited. If reliability is needed by a routing solution, only epidemic routing or message coding can be employed.

Table 2.2 aims at summarizing the correspondence between network/application characteristics and DTN routing solutions. The rows in the table represent the properties of applications (networks), whereas each column provides a different routing solution. If read line-by-line (horizontally), it states which routing modules may be *useful* or *necessary* to cope with the given characteristic (one per line). If read column-by-column (vertically), then it describes particular scenarios where the given protocol (one per column) is the best choice (or simply a good choice). We do not intend that this table is all-inclusive or without exceptions. It is only rather indicative of which routing strategies might match better to which DTN environments. It is also important to note that this table characterizes the suitability of a routing solution according to the set of network/application characteristics that we have presented in Section 2.4.

Table 2.2: Routing Module Applicability

		Epidemic	Replicate	Smart Replicate	Focus	MANET	Code
Connectivity	low	✓	✓	✓	✓		
	high				✓	✓	
Mobility	low	✓			✓	✓	✓
	high		✓	✓			
Mob. Model	homog.		✓		✓		
	heterog.			✓	✓		
	correl.			✓	✓		
Resources	low		✓	✓			✓
	high	✓			✓	✓	
Priority		✓				✓	
Reliability		✓					✓

2.6 Case Studies

In this section we will present simulation results to support/demonstrate our claims (design principles) from the previous section. We could actually include here those scenarios that we addressed in detail in the previous section. Our goal here is not to provide extensive simulation results or argue for specific protocols, but rather to demonstrate the validity of our analysis of the routing solution space.

In the following sections, we describe three case studies.

2.6.1   Pocket Switched Networks

Pocket Switched Networks have recently been proposed [18] as a special type of DTN networks. The idea is to extend Internet connectivity beyond access points, by taking advantage of all possible means of "communication," including peer-to-peer links (e.g., Bluetooth), ephemeral access to a connected infrastructure (e.g., wireless Infostations [48]), as well as physical mobility.

In this paradigm, nodes are assumed to be carried in the users' "pockets" during their daily life activities. This implies that patterns existing in the daily movement of different nodes (e.g., time of commuting to work and means of transportation used, time spent in the office or in other job locations, etc.), as well as interaction and social patterns between different users, are expected to affect considerably the transmission opportunities "seen" by the nodes.

There have been a lot of experimental studies recently trying to discover and quantify these mobility and "inter-meeting" patterns between users/nodes [50]. Some key findings include the following [49]: (i) nodes tend to show strong location or peer preference; that is, each node has a number of access points (peers) that it visits (sees) more often than others; (ii) nodes are rather heterogeneous in their mobility and interaction behavior; some nodes tend to see all other nodes often, while others only see a small set of peers throughout the measurement periods; (iii) inter-contact times between nodes have "heavy-tailed" behavior.

The above create a scenario where the respective transmissions opportunities have detailed structure. As we mentioned in the previous section, utility-based routing protocols that take into account, for example, the age of last encounter between nodes, are capable of discovering and taking advantage of such structure. In Figure 2.2 we compare the performance of 3 protocols for the traces collected at the Infocom 2005 conference scenario [49]: (i) epidemic routing, (ii) a controlled replication protocol that blindly hands over copies [27], and (iii) one that maintains last encounter information between nodes, and may forward a message copy further to another node with a more correlated mobility/encounter pattern with the destination.

As can be seen there, using controlled replication rather than epidemic routing can utilize the available bandwidth much better. Furthermore, using a utility function to discover better relays for a given destination can improve performance even more. (Note that this is a small scenario with only 41 nodes, so performance improvement is more modest than in other scenarios; we expect that a similar mobility pattern in a larger setting — more nodes and more locations — would further demonstrate the importance of using the "right" routing modules).

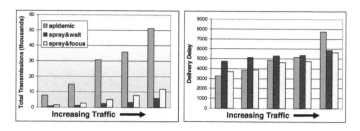

Figure 2.2: Performance of different routing modules for trace-based mobility: Infocom 2005 traces collected by the HAGGLE project.

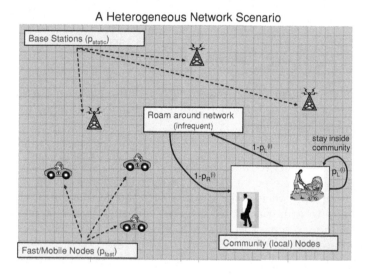

Figure 2.3: Example scenario with heterogeneous wireless nodes.

2.6.2   Metropolitan Networks with Heterogeneous Nodes

Even though nodes in the previous scenario exhibit different social and movement behavior, they all still correspond to humans, and specifically pedestrians. However, there are situations where a larger variety of nodes may collaborate/coexist to enable intermittent connectivity in a larger/metropolitan scale. Such a scenario might include, for example, nodes carried by pedestrians, other nodes mounted on vehicles, static nodes corresponding to base stations, sensors, or throwboxes [17], etc., as shown in Figure 2.3.

Scenarios like the one just described, involve a larger amount of heterogeneity. In addition to different social interactions, nodes in this case might also have largely varying amounts of resources as well as mobility ranges and speeds. For example, a node mounted on a car or a bus may cover a much larger network area than a node carried in the pocket of a pedestrian, and

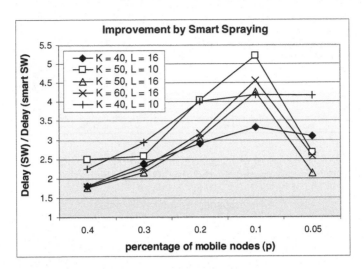

Figure 2.4: Performance improvement of Smart Spraying over greedy Spray and Wait, as a function of the percentage p of "mobile" (useful) nodes; K is a node's transmission range.

also may have no energy considerations. This implies the following: In the previous scenario some nodes may be better relays *for a specific destination* due, for example, to their social relation with the destination or their physical proximity to it; in this scenario some nodes may be better relays *for all destinations* due to some special capabilities of theirs like more resources or more peer encounters.

Imagine an example scenario where a percentage of nodes are mobile (e.g. cars, buses) and often perform long trips around the network, while the rest of the nodes each move inside its own local community only (e.g. campus, office building, etc.), which is much smaller than the total network area. In this network, in order to route messages between nodes that lie in different communities, it is crucial to discover and take advantage of the few "mobile" nodes in the network. The rest of the nodes are useless for inter-community traffic.

In Figure 2.4 we compare the delivery delay for two different routing strategies: (*Greedy Spraying*) In the first scheme, controlled replication is performed using a greedy distribution of the copies; all L copies of a message are handed over to the first L nodes encountered; (*Smart Spraying*) In the second scheme, we assume that each node carries a label that indicates what type of node it is (e.g., "Vehicle," "Pedestrian," "Base Station," etc.).[3] Copies of the messages

---

[3]Although such labeling could be possible in some cases, we do not assume that a node must necessarily know its type; Instead, each node could maintain statistics of the "intensity of its mobility" by maintaining a running average of the number of different nodes it encounters within a given window of time; this could then be used as an estimate of how mobile a node is.

are handed over only to nodes that carry a given label (e.g., "Vehicle"), and that can travel outside the source's local community.

As can be seen in Figure 2.4, blindly choosing relays could result in significant performance degradation in such a scenario. Although a few copies might happen to be handed over to "mobile" nodes that may eventually see a destination in a different community, most copies are wasted on nodes that rarely or never see the destination. On the other hand, a very simple optimization that tries to "read" a bit further into the structure of the surrounding network could result in up to 5× improvement. Specifically, the fewer the correct choices (i.e., the ratio of "good" over "bad" relays), the higher the potential improvement by trying to identify the good ones. Nevertheless, if the choices become too few, even "smart replication" is not powerful enough to discover the very few existing "paths-over-time," as evident in the plots as well. In that case, additional or different routing modules might be necessary to tackle the problem (e.g., flooding or utility-based forwarding).

2.6.3   Applications with Priorities

Despite the inherent delay-tolerance of the networks discussed, there can be situations where some messages may be more important than others. For example, in a VANET network it is reasonable to assume that an accident notification message will have higher priority than a chat message, or advertisements of nearby shops. Consequently, it would be useful to be able to treat priority messages preferentially, and ensure that *they get the best possible service, given the network limitations*. The questions is then, which routing strategy should be used for the priority messages and which for the non-priority ones, to satisfy the demands and semantics of both services?

Let us look at an example scenario where a $p\%$ of the messages have higher priority. In order to ensure the best possible service to these messages, we can use epidemic routing to route these messages only. Epidemic routing is guaranteed, under no buffer and bandwidth limitations, to find the optimal paths in *any* scenario. Thus, it provides the *best effort priority service* necessary in this context. The rest of the messages can be routed using a scheme like Spray and Wait. Spray and Wait: (i) generates very little traffic, which is important to not interfere significantly with the priority service; (ii) is robust enough to deliver good performance in a number of scenarios.[4] We have used simulations to answer the following two questions: What is the performance degradation to each service type, by the cross-traffic interference? How do these two services behave when the network becomes congested?

In Figure 2.5 we assume there are 100 nodes that move according to the Random Waypoint mobility model in a 500×500 network. We also assume that 10% of the messages (chosen randomly) have priority and are routed

---

[4]We intend to look into using Spray and Focus rather than Spray and Wait for the non-priority service.

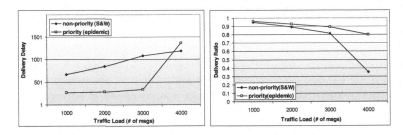

Figure 2.5: Delivery Delay (left) and Delivery Ratio (right) for traffic classes with different priorities, as a function of total traffic.

using epidemic routing, and the rest of the messages are routed using Spray and Wait with $L = 16$ copies. We look first at how congestion affects the two traffic classes. As can be seen there, when traffic is not too high, both traffic classes coexist smoothly. Priority messages get the best possible services with at most $10 - 20\%$ degradation, while the delay of non-priority messages gets increased a bit, and still remains competitive. On the other hand, if the network reaches congestion, it is important to note that it is the non-priority traffic class whose performance degrades the most and the fastest. This is very important, as it satisfies the semantics of a priority class, which is supposed to get the best service available.

Finally, in Figure 2.6 we again depict the delivery delay and delivery ratio for the two traffic classes, as a function of the percentage of total messages that have high priority (we assume a fixed traffic load of 800 messages). We also include the delivery delay for the case where all messages are routed using epidemic routing, and for the case where all messages are routed using Spray and Wait. As is evident by these plots, using a different routing strategy for the two classes achieves a much better trade-off than using the same routing protocol for all traffic, if the priority messages are only a fraction of the total messages (this is the *desirable* case in all priority services — see for example the air industry). Priority messages get better service than using Spraying for

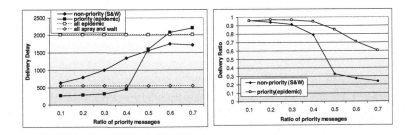

Figure 2.6: Delivery Delay (left) and Delivery Ratio (right) for traffic classes with different priorities, as a function of the ratio of priority messages.

all traffic, which is the desired semantic. Furthermore, *both traffic classes get better service than if all messages were treated as priority!* Finally, if for some reason the priority traffic increases (e.g., a major accident, natural disaster, etc.), it is the performance of the non-priority class that degrades first, with the priority traffic being again able to capture all available resources.

Bibliography

[1] T. Clausen and P. Jacquet, "Optimized Link State Routing Protocol, draft-ietfmanet-olsr-11.txt," 2003.

[2] J. Broch, D. A. Maltz, D. B. Johnson, Y.-C. Hu, and J. Jetcheva, "A Performance Comparison of Multi-Hop Wireless Ad Hoc Network Routing Protocols," In Proceedings of Mobile Computing and Networking, 1998.

[3] Sushant Jain, Kevin Fall, and Rabin Patra, "Routing in a Delay Tolerant Network," In Proceedings of ACM SIGCOMM, 2004.

[4] W. Zhao, M. Ammar, and E. Zegura, "A Message Ferrying Approach for Data Delivery in Sparse Mobile Ad Hoc Networks," In Proceedings of ACM/IEEE MOBIHOC, 2004.

[5] R. Shah, S. Roy, S. Jain, and W. Brunette, "Data MULEs: Modeling a Three-tier Architecture for Sparse Sensor Networks," IEEE SNPA Workshop, May 2003.

[6] A. Vahdat and D. Becker, "Epidemic Routing for Partially Connected Ad Hoc Networks," Technical Report Number CS-200006, Duke University, 2000.

[7] D. Yu and H. Li, "On the Definition of Ad Hoc Network Connectivity," In Proceedings of International Conference on Communications Technologies (ICCT), 2003, pages 990–994.

[8] B. Krishnamachari, S.B. Wicker, and R. Bejar, "Phase Transition Phenomena in Wireless Ad Hoc Networks," In Proceedings of IEEE Globecom, 2001, pages 2921–2925.

[9] B. Krishnamachari, S.B. Wicker, R. Bejar, and M. Pearlman, "Critical Density Thresholds in Distributed Wireless Networks," In Proceedings of Communications, Information and Network Security, 2002, pages 1–15.

[10] T. Spyropoulos, T. Turletti, and K. Obraczka, "Utility-based Message Replication for Intermittently Connected Heterogeneou Networks," In Proceedings of The First International IEEE WoWMoM Workshop on Autonomic and Opportunistic Communications (AOC), June 2007.

[11] S. Katti, H. Rahul, W. Hu, D. Katabi, M. Medard, and J. Crowcroft, "XORs In The Air: Practical Wireless Network Coding," In Proceedings of ACM SIGCOMM, 2006.

[12] N. Sarafijanovic-Djukic, M. Piorkowski, and M. Grossglauser, "Island Hopping: Efficient Mobility-Assisted Forwarding in Partitioned Networks," In Proceedings of IEEE SECON, September 2006.

[13] T. Henderson, D. Kotz, and I. Abyzov, "The Changing Usage of a Mature Campus-wide Wireless Network," In Proceedings of the 10th Annual International Conference on Mobile Computing and Networking, 2004.

[14] P. Juang, H. Oki, Y. Wang, M. Martonosi, L. Shiuan Peh, and D. Rubenstein, "Energy-efficient computing for wildlife tracking: design tradeoffs and early experiences with ZebraNet," In Proceedings of ACM ASPLOS, 2002.

[15] J. Ghosh, S. J. Philip, and C. Qiao, "Sociological Orbit Aware Location Approximation and Routing in MANET," In Proceedings of 2nd International Conference on Broadband Networks, 2005.

[16] M. Musolesi and C. Mascolo, "A Community Based Mobility Model for Ad Hoc Network Research," In Proceedings of ACM REALMAN, 2006.

[17] W. Zhao, Y. Chen, M. Ammar, M. Corner, B. Levine, and E. Zegura, "Capacity Enhancement using Throwboxes in DTNs," In Proceedings of IEEE MASS, 2006.

[18] J. Scott, P. Hui, J. Crowcroft, and C. Diot, "Haggle: A Networking Architecture Designed Around Mobile Users," In Proceedings of IFIP Conference on Wireless On-Demand Network Systems and Services (WONS), 2006.

[19] J. Leguay, T. Friedman, and V. Conan, "DTN Routing in a Mobility Pattern Space," In Proceedings of ACM SIGCOMM Workshop on Delay Tolerant Networking (WDTN), 2005.

[20] A. Lindgren, A. Doria, and O. Schelen, "Probabilistic routing in intermittently connected networks," SIGMOBILE Mobile Computing and Communication Review, Vol. 7, No. 3, 2003.

[21] J. Boice, J.J. Garcia-Luna-Aceves, and K. Obraczka, "Disruption-Tolerant Routing with Scoped Propagation of Control Information," In Proceedings of International Conference on Communications (ICC), 2007, pages 3114–3121.

[22] X. Zhang, G. Neglia, J. Kurose, and D. Towsley, "Performance Modeling of Epidemic Routing," In Proceedings of IFIP Networking, 2006.

[23] T. Small and Z. Haas, "Resource and Performance Tradeoffs in Delay-Tolerant Wireless Networks," In Proceedings of ACM SIGCOMM workshop on Delay Tolerant Networking (WDTN), 2005.

[24] P. Ramanathan and A. Singh, "Delay Differentiated Gossiping in Delay Tolerant Networks," In Proceedings of International Conference on Communications (ICC), 2008, pages 3291–3295.

[25] J. Burgess, B. Gallagher, D. Jensen, and B. N. Levine, "MaxProp: Routing for Vehicle-Based Disruption-Tolerant Networks," In Proceedings of IEEE INFOCOM, April 2006.

[26] M. Grossglauser and D. Tse, "Mobility Increases the Capacity of Ad Hoc Wireless Networks," IEEE/ACM Transactions on Networking, Vol. 10, No. 4, August 2002.

[27] T. Spyropoulos, K. Psounis, and C. S. Raghavendra, "Spray and Wait: Efficient Routing in Intermittently Connected Mobile Networks," In Proceedings of ACM SIGCOMM Workshop on Delay Tolerant Networking (WDTN), 2005.

[28] G. Neglia, and X. Zhang, "Optimal Delay-Power Tradeoff in Sparse Delay Tolerant Networks: A Preliminary Study," In Proceedings of ACM SIGCOMM Workshop on Challenged Networks (CHANTS'06), 2006.

[29] Y. Wang, S. Jain, M. Martonosi, and K. Fall, "Erasure Coding Based Routing for Opportunistic Networks," In Proceedings of ACM SIGCOMM Workshop on Delay Tolerant Networking (WDTN), 2005.

[30] X. Zixiang, A.D. Liveris, and S. Cheng, "Distributed Source Coding for Sensor Networks," IEEE Signal Processing Magazine, Vol. 21, Issue 5, September 2004.

[31] I. Solis and K. Obraczka, "Efficient Continuous Mapping in Sensor Networks Using Isolines," In Proceedings of MobiQuitous, 2005, pages 325–332.

[32] A. Kamra, V. Misra, J. Feldman, and D. Rubenstein, "Growth Codes: Maximizing Sensor Network Data Persistence," In Proceedings of ACM SIGCOMM, 2006.

[33] J. Widmer and J.-Y. Le Boudec, "Network Coding for Efficient Communication in Extreme Networks," In Proceedings of ACM SIGCOMM Workshop on Delay Tolerant Networking (WDTN), 2005.

[34] S.-Y. R. Li, R. W. Yeung, and N. Cai, "Linear Network Coding," IEEE Transactions on Information Theory, February 2003.

[35] S. Deb, C. Choute, M. Medard, and R. Koetter, "Data Harvesting: A Random Coding Approach to Rapid Dissemination and Efficient Storage of Data," In Proceedings of the IEEE INFOCOM, March 2005.

[36] R. N. B. Rais, T. Turletti, and K. Obraczka, *Coping with Episodic Connectivity in Heterogeneous Networks*. In Proceedings of the 11th International Symposium on Modeling, Analysis and Simulation of Wireless and Mobile Systems (MSWiM), pages 211–219, Canada, 2008.

[37] T. Spyropoulos, T. Turletti, and K. Obraczka, *Routing in Delay Tolerant Networks Comprising Heterogeneous Node Populations*, IEEE Transactions on Mobile Computing (TMC) Vol. 8, No. 8, pages 1132–1147, August 2009.

[38] H. Dubois-Ferriere, M. Grossglauser, and M. Vetterli, "Age Matters: Efficient Route Discovery in Mobile Ad Hoc Networks Using Encounter Ages," In Proceedings of ACM MobiHoc, 2003.

[39] M. Grossglauser and M. Vetterli, "Locating Mobile Nodes with EASE: Learning Efficient Routes from Encounter Histories Alone," IEEE/ACM Transactions on Networking, Vol. 14, No. 3, June 2006.

[40] First Mile Solutions, "http://www.firstmilesolutions.com."

[41] KioskNet (VLINK), "http://blizzard.cs.uwaterloo.ca/tetherless/index.php/VLink," University of Waterloo, Canada.

[42] J. Leguay, V. Conan, and T. Friedman, "Evaluating MobySpace-Based Routing Strategies in Delay-Tolerant Networks," Wireless Communications and Mobile Computing, May 2007.

[43] S. Capkun, L. Buttyan, and J. Hubaux, "Self-Organized Public Key Management for Mobile Ad Hoc Networks," IEEE Transactions on Mobile Computing, Vol. 1, No. 1, 2002.

[44] P. Resnick, R. Zeckhauser, E. Friedman, and K. Kuwabara, "Reputation Systems, Facilitating Trust in Internet Interactions," In Proceedings of Communications of the ACM, December 2000, pages 45–48.

[45] T. Spyropoulos, K. Psounis, and C. Raghavendra, "Spray and Focus: Efficient Mobility-Assisted Routing for Heterogeneous and Correlated Mobility," In Proceedings of IEEE PERCOM, on the International Workshop on Intermittently Connected Mobile Ad Hoc Networks (ICMAN), March 2007.

[46] A. Balasubramanian, B. N. Levine, and A. Venkataramani, "DTN Routing as a Resource Allocation Problem," In Proceedings of the ACM SIGCOMM, August 2007.

[47] A. Balasubramanian, B. N. Levine, and A. Venkataramani, "Replication Routing in DTNs: A Resource Allocation Approach," To appear in IEEE/ACM Transactions on Networking, 2010.

[48] T. Small and Z. J. Haas, "The shared wireless infostation model: a new ad hoc networking paradigm (or where there is a whale, there is a way)," In Proceedings of the 4th ACM International Symposium on Mobile Ad Hoc Networking and Computing, June 2003.

[49] P. Hui, A. Chaintreau, J. Scott, R. Gass, J. Crowcroft, and C. Diot, "Pocket Switched Networks and Human Mobility in Conference Environments," In Proceedings of ACM SIGCOMM Workshop on Delay Tolerant Networking (WDTN), 2005.

[50] M. McNett and G. M. Voelker, "Access and Mobility of Wireless PDA Users," ACM Mobile Computing and Communication Review, 2003.

[51] B. Burns, O. Brock, and B. Neil Levine, "MV Routing and Capacity Building in Disruption Tolerant Networks," In Proceedings of the IEEE INFOCOM, March 2005.

[52] M. Shiny, S. Hongyy, and I. Rhee, "DTN Routing Strategies Using Optimal Search Patterns," In Proceedings of the Third ACM Workshop on Challenged Networks (CHANTS), September 15, 2008, San Francisco, California, USA.

[53] S. C. Nelson, M. Bakht, and R. Kravets, "Encounter-Based Routing in DTNs," In Proceedings of the IEEE INFOCOM, Rio de Janerio, Brazil, April 2009.

[54] R. Ramanathan, R. Hansen, P. Basu, R. Rosales-Hain, and R. Krishnan, "Prioritized Epidemic Routing for Opportunistic Networks," In Proceedings of the 1st International MobiSys Workshop on Mobile Opportunistic Networking, Puerto Rico, 2007.

[55] V. Erramilli and M. Crovella, "Forwarding in Opportunistic Networks with Resource Constraints," In Proceedings of Third ACM Workshop on Challenged Networks (CHANTS), September 15, 2008, San Francisco, California, USA.

[56] G. Sandulescu and Simin Nadjm-Tehrani, "Opportunistic DTN Routing with Window-Aware Adaptive Replication," In Proceedings of the ACM 4th Asian Conference on Internet Engineering (AINTEC), Bangkok, Thailand, November 2008.

[57] P. Hui, J. Crowcroft, and E. Yoneki, "BUBBLE Rap: Social-Based Forwarding in Delay Tolerant Networks," In Proceedings of ACM Mobi-Hoc'08, Hong Kong, May 2008.

[58] V. Erramilli, M. Crovella, A. Chaintreau, and C. Diot, "Delegation For-warding," In Proceedings of ACM MobiHoc'08, Hong Kong, May 2008.

[59] H. Jun, M. H. Ammar, and E. W. Zegura, "Power Management in Delay Tolerant Networks: A Framework and Knowledge-Based Mechanisms," In Proceedings of IEEE SECON, 2005.

[60] W. Wang, V. Srinivasan, and M. Motani, "Adaptive Contact Probing Mechanisms for Delay Tolerant Applications," In Proceedings of ACM Mobicom, 2007.

[61] E. Altman, A. Prakash Azad, T. Basar, and F. De Pellegrini, "Optimal Activation and Transmission Control in Delay Tolerant Networks," In Proceedings of IEEE Infocom, 2010.

[62] Y. Xi, M. Chuah, and K. Chang, "Performance Evaluation of a Power Management Scheme for Disruption Tolerant Network," Lecture Notes in Computer Science, Vol. 12, No. 5.6, pages 370–380, December 2007.

[63] B. Jun Choi and X. Shen, "Adaptive Asynchronous Clock-Based Power Saving Protocols for Delay Tolerant Networks," In Proceedings of IEEE Globecom'09, Honolulu, Hawaii, USA, 2009.

[64] N. Banerjee, M. D. Corner, and B. N. Levine, "Design and Field Experimentation of an Energy-Efficient Architecture for DTN Throwboxes," To appear in IEEE/ACM Transactions on Networking, 2010.

Chapter 3

Energy-Aware Routing Protocol for Delay Tolerant Networks

Seung-Keun Yoon and Zygmunt J. Haas

3.1	Introduction	70
3.2	Routing in Sparse Networks	70
3.3	The Epidemic Routing Protocol for Delay Tolerant Network	74
	3.3.1 The Properties of the Epidemic Routing Protocol	74
	3.3.2 The Network Lifetime of the Epidemic Routing Protocol	74
3.4	The Analytical Model of the Epidemic Routing Protocol	77
	3.4.1 The Statistics of Encounter Times	77
	3.4.2 The Transitional Epidemic Routing Model	77
	3.4.3 The Solution of the Epidemic Routing Model	79
3.5	The Restricted Epidemic Routing (RER) Protocols	82
	3.5.1 The Exclusion (EX) Scheme	82
	3.5.2 The Limited Time (LT) Scheme	83
	3.5.3 The Limited Number of Copies (LC) Scheme	84
3.6	The Tradeoff Function of the Restricted Epidemic Routing	86
	3.6.1 The Number of Copies and the Time Delay	86
	3.6.2 Evaluation of the Tradeoff Function	88
	3.6.3 The Network Lifetime of the RER Schemes	88
	3.6.3.1 The Lifetime of the EX Scheme	90
	3.6.3.2 The Lifetime of the LT Scheme	90
	3.6.3.3 The Lifetime of the LC Scheme	91
3.7	The Residual-Energy (RE) Scheme	91
	3.7.1 The LC Scheme with Residual-Energy Information	92

3.8   Maximizing the Lifetime of the Delay Tolerant Networks...........93
        3.8.1   The Residual-Battery Information........................93
        3.8.2   The Comparison of Lifetime Performance.................94
3.9   Summary and Concluding Remarks...............................95

3.1   Introduction

As mobile nodes are typically limited in their energy capacity, one of the main considerations in the design of wireless mobile networks is energy consumption, with the goal of extending the lifetime of the network. One approach which reduces energy consumption is to limit the transmission power, and consequently the transmission range, of the network nodes. However, unavoidably, this approach leads to sparse network topologies, in which the average number of neighbors of a network node is less than one. In sparse networks, nodes remain often disconnected from the rest of the network nodes. Routing in such network becomes a challenge, as frequently an end-to-end path does not exist when a packet is to be routed in the network. A possible approach to routing in such networks is for packets to be *carried* by network nodes until such time as the mobile node is able to create a forwarding link to another node in the network. This new networking paradigm is referred to as *store-carry-forward*, to distinguish it from the traditional *store-and-forward* paradigm. Store-carry-forward routing increases, often quite significantly, the delay and the delay jitter relative to the store-and-forward routing protocols. Consequently not all applications are capable of using the store-carry-forward networking paradigm. Rather, only those applications that can tolerate the increased delay are good candidates for this type of network. Networks which support such delay-insensitive applications are referred to as *Delay Tolerant Networks (DTN)*.

Most of the DTN routing protocols are derivatives of the *Epidemic Routing Protocol (ERP)*, in which a packet is replicated in every node that comes in contact with the node that carries the packet. On the one hand, replication increases the probability of packet delivery, which, as a result, reduces the time at which a copy of the packet is delivered to the destination. However, on the other hand, replication requires extra transmissions, increasing the energy consumption, which, in turn, reduces the network lifetime. Thus, ideally, packet replication should be limited. Therefore, there exists a tradeoff between energy consumption and delivery delay in the ERP. This chapter examines this tradeoff as a function of the network parameters and proposes an efficient method to extend the network lifetime with ERP.

3.2   Routing in Sparse Networks

A Mobile Ad Hoc Network (MANET) is composed of mobile nodes, where each node is capable of communicating with other network nodes when the nodes are within communication range of each other. Data packets in a MANET are routed from the source node to the destination node utilizing the *multihop* approach; i.e., by being relayed through other mobile nodes of the network. However, due to the mobility of the nodes and the constantly changing wireless communication conditions, the topology of the network frequently changes. More specifically, when two communicating nodes move out of each other's transmission range, the communication link between the two nodes breaks and the end-to-end path becomes disconnected. Thus routing in MANETs has been an intricate problem. OLSR [1],[2], AODV [3], DSR [4], ZRP [5], and many other protocols [6]–[9] have been proposed as MANET routing protocols. These protocols cope with the frequent path disconnections by discovering a new routing path when the old path fails.

The traditional routing protocols, which operate based on the *store-and-forward* communication paradigm, assume that a network is fully connected and that there always exists an end-to-end routing path between any pair of source and destination nodes. However, this is not always the case; in some communication scenarios, when the transmission range is too short or the node density is too low, connections between the nodes in MANET can become intermittent. When loss of connectivity becomes a frequent occurrence, so that a node sees, on the average, less than one neighbor, we refer to such communication scenarios as *sparse networks*. A possible approach to routing in sparse networks is to rely on mobility of the nodes and the likelihood of creation of temporary links among the network nodes. In this new approach, while packets are forwarded from one node to another, the receiving node may not have the ability to forward the packet on immediately after the packet reception. Thus the receiving node *carries* the packet "for a while" before it encounters another node, which the packet could be forwarded to. This leads to the new *store-carry-forward* communication paradigm. The *store-carry-forward* routing paradigm leads to an increased end-to-end packet delivery delay and may not be a suitable approach when delivery delay and delay jitter are of concern. However, some applications are not particularly sensitive to delay performance figures, such as is the case in long-term monitoring or long-term data collection applications. A network that supports applications which can tolerate a significant degree of end-to-end delay is referred to as a *Delay Tolerant Network (DTN)*. Potentially, DTNs could be used for wildlife monitoring [11],[12], for underwater surveillance [13], for interplanetary communications [10], and for remote village messaging [14]. These, and other applications of DTN, are characterized by sparse network topologies, often as a result of short transmission range relative to the diameter of the networks, where the limited transmission range is a result of the restricted capacity of the energy sources of the network nodes and the necessity of preserving energy to extend the network lifetime.

DTN has recently been the subject of extensive research efforts [15]–[21]. The majority of the research into routing protocols for DTN focuses either on finding the shortest (in hops) end-to-end routing path or the path with the shortest routing delay, or on reducing the total energy consumed for routing at the expense of increased end-to-end delay. Since nodes in DTN rely on their mobility to relay data packets to other nodes, the likelihood of encountering other nodes is an important factor in performance of any DTN routing protocol. If the behavior of the mobile nodes is predictable, future transmissions between nodes can be scheduled ahead of time using routing protocols that exploit this predictive behavior [22]–[25]. Assuming that the future network characteristic can be known, [22] proposes six types of routing algorithms depending on the amount of knowledge of the network. This knowledge includes information about contacts between nodes, their queuing occupancies, and future traffic demands. Assuming that the network characteristic can be predicted over some time into the future, [23] proposed a routing algorithm that finds the best path by looking ahead over a fixed time interval. These deterministic DTN routing algorithms compute the end-to-end path before the source node transmits its data packet using dynamic programming and shortest path algorithms.

However, in many cases, the behavior of the mobile nodes is random and, therefore, unpredictable. In these cases, the network requires a stochastic routing protocol in which each relaying node *dynamically* decides on its next recipient at the forwarding times, rather than before the packet leaves the source node. A relaying node can also choose its recipient based on mobility pattern, encounter history, or other information. Algorithms proposed in [26]–[30] use one-hop information, while [31]–[33] accumulate end-to-end information. Usage of special nodes with high mobility and high storage capacity has also been proposed [34]–[38]. To improve reliability, forwarding data packets to these special nodes may be preferable.

The *Epidemic Routing Protocol (ERP)* [39] is a routing protocol which exhibits the shortest end-to-end delay in intermittently connected mobile ad hoc networks. In ERP, a packet-carrying node forwards a replicated packet to every encountered node, without any other consideration. The basic idea is referred to as *packet flooding* and resembles the model of disease spread in epidemiology [40]–[44]. Replication of packets in ERP increases the probability of packet delivery by maximizing the number of nodes carrying the packet. However, ERP also increases the use of network resources, such as the energy consumed on the multiple transmissions, most of which will eventually not be useful in delivering the packet to the destination node. Furthermore, ERP also reduces the network capacity, which may become a bottleneck, as the encounter times are typically short and a link capacity is finite. In practical communication scenarios, the limited amount of energy stored in the batteries of a node becomes a major drawback of ERP.

Several works proposed efficient ways to overcome this ERP drawback [12],[45]–[60]. SWIM [12],[48] uses a small-sized *anti-packet* to restrict the packet replication. The anti-packet would typically contain just the ID (e.g., the sequence number) of the delivered data packet. After the destination receives a data packet, it starts back-propagating an anti-packet into the network. Anti-packets propagate in the network using ERP as well. An anti-packet signifies the delivery of the data packet to the destination node, allowing the nodes which still store the data packet to dispose of the stored packet. More advanced variations of the anti-packet scheme allow a node that receives an anti-packet to become "immune" from receiving the same data packet in the future, or even to erase the packets from nodes that it encounters. An anti-packet is deleted from the network at approximately the same time that the corresponding data packet expires.

In the *Spray and Wait* routing protocol [50], the source node restricts the total number of ERP transmissions during the propagation of a data packet in the network. In this protocol there are two periods: the *Spray* period and the *Wait* period. During the *Spray* period packets are replicated using ERP. When the total number of copies reaches the limit, the state changes to the *Wait* period and a node carrying a packet is allowed to transmit the packet only to the destination node.

Controlling the packet flow in mobile ad hoc networks can also restrict the ERP packet replication [39],[52]–[54]. In [39], a "limited-hop" algorithm is used to reduce the number of packet copies. In this algorithm, a packet can only be forwarded a limited number of times. When the ERP is limited to 2 hops, the source node is the only node that can transmit a replication of its packet to a peer node, while the other nodes can only transmit the packet to the destination node. A gossip-based algorithm, proposed in [53], can also control the packet flow. Transmitting a packet only to a fraction of encountered nodes based on certain probability can conserve energy, while the packet can still be delivered to the destination with high probability.

Controlling the packet flow based on the energy information has been shown to be an effective method for conserving energy in ad hoc networks [55]–[57]. In particular, controlling the packet flow based on the node's residual battery energy results in increased network lifetime [54]. Coding-based protocols [58]–[60] also reduce the energy consumption of packet routing in DTN. In erasure-coding [59], "abbreviated packets" carry only partial information of the original data packet to reduce energy consumption. Network-coding [60] was also used to combine multiple different packets to reduce the number of transmissions. In both cases, the destination has to receive more than one packet to recover the original data packet.

In this chapter, we focus on the question of how to extend the network lifetime by controlling the packet flow in ERP. It is important to notice that maximizing the network lifetime is not necessarily equivalent to minimizing the number of copies of a packet in the network. For example, creating abundance of copies by energy-rich nodes, while even minimally reducing the number of copies produced by energy-poor nodes, would still extend the network lifetime, amid an overall increase in the number of copies.

3.3   The Epidemic Routing Protocol for Delay Tolerant Network

3.3.1   The Properties of the Epidemic Routing Protocol

The basic idea behind Epidemic Routing (also referred to as packet flooding) is for a node carrying a copy of a data packet to replicate the packet onto every encountered node, unless the encountered node already has a copy of the packet. From the time a packet is generated at the source node, the number of nodes carrying a copy of the packet increases in time. Consequently, the probability of the destination node encountering a node that carries a packet copy increases as well.

In order to remove the copies of already delivered packets, [48] used the Time-to-Live (TTL) concept. The TTL, set by the source node, is a timer which limits the lifetime of a packet in the network. When a packet is transmitted from a node to another node, only the residual TTL is included in the transmission.[1] When the TTL expires, all the copies stored in the network nodes are erased. Short TTL can actually stop the Epidemic Routing before the number of copies in the network increases excessively. However, short TTL also reduces the probability that a copy of the packet will be delivered to the destination node. Since the number of copies in the network refers to the total energy consumption, the initial value of the TTL trades off the energy consumption with the packet delivery probability.

3.3.2   The Network Lifetime of the Epidemic Routing Protocol

Figure 3.1 depicts the delivery probability of a sequence of 100 consecutive routings in the network using Epidemic Routing. The network consists of 50 mobile nodes, in addition to one mobile destination node. The energy consumed in one transmission of a copy of a packet is 1 energy unit [EU], and we neglected the energy of packet reception. At the beginning, each node is equipped with a battery charged with 40 [EU] and the battery is used only for transmission. A data packet is created periodically at packet-time intervals of TP = 150 [sec] at a randomly selected source from among the 51 network nodes. The TTL was set to 150 [sec] in Figure 3.1(a), to 120 [sec] in Figure 3.1(b), and to 75 [sec] in Figure 3.1(c).

The graphs in Figure 3.1 show gradual decrease in the packet delivery probability. Since the batteries of the DTN nodes have a limited amount of energy, the nodes with depleted batteries are effectively removed from the network. Thus, as time goes by, the number of active nodes gradually decreases to zero, and so does the packet delivery probability. One can postulate that

---

[1]Note that global clock synchronization is not required, as the nodes only need to clock the residual packet lifetime.

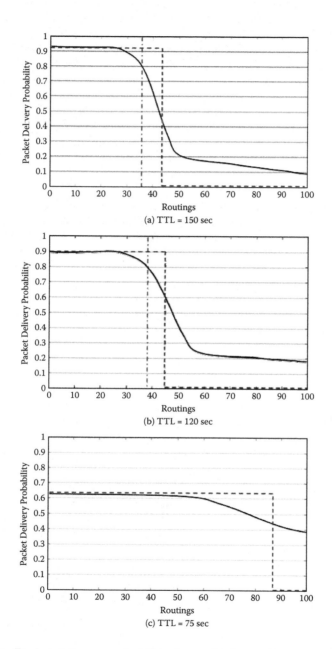

Figure 3.1: Packet delivery probability for multiple routing attempts of ER.

the network continues to be useful only as long as the packet delivery probability is above some minimum level, which we define as the *Minimum Delivery Probability (MDP)*.[2] Consequently, the *MDP lifetime* of a network is defined as the time when the packet delivery probability crosses the MDP level.

With MDP set to 80% and for TTL of 150 [sec], the MDP lifetime is 36 packet intervals (TP). The MDP lifetime extends to 39 packet intervals for TTL of 120 [sec]. This lifetime extension is facilitated by reducing the maximum achievable packet delivery probability (when all the nodes are active). That is, when the value of TTL is reduced from 150 [sec] to 120 [sec], the maximum achievable packet delivery probability decreases from 93% to 90%. This example demonstrates the tradeoff between maximum packet delivery probability and the lifetime of the network; i.e., the network lifetime can be extended at the expense of packet delivery probability. However, if we reduce the TTL down to 75 [sec], the packet delivery probability never reaches the MDP level, even though it decreases less rapidly as a function of time.

Through appropriate reduction in the value of TTL, the MDP lifetime can be extended by decreasing the maximum packet delivery probability to the MDP level. However, this may not always be desirable, as one would often prefer the network to achieve packet delivery probability higher than the MDP level. In other words, typically the network should achieve packet delivery probability of some *Target Delivery Probability (TDP)* most of its lifetime, and only at the end of the network lifetime the packet delivery probability should degrade to the MDP level, where TDP $\geq$ MDP. As we have seen above, the MDP lifetime of the network can be extended at the cost of reducing the maximum packet delivery probability to TDP.

Ideally, if the batteries of all the nodes are depleted at the same time, the packet delivery probability will instantaneously decrease to zero, as shown by the step-function dotted line in Figure 3.1. In Figure 3.1(a), for example, while the packet delivery probability is maintained at 93%, there were on the average 46.2 packets copied per packet routing. Hence, in an ideal case, the duration of the network lifetime should be $50 \cdot 40/46.2 = 43.29$ [routings]. Even after the ideal lifetime expires, there still exist some active nodes capable of delivering packets to the destination. The *ideal lifetime* can be calculated simply by dividing the total energy in the network by the average energy consumed during a packet routing. This ideal lifetime can be considered as the maximal possible extension of the MDP lifetime.

---

[2]This assumption should be considered valid for most DTN applications; e.g., for the monitoring application, one could define the level of reliability, which translates into the probability with which a single alarm message has to reach the destination.

3.4 The Analytical Model of the Epidemic Routing Protocol

3.4.1 The Statistics of Encounter Times

The process of Epidemic Routing is similar to the process of a virus spreading in epidemiology [40–43]. During epidemic routing, the number of nodes carrying a packet (referred to as "infected nodes") increases every time an infected node encounters an uninfected node (referred to as a "susceptible node"). If there were infinitely many nodes in the network, the rate of infection would increase linearly as the number of infected nodes increases. However, in a network with finite number of nodes, as the number of infected nodes increases, the number of susceptible nodes decreases, so that the infection rate saturates when the number of infected nodes is (approximately) equal to the number of susceptible nodes. From that point, the number of infected nodes continues to increase, albeit at an infection rate that decreases with time. Eventually, all the nods become infected.

In order to estimate the number of packet copies in the network at a certain time during the epidemic routing, we first analyze the packet flooding mechanism using the time interval between nodes encounter. The basic assumption in this chapter is that the number of encounters between two particular network nodes follows the Poisson distribution, so that the time interval between two consecutive encounters has exponential distribution with rate λ.

In a mobile network with total of N nodes, when the number of infected nodes is k and the number of susceptible nodes is N-k, the encounter rate between infected nodes and susceptible nodes is $k(N-k)\lambda$. Whenever an infected node encounters a susceptible node, the number of infected nodes increases by 1 and the number of susceptible nodes decreases by 1. Hence we can see that, as the number of infected nodes increases, the encounter rate between the infected nodes and the susceptible nodes increases until k reaches the value of N/2 (or (N-1)/2), and then decreases symmetrically down to $(N-1)\lambda$.

Since the time interval between encounters follows the exponential distribution, the next encounter time between an infected node and a susceptible node is totally independent from the past encounter times. If there are k nodes carrying a copy of a packet, the rate of the destination node encountering one of these nodes is $k\lambda$. Hence, as the number of propagating nodes increases, the probability of the destination encountering one of the propagating nodes increases as well.

3.4.2 The Transitional Epidemic Routing Model

Using the encounter rate between the infected nodes and the susceptible nodes, we derive a Markov chain model, as shown in Figure 3.2, where the states in the Markov chain denote the number of copies in the network. Using this Markov chain model, we derive the probability of having k copies of the packet

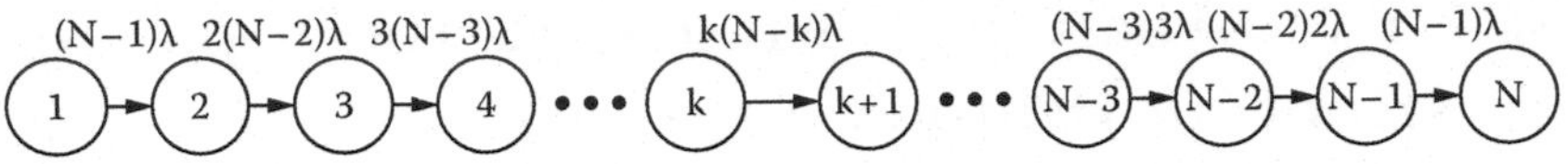

![Transition diagram with states 1, 2, 3, 4, ..., k, k+1, ..., N−3, N−2, N−1, N with rates (N−1)λ, 2(N−2)λ, 3(N−3)λ, k(N−k)λ, (N−3)3λ, (N−2)2λ, (N−1)λ]

Figure 3.2: Transition diagram of Markov chain model for number of copies.

among N network nodes at time t, $P_k(t)$:

$$P_k(t) = \int_0^t P_{k-1}(x) \cdot (k-1)\{N - (k-1)\}\lambda e^{-k(N-k)\lambda(t-x)}dx \quad (k > 1);$$
$$P_1(t) = e^{-(N-1)\lambda t}$$

$$(3.1)$$

Using (3.1) we derive the average number of copies as a function of time t.

$$E_n(t) = \sum_{k=1}^{N} kP_k(t)$$

$$= e^{-(N-1)\lambda t} + \sum_{k=2}^{N} k \int_0^t P_{k-1}(x) \cdot (k-1)\{N - (k-1)\}\lambda e^{-k(N-k)\lambda(t-x)}dx$$

$$(3.2)$$

Figure 3.4(a) compares the number of copies in the network as a function of time obtained by equation (1.2) (labeled "Analytical result") with the number of copies obtained by the discrete event simulation. As expected, the figure shows that the rate of increase of the number of copies is largest when the number of copies reaches 25; i.e., half of the total number of nodes.

In our network model, there are N=50 mobile nodes, in addition to one destination node. The transmission range of all the nodes is 25[m]. The network is a closed square of 1000[m] by 1000[m], a torus-like area. Each node determines its velocity independently. The time between velocity changes is an exponentially distributed random variable with average of

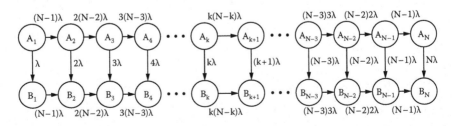

Figure 3.3: Transition diagram of Markov chain model for number of copies with destination.

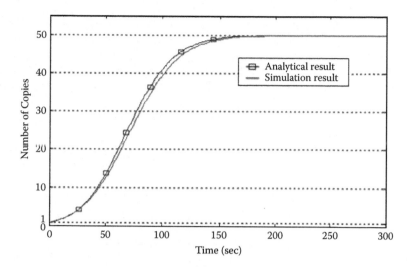

Figure 3.4: Result of average number of copies vs. time.

0.2 [sec]. The direction is uniformly distributed in $[0, 2\pi]$ and the speed is uniformly distributed in $(20, 70)$ [m/s]. As a result, the encounter rate is, $\lambda = 1/E[T_E|T_E > 0] = 0.00127/sec$, where T_E is the encounter time between two particular nodes. This random mobility model and the particular encounter rate will be used in the rest of this chapter.

3.4.3 The Solution of the Epidemic Routing Model

Although up to now we discussed DTN in the context of MANET, *Sensor Networks* could also operate as DTN. In a sensor network [12],[13], there is usually a single (or a small number of) special nodes to which all the packets are to be delivered. These nodes are referred to as *sink nodes*. If multiple sink nodes are present, it suffices that a copy of a packet is received by only one sink node. Sink nodes can be stationary, such as the offloading stations in SWIM [12],[48], or mobile, such as the data *MULEs* in [26]. From the analysis point of view, the only difference between these two cases is the encounter rate between the sink node and the other network nodes. Sink nodes do not transmit data packets, but only receive packets from other network nodes. With the exception of the sink node, all other network nodes do not know whether a data packet was received by the sink node.[3] Hence, the data packets are erased from the nodes only based on the value of the TTL. The initial TTL value is determined according to the calculated probability of a sink receiving a copy of the packet by some time t and some agreed upon level of confidence that the packet is to reach a sink before it is erased from the

---

[3]Unless anti-packets are implemented in the network.

network nodes. In this article, we consider a single mobile sink node, with the same transmission range and mobility pattern as the other network nodes.

In order to evaluate the Epidemic Routing performance, we need to consider three metrics: the number of packet copies in the network, the delay of the data packet,[4] and the probability of the sink node receiving a copy of the packet *(packet delivery probability or delivery confidence level)*. As depicted in Figure 3.4, the number of copies and the packet delivery probability are both non-decreasing function of time.

In Figure 3.3, when the network is in a state A_k, the sink node has not yet received a copy of the packet, and when the network is in a state B_k, the sink node has received at least one copy of the packet. The index k in the states A_k and B_k denotes the number of packet copies present in the network. While in a state A_k, the transition rate from state A_k to state B_k is $k\lambda$, which is the encounter rate between any of the k nodes and the sink node.

The probability of having k packet copies in the network at time t, $P_k(t)$ is calculated using equation (1.1). From the model in Figure 3.3, we calculate the probability $P_{A,k}(t)$ of the network being in a state A_k at time t. The probability of the network being in state B_k at time t, $P_{B,k}(t)$, can then be calculated by subtracting $P_{A,k}(t)$ from $P_k(t)$:

$$P_{A,k}(t) = \int_0^t P_{A,k-1}(x) \cdot (k-1)\{N-(k-1)\}\lambda e^{-k(N-k+1)\lambda(t-x)}dx \ (k > 1);$$

$$P_{A,1}(t) = e^{-N\lambda t}$$

$$\text{(3.3)}$$

$$P_{B,1}(t) = P_k(t) - P_{A,1}(t) \qquad \text{(3.4)}$$

Using these probabilities, we calculate the packet delivery probability at time t, $D(t)$:

$$D(t) = \sum_{k=1}^{N} P_{B,k}(t) \qquad \text{(3.5)}$$

The expected number of packet copies at time t is calculated by equation (3.2), and since the infected nodes do not have the knowledge whether the sink node received the data packet or not, the average number of packet copies is independent of the packet delivery probability in equation (3.5). Based on equations (3.2) and (3.5), the number of copies and the packet delivery probability are plotted as a function of time in Figures 3.4 and 3.5, and further combined in Figure 3.6.

These results show that if the packet propagation is left uninterrupted, the network will eventually transition to one of the B_k states. The transition to the B-states usually occurs when the value of k is close to N. Thus, typically,

---

[4]The time duration from when the packet is generated until when a copy is received by the sink node.

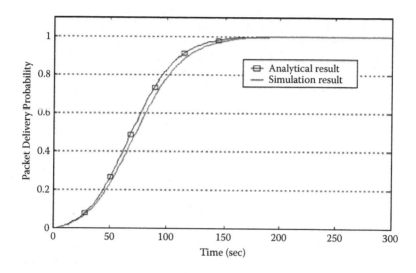

Figure 3.5: Result of packet delivery probability at destination vs. time.

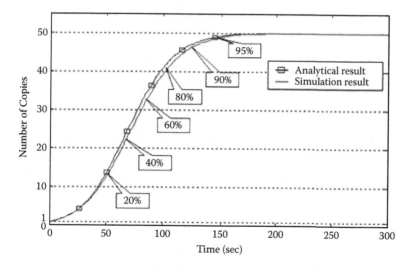

Figure 3.6: Packet delivery probability combined with number of copies vs. time.

there are many redundant copies in the network at the time that the sink node receives the packet. Of course, considering the energy consumption, one would like to have as few packet copies as possible at the time a copy is delivered to the sink node. Notice in Figure 3.3 that the rates of transition from the A-states to the B-states increase with k. Therefore, if this transition rate could be made larger for smaller k, this would allow us to terminate the packet propagation sooner, reducing the number of packet copies in the network. This is the idea behind selective transmissions, where the packet is replicated primarily on nodes that have larger probability of encountering the sink node.

3.5   The Restricted Epidemic Routing (RER) Protocols

The purpose of the Restricted Epidemic Routing is to reduce the number of copies of a packet in the network needed to deliver a copy to the destination node. Instead of transmitting the data packet to all the encountered nodes, a node may transmit only to the nodes with larger priority, where a node's priority could be based on the destination encounter history, the node's mobility pattern, or some other information [18]–[25]. If the mobility of the nodes is totally random, all the nodes would have the same priority [26]–[33]. In this case, reduction of the number of copies could be achieved by a control algorithm, which is independent of the information of a particular encountered node. For example, the source node could include a replication restriction in the propagating data packet [48]–[52]. Of course, restricting packet replication may lead to worse routing performance, such as increased delay at a particular delivery probability. (An alternative would be for the destination node to broadcast acknowledgements of the received packets (e.g., in the form of an anti-packet) [48].)

In what follows, we present three different schemes to restrict the number of copies in Epidemic Routing, which we refer to as the Restricted Epidemic Routing (RER) schemes. Those schemes rely on limiting the number of nodes that participate in the packet propagation (the Exclusion scheme), limiting the time allowed for propagation (the Limited Time scheme), and directly controlling the number of copies (the Limited Number of Copies scheme).

3.5.1   The Exclusion (EX) Scheme

In the Exclusion scheme (EX scheme) some of the network nodes are prevented from participation in routing the data packet. Before the source node encounters another node and propagates the data packet, the source node determines which nodes are excluded from the packet propagation. If we label the number of non-excluded nodes as M, where M < N, the EX scheme is effectively the ERP, but in a "smaller" network which includes only the

M non-excluded nodes. Of course, the number of copies in the EX scheme increases more slowly than in ERP. Since the EX scheme could be seen as simply reducing the total number of network nodes from N to M, the Markov Chain model in Figure 3.3 remains valid, with the exception that N should now be replaced by M. Consequently, equations (1.2) and (1.5) remain valid for this case as well, with the simple substitution of M in place of N:

$$P_k(t) = \int_0^t P_{k-1}(x) \cdot (k-1) \{M - (k-1)\} \lambda e^{-k(M-k)\lambda(t-x)} dx \quad (k > 1)$$

$$(3.6)$$

$$P_{A,k}(t) = \int_0^t P_{A,k-1}(x) \cdot (k-1) \{M - (k-1)\} \lambda e^{-k(M-k+1)\lambda(t-x)} dx \quad (k > 1)$$

$$(3.7)$$

3.5.2 The Limited Time (LT) Scheme

In the Limited Time scheme (LT scheme), packet propagation is the same as in ERP, except that there is a time limit allowed for replication. To accomplish this, a timer is introduced – the *Replication Time Limit (RTL)*, which is included within the data packet format, very much like the TTL is. The nodes can replicate the data packet until the RTL expires, at which time the copies persist in the network, albeit without further replication. Thus, RTL≤TTL. Assuming that no copies of the packet were delivered prior to RTL, the packet delivery probability depends on the number of copies in the network at the time RTL. The replication rates remain the same as in the case of ERP, and the Markov Chain model of LT scheme is the same as that of the ERP in Figure 3.3 for 0≤t≤RTL. For t>RTL, the network follows a simple two-state Markov Chain (Figure 3.7), where the network starts either in state $B_{k'}$ or $A_{k'}$, depending on whether at time t=RTL the destination received a copy of the packet or not, respectively, and where k' is the number of copies in the network at the time RTL. Consequently:

for $0 < t < PTL$

$$P_k(t) = \int_0^t P_{k-1}(x) \cdot (k-1) \{N - (k-1)\} \lambda e^{-k(N-k)\lambda(t-x)} dx \quad (k > 1)$$

$$(3.8)$$

$$P_{A,k}(t) = \int_0^t P_{A,k-1}(x) \cdot (k-1) \{N - (k-1)\} \lambda e^{-k(N-k+1)\lambda(t-x)} dx \quad (k > 1)$$

$$(3.9)$$

for $t \geq PTL$

$$P_k(t) = P_k(PTL) \quad (k > 1) \tag{3.10}$$

$$P_{A,k}(t) = P_{A,k}(PTL) \cdot e^{k\lambda(t-PTL)} \quad (k > 1) \tag{3.11}$$

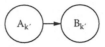

Figure 3.7: Transition diagram of Markov chain model for LT-Scheme
(t>RTL).

3.5.3 The Limited Number of Copies (LC) Scheme

In the Limited Number of Copies scheme (LC scheme), there is an upper
bound on the number of copies that the network is allowed to replicate. Of
course, disseminating the information about the current number of copies in
the network to all the network nodes would be prohibitively expensive in
network resources. Alternatively, one way to limit the total number of copies
is to determine a priori how many copies a node can create at the time when
the node receives a copy of the packet. Another way of looking at the scheme
is to assume that upon packet creation, all the possible replicas are created
as well and from then on, the replicas are forwarded (as batches larger than
one), but not replicated anymore. Nodes that receive a single replica cannot
either forward the single replica anymore, or create any new replicas.

Assume we want to propagate up to 12 copies of a packet in the network.
Upon an encounter, a node carrying more than one copy splits its load with
an encountered susceptible node. (When two nodes that already carry one or
more copies meet, no exchange takes place.) Figure 3.8 shows the evolution
of the network, where the numbers in a state indicates the load carried by
the infected nodes. For example, in the state [6/2/2/1/1], there are 5 infected

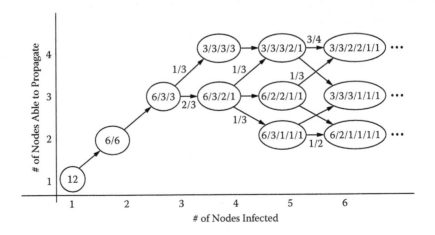

Figure 3.8: 2-D transition diagram of Markov chain model for LC scheme
(max. 12 copies).

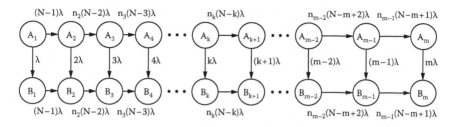

Figure 3.9: Transition diagram of Markov chain model for LC scheme (max. m copies).

nodes: one node with the load of 6 copies, two nodes with 2 copies, and two other nodes with one copy each. The next state is determined by the number of nodes that can propagate the copies, with the assumption that each node has the same probability of encountering another node. Hence, we can calculate the probability of the network being in a state for the 2-D Markov chain in Figure 3.8.

Table 3.1: Average Number of Nodes in LC Scheme That Can Propagate in State $k(n_k)$

k	1	2	3	4	5	6	7	8	9	10	11	12	13	14	15	...
m = 2	1	0	0	0	0	0	0	0	0	0	0	0	0	0	0	...
m = 3	1	1	0	0	0	0	0	0	0	0	0	0	0	0	0	...
m = 4	1	2	1	0	0	0	0	0	0	0	0	0	0	0	0	...
m = 5	1	2	1.5	1	0	0	0	0	0	0	0	0	0	0	0	...
m = 6	1	2	2	1.5	1	0	0	0	0	0	0	0	0	0	0	...
m = 7	1	2	2.5	1.667	1.833	1	0	0	0	0	0	0	0	0	0	...
m = 8	1	2	3	2.667	2.333	2	1	0	0	0	0	0	0	0	0	...
m = 9	1	2	3	2.833	2.611	2.306	1.75	1	0	0	0	0	0	0	0	...
m = 10	1	2	3	3	2.889	2.569	2.278	1.639	1	0	0	0	0	0	0	...
m = 11	1	2	3	3.167	3.111	2.921	2.659	2.262	1.631	1	0	0	0	0	0	...
m = 12	1	2	3	3.333	3.333	3.231	2.986	2.723	2.262	1.631	1	0	0	0	0	...
m =13	1	2	3	3.5	3.597	3.573	3.403	3.18	2.875	2.261	1.723	1	0	0	0	...
m = 14	1	2	3	3.667	3.861	3.894	3.772	3.554	3.344	2.979	2.429	1.874	1	0	0	...
m = 15	1	2	3	3.833	4.097	4.179	4.141	4.018	3.772	3.52	3.048	2.496	1.944	1	0	...
m = 16	1	2	3	4	4.333	4.444	4.478	4.349	4.132	3.951	3.657	3.104	2.552	2	1	...

In Figure 3.9, n_k is the average number of nodes that can still propagate copies when there are k copies in the network, and m is the maximum copies allowed in the network. Each n_k can be calculated using the 2-D transition Markov chain; the values for n_k are listed in Table 3.1. Using the Markov Chain model in Figure 3.9, the probability of the network being in a state with k infected nodes, and the probability of k infected nodes without having a copy delivered to the destination are:

$$P_k(t) = \int_0^t P_{k-1}(x) \cdot n_{k-1} \left\{ N - (k-1) \right\} \lambda e^{-n_k(N-k)\lambda(t-x)} dx \quad (k > 1)$$

$$(3.12)$$

$$P_{A,k}(t) = \int_0^t P_{A,k-1}(x) \cdot n_{k-1} \left\{ N - (k-1) \right\} \lambda e^{-(n_k(N-k)+k)\lambda(t-x)} dx \quad (k > 1)$$

$$(3.13)$$

The expected number of copies in the network and the packet delivery probability for a given time t can be calculated by applying equations (3.12) and (3.13) to (3.2) and (3.5).

Due to the limited number of propagating nodes, the propagation rate in the LC scheme is slightly smaller than the propagation rate in the LT scheme, but still larger than the propagation rate in the EX scheme, since the number of susceptible nodes in the LC scheme is not limited.

3.6 The Tradeoff Function of the Restricted Epidemic Routing

In this section we evaluate and verify the performance of the restricted Epidemic Routing schemes by plotting the average number of copies at time t, E[n(t)]. The graphs were obtained using the formulas developed in the previous sections and present the results for various degrees of restriction.

3.6.1 The Number of Copies and the Time Delay

Figure 3.10(a) shows the average number of copies in the network as a function of time with the number of included nodes (M) as a parameter. For larger values of M, the protocol behaves closer to ERP, and for small M, more like a single hop routing. Using Figure 3.10(a) and for each one of the M-curves, marking the points that correspond to delivery probability of 95% (with a triangle) and of 60% (with a square), we obtain the Tradeoff function (T-function) in Figure 3.11. The T-function of a particular scheme is specific to a particular level of packet delivery probability and presents the tradeoff between the number of copies (an indication of energy consumption) and the packet delivery delay. This tradeoff can be controlled for the EX scheme by the value of the parameter M; smaller M leads to a smaller number of copies but with longer delivery delay, and vice versa.

Similarly, a T-function could be obtained for the two other schemes using Figure 3.10(b) and Figure 3.10(c), where the controlling parameter is the value of RTL for the LT scheme and the maximal number of copies (m) for the LC scheme.

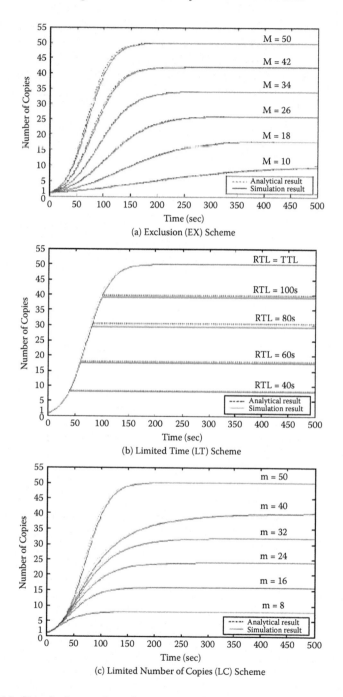

Figure 3.10: Simulation and analysis results for different methods and degrees of restriction.

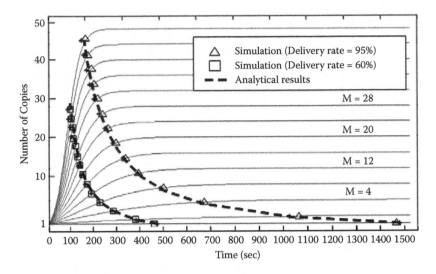

Figure 3.11: T-function for packet delivery probability 60% and 95% (EX scheme).

3.6.2   Evaluation of the Tradeoff Function

Figure 3.12(a) shows the comparison of the T-functions for the three RER schemes and for 90% packet delivery probability. Notice that all the tradeoff functions start and end at the same point. For any delay time, the T-function of the LC scheme is closer to the origin than the T-functions of the other two schemes. Consequently, we can conclude that the LC scheme outperforms the other two RER schemes in terms of the energy-delay tradeoff.

Figure 3.12(b) compares the efficiency of the RER schemes in a different way. It shows the packet delivery probability as a function of the number of packet copies at a fixed time t=150 [sec]. The figure demonstrates that the LC scheme has higher packet delivery probability with any particular number of copies, as compared to the other two schemes. Looking at these results from a different perspective, for the TTL equal to 150 [sec] and for the packet delivery probability equal to 90%, the LC scheme produces on the average approximately 31 copies in the network, while the LT scheme and the EX scheme produce 33 and 37 copies, respectively.

3.6.3   The Network Lifetime of the RER Schemes

The parameters used in this section are the same as those used in section 3.5.2; in particular, the encounter rate between two nodes is $\lambda = 0.00127/sec$. Additionally, we define the MDP lifetime by setting the TDP to 90% and the MDP to 80%. A packet is created from an arbitrarily selected node every 150 [sec], and the TTL in each packet is set to 150 [sec]. All 50 nodes carry

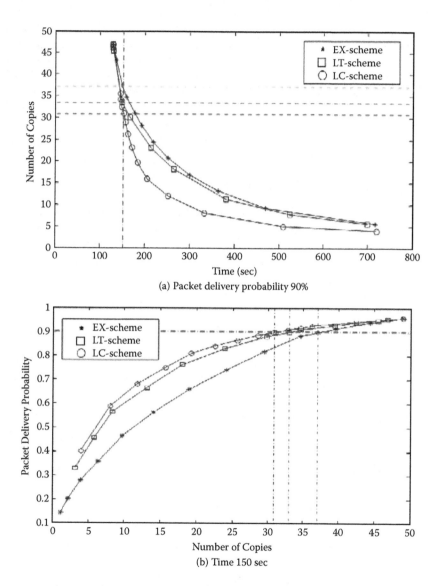

Figure 3.12: Comparison of efficiency between different methods of restriction.

batteries with 40 [EU] each.

3.6.3.1   The Lifetime of the EX Scheme

Figure 3.12 shows that the EX scheme requires 37 nodes in order to obtain 90% TDP, while Figure 3.13 demonstrates that, when the total number of nodes is limited to M = 40, the average number of copies in the network grows to 37 at 150 [sec]. Hence in order for the EX scheme to obtain the TDP of 90%, the total number of participating nodes should be set to 40. Assuming that every node starts with battery of 40 [EU], since the average number of copies at 150 [sec] is 37, the expected lifetime of the EX scheme would be approximately 8108 [sec] (i.e., transmission of $(40 \cdot 50/37) = 54$ data packets, with each packet remaining in the network for 150 [sec]).

3.6.3.2   The Lifetime of the LT Scheme

Figure 3.12 shows that the LT scheme requires 33 nodes to obtain the 90% TDP. As mentioned in Section 3.5.2, the Markov Chain model of the LT scheme is the same as that of the ERP and the number of copies in the LT scheme does not increase after the RTL expiration. Hence, we can easily calculate from Figure 3.10(b) the time when the number of copies in the network reaches 33. From Figure 3.14, when the RTL is set to 85 [sec], the average number of copies in the network grows to 33. Assuming that every node has battery energy of 40 [EU], and since the average number of copies at 150 [sec] for the LT scheme is 33, the expected lifetime of the LT scheme

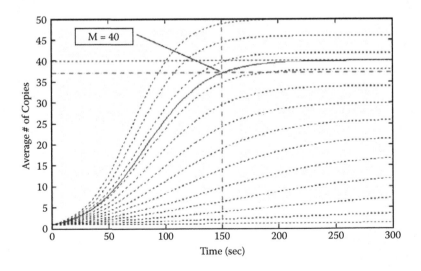

Figure 3.13: Average number of copies for EX scheme with different values of M.

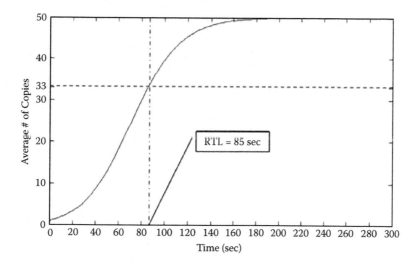

Figure 3.14: Average number of copies for LT scheme.

would be approximately 9091 [sec] (i.e., transmission of $(40 \cdot 50/33) = 60.6$ data packets, with each packet remaining in the network for 150 [sec]).

3.6.3.3 The Lifetime of the LC Scheme

The LC scheme requires 31 nodes for TDP of 90%. Figure 3.15 shows that when the total number of copies is limited to m = 36, the average number of copies in the network grows up to 31 at 150 [sec]. Hence in order for the LC scheme to satisfy the TDP of 90%, the total number of copies should be upper bounded by 36. Assuming that every node has battery energy of 40 [EU], the expected lifetime of the LC scheme would be approximately 9677 [sec] (i.e., transmission of $(40 \cdot 50/31) = 64.5$ data packets, with each packet remaining in the network for 150 [sec]).

3.7 The Residual-Energy (RE) Scheme

The performance of the Epidemic Routing scheme mostly depends on the number of active nodes in the network. To obtain the Ideal Lifetime, the batteries of all the nodes should be depleted at the same time. Of course, without a special mechanism implemented in the network, practically, this would be impossible. However, using residual energy information, it is possible to implement a control algorithm to manage the energy consumption of each node [54]. Such a scheme requires each node to maintain its own residual energy information and transmit the information to the encountered node prior to transmitting a data packet [54]–[57].

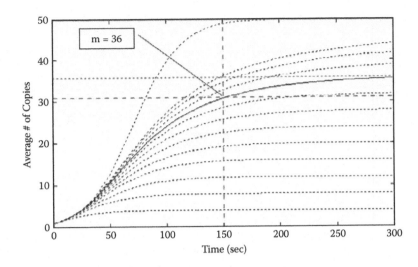

Figure 3.15: Average number of copies for LC scheme with different values of m.

Suppose at the encounter time of two nodes, the nodes share their residual energy information. A prudent thing to do would be to allow the node with larger residual energy to create more copies in future encounters, as compared with the node with less residual energy. Furthermore, the LC scheme is designed to allow controlling the number of copies that a node could spin off in its future encounters. This suggests that the LC scheme, combined with the above energy control mechanism, could result in more uniform energy depletion among the network nodes. This scheme is explored in the next section.

3.7.1   The LC Scheme with Residual-Energy Information

Normally in the LC scheme, when a node transmits a copy to another node, it divides its load (the number of copies that could be created in future encounters) and passes half of the load onto the receiving node. This way the two nodes have the same amount of copies to propagate. However, if the nodes share their residual energy information, they should divide the number of copies in some relation to the residual battery energy. The simplest way to divide the load is to split the copies in proportion to the residual energies of the nodes. We refer to this scheme as the LE scheme.

In order to find the value of the parameter m that limits the total number of copies for the LE scheme, we use the same method that we used for the LC scheme. Hence, for our example, the total number of copies that can be created in the network should be limited to 36 copies. Of course, the Ideal Lifetime of the LE scheme is the same as that of the LC scheme, which is approximately 9677 [sec] (see section 3.7.3.3).

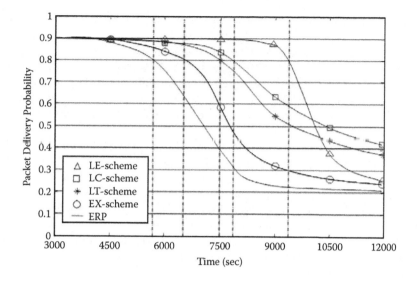

Figure 3.16: MDP lifetime of ERP and RER schemes.

The method that we used to extend the LC scheme to the LE scheme is difficult (although not impossible) to apply to the EX and the LT schemes, since these two schemes cannot naturally control the number of transmissions of nodes in the future. To clarify this further, suppose that we extend the EX scheme or the LT scheme by allowing a node to transmit to another node only when the first node has more residual battery energy than the second node. Then, if the source node has less residual battery energy than any nodes it encounters, the packet can be still propagated with the LE scheme. However, with the EX scheme or the LT scheme, the only opportunity for the packet to reach the destination is for the source to deliver the packet to the destination itself.

Similarly, if we allow a node to transmit based on some probabilistic function of its residual energy, this would reduce the propagation rate, decreasing the packet delivery probability below the TDP. In order to satisfy the TDP, the number of copies would have to be increased, reducing the effectiveness gained by extending the scheme to rely on the residual energy information.

3.8 Maximizing the Lifetime of the Delay Tolerant Networks

3.8.1 The Residual-Battery Information

Figure 3.16 shows the comparison of the 80% MDP lifetime for all the schemes we have discussed in this chapter and for our example network scenario. After

4500 [sec], the packet delivery probability of the ERP scheme starts to decrease and the rate drops below the MDP at approximate time of 5850 [sec]. The EX scheme has the shortest MDP lifetime among the RER schemes, while the LE scheme has the longest MDP lifetime. The LC scheme has a longer MDP lifetime than the LT scheme or the EX scheme, since the LC scheme requires the smaller number of copies. This conserves energy of a single packet routing, allowing more packet routings before the performance drops below the MDP level.

The extension of the LC scheme to the LE scheme by including the residual energy information resulted in a significant MDP lifetime extension. An interesting observation is that the packet delivery probability of the LE scheme decreases rapidly after it drops below the MDP level, at which time the packet delivery probability becomes lower than that of the other schemes. This suggests that for the LE scheme, most of the batteries of the nodes become depleted closer to the MDP lifetime and after this time most of the nodes become inactive. In this sense, as compared with the other schemes, the LE scheme lifetime better resembles the Ideal Lifetime curve.

Table 3.2: Ideal Lifetime and MDP Lifetime for Each Scheme

Scheme	Ideal lifetime(α)	MDP(80%) lifetime (β)	$\alpha - \beta$	$E(n)$	$\sigma(n)$	$\sigma(n)/E(n)$
ERP	6667 s	5850 s	817 s	0.90	1.360	1.511
EX	8108 s	6450 s	1658 s	0.74	1.206	1.630
LT	9091 s	7500 s	1591 s	0.66	1.187	1.798
LC	9677 s	7800 s	1877 s	0.62	1.213	1.956
LE	9677 s	9250 s	427 s	0.62	0.904	1.458

3.8.2   The Comparison of Lifetime Performance

The MDP lifetime for the various schemes are summarized in Table 3.2 and compared with the corresponding expected Ideal Lifetimes. $E(n)$ is the average number of transmissions per node per routing and $\sigma(n)$ is the standard deviation of the number of transmissions per node per routing for packet delivery probability of TDP. The last column represents the coefficient of variation of the number of transmissions per node, calculated as $\sigma(n)/E(n)$.

Indeed, the results in the Table demonstrate again that the LE scheme has the longest lifetime and that it is closer to the Ideal Lifetime than the other schemes. Although all the RER schemes have longer MDP lifetime than the ERP lifetime, the difference between the Ideal Lifetime and the MDP lifetime is larger for the RER schemes, compared with the ERP scheme. The only exception is the LE scheme, where the MDP lifetime is relatively close to the Ideal Lifetime. The results in the Table show that a large value of the coefficient of variation suggests a large difference between the MDP lifetime and the Ideal Lifetime.

3.9   Summary and Concluding Remarks

In this chapter, we have studied a number of variations of the Epidemic Routing Protocol (ERP) in an attempt to extend the lifetime of the network. Our study started by presenting an analytical model for the ERP, followed by the presentation of the ERP variations, for which we coined the name Restricted Epidemic Routing (RER) schemes. We derived the energy-delay tradeoff function (the T-function) and we showed how the T-function could be used for comparison of the RER schemes. It is worth noticing that this comparison approach could be used for other DTN routing schemes as well.

We then presented our definition of the network lifetime, by introducing the Target Delivery Probability (TDP) and the Minimum Delivery Probability (MDP) concepts. Using these concepts, we proposed a new RER scheme, the LE scheme based on the residual battery information, which outperforms the other RER schemes in terms of extending the network lifetime. Indeed, the LE scheme better approximates the Ideal Lifetime profile, as compared with the other RER schemes.

In our quest for a DTN scheme which maximizes the network lifetime, a basic observation that we made was that merely reducing the number of packet copies does not, necessarily, lead to a scheme that maximally extends the network lifetime. Rather, a scheme that minimizes the number of copies and that shifts the production of the copies to the nodes with larger residual energy would be most effective in extending the lifetime of the network. Indeed, this is the principle of the LE scheme proposed here.

Bibliography

[1] P. Jackquet, P. Muhlethaler, T. Clausen, A. Laouiti, A. Qayyum, and L. Viennot, "Optimized Link State Routing Protocol for Ad Hoc Networks," IEEE International Multi-Topic Conference, Lahore, Pakistan, Dec. 28–30, 2001.

[2] Y. Ge, T. Kunz, and L. Lamont, "Quality of Service Routing in Ad Hoc Networks Using OLSR," 36th HICSS, Big Island, Hawaii, Jan. 6–9, 2003.

[3] C. Perkins and E. Royer, "Ad Hoc On-Demand Distance Vector Routing," Proceedings of the 2nd IEEE Workshop on Mobile Computing Systems and Applications, New Orleans, Louisiana, Feb. 1999.

[4] D. Johnson, D. Maltz, and J. Broch, "DSR: The Dynamic Source Routing Protocol for Multi-Hop Wireless Ad Hoc Networks," Ad Hoc Networking, 2001.

[5] Z. J. Haas and M. R. Pearlman, "The Performance of Query Control Schemes for the Zone Routing Protocol," IEEE/ACM Transactions on Networking, Vol. 9, No. 4, August 2001.

[6] Y. Ko and N. Vaidy, "Location-Aided Routing in Mobile Ad Hoc Networks," ACM Wireless Networks Journal, June 2000.

[7] M. Grossglauser and M. Vetterli, "Nodes with EASE: Mobility Diffusion of Last Encounters in Ad Hoc Networks," Proceedings of INFOCOM, San Francisco, California, Mar. 30–Apr. 3, 2003.

[8] H. Dubois-Ferriere, M. Grossglauser, and M. Vetterli, "Age Matters: Efficient Route Discovery in Mobile Ad Hoc Networks Using Encounter Age," Mobihoc, Annapolis, Maryland, June 1–3, 2003.

[9] J. J. Garcia-Luna-Aceves, M. Mosko, and C. Perkins, "A New Approach to On-Demand Loop-Free Routing in Ad Hoc Networks," Proc. 22nd ACM Symp. PODC, Boston, Massachusetts, July 13–16, 2003.

[10] S. Burleigh, A. Hooke, L. Torgerson, K. Fall, V. Cerf, B. Durst, K. Scott, and H. Weiss, "Delay Tolerant Networking: An Approach to Interplanetary Internet," *IEEE Communications Magazine*, Vol. 41, 2003.

[11] P. Juang, H. Oki, Y. Wang, M. Martonosi, L. S. Peh, and D. Rebenstein, "Energy Efficient Computing for Wildlife Tracking: Design Tradeoffs and Early Experiences with ZebraNet," ASPLOS-X, San Jose, California, Oct. 2002.

[12] T. Small and Z. J. Haas, "The Shared Wireless Infostation Model: A New Ad Hoc Networking Paradigm (or Where There is a Whale There is a Way)," MobiHoc, Annapolis, Maryland, June 1–3, 2003.

[13] S. M. Nazrul Alam and Z. J. Haas, "Coverage and Connectivity in Three-Dimensional Underwater Sensor Networks," *Wireless Communication and Mobile Computing (WCMC) Journal*, Vol. 8, Issue 8, October, 2008.

[14] A. Pentland, R. Fletcher, and A. Hasson, "A Road to Universal Broadband Connectivity," 2nd International Conference of Open Collaborative Design for Sustainable Innovation; Development by Design, Dec. 2002.

[15] M. Grossglauser and D. Tse, "Mobility Increases the Capacity of Ad Hoc Wireless Networks," INFOCOM, Anchorage, Alaska, Apr. 22–26, 2001.

[16] A. L. Iacono and C. Rose, "Infostation: New Perspectives on Wireless Data Networks," WINLAB, Technical Document, Rutgers University, 2000.

[17] Q. Li and D. Rus, "Sending messages to mobile users in disconnected ad-hoc wireless networks," ACM MobiCom, The Sixth Annual International Conference on Mobile Computing and Networking, Boston, Massachusetts, Aug. 6–11, 2000.

[18] Z. Zhang, "Routing in Intermittently Connected Mobile Ad Hoc Networks and Delay Tolerant Networks: Overview and Challenges," IEEE Communications Surveys Tutorials, Vol. 8, No. 1, pages 24–37, 2006.

[19] A. Jindal and K. Psounis, "Fundamental Mobility Properties for Realistic Performance Analysis of Intermittently Connected Mobile Networks," PerCom Workshop on Intermittently Connected Mobile Ad Hoc Networks (ICMAN), White Plains, New York, Mar. 2007.

[20] Q. Zheng, X. Hong, P. Wang, L. Tang, and J. Liu, "Delay Management in Delay Tolerant Network," *International Journal of Network Management*, 2008.

[21] G. Karlsson, V. Lenders, and M. May, "Delay-Tolerant Broadcasting," ACM SIGCOMM Workshops, Pisa, Italy, Sept. 11–15, 2006.

[22] S. Jain, K. Fall, and R. Patra, "Routing in a Delay Tolerant Network," Proceedings of ACM SIGCOMM Computer Communication Review, Vol. 34, No. 4, pages 145–158, Portland, Oregan, Aug. 30–Sept. 3, 2004.

[23] S. Merugu, M. Ammar, and E. Zegura, "Routing in Space and Time in Networks with Predicable Mobility," Georgia Institute of Technology, Technical Report, GIT-CC-04-07, 2004.

[24] R. Handorean, C. Gill, and G. Roman, "Accommodating Transient Connectivity in Ad Hoc and Mobile Settings," Pervasive, Linz/Vienna, Austria, Apr. 18–23, 2004.

[25] V. Cerf, S. Burleigh, A. Hooke, L. Torgerson, R. Durst, K. Scott, K. Fall, and H. Weiss, "RFC 4838, Delay-Tolerant Networking Architecture," IRTF DTN Research Group, Apr. 2007.

[26] A. Lindgren, A. Doria, and O. Scheln, "Probabilistic routing in intermittently connected networks," Proceedings of the Fourth ACM International Symposium on Mobile Ad Hoc Networking and Computing, 2003.

[27] J. Burgess, B. Gallagher, D. Jensen, and B. N. Levine, "MaxProp: Routing for vehicle based disruption-tolerant network," Proc. IEEE INFOCOM, Barcelona, Catalunya, Spain, Apr. 23–29, 2006.

[28] J. Ghosh, H. Q. Ngo, and C. Qiao, "Mobility Profile-Based Routing Within Intermittently Connected Mobile Ad Hoc Networks (ICMAN)," International Conference on Wireless Communications and Mobile Computing (IWCMC), Vancouver, Canada, July 3–6, 2006.

[29] Y. Wang and H. Wu, "DFT-MSN: The Delay Fault Tolerant Mobile Sensor Network for Pervasive Information Gathering," Proc. IEEE INFOCOM, Barcelona, Catalunya, Spain, Apr. 23–29, 2006.

[30] A. Balasubramanian, B. N. Levine, and A. Venkataramani, "DTN routing as a resource allocation problem," Proc. ACM SIGCOMM, Kyoto, Japan, Aug. 27–31, 2007.

[31] K. Tan, Q. Zhang, and W. Zhu, "Shortest Path Routing in Partially Connected Ad Hoc Networks," IEEE Globecom, San Francisco, Dec. 1–5, 2003.

[32] E. P. C. Jones, L. Li, J. K. Schmidtke, and P. A. S. Ward, "Practical Routing in Delay Tolerant Networks," IEEE Transactions on Mobile Computing, Aug. 2007.

[33] Y. Gong, Y. Xiong, Q. Zhang, Z. Zhang, W. Wang, and Z. Xu, "Anycast Routing in Delay Tolerant Networks," Technical Report MSR-TR-2006-04, Microsoft Research, Jan. 2006.

[34] R. Shah, S. Roy, S. Jain, and W. Brunette, "Data MULEs: Modeling a Three-Tier Architecture for Sparse Sensor Networks," IEEE SNPA Workshop, Egan Convention Center, Anchorage, Alaska, May 11, 2003.

[35] W. Zhao, M. Ammar, and E. Zegura, "A Message Ferrying Approach for Data Delivery in Sparse Mobile Ad Hoc Networks," MobiHoc, Ropping Hills, Tokyo, Japan, May 24–26, 2004.

[36] W. Zhao, M. Ammar, and E. Zegura, "Controlling the Mobility of Multiple Data Transpor Ferries in a Delay-Tolerant Network," INFOCOM, Miami, Florida, Mar. 13–17, 2005.

[37] Q. Li and D. Rus, "Communicating in Disconnected Ad Hoc Networks Using Message Relay," *Journal of Parallel and Distributed Computing*, 63, 2003.

[38] B. Burns, O. Brock, and B. Levine, "MV Routing and Capacity Building in Disruption Tolerant Networks," IEEE INFOCOM, Miami, Florida, Mar. 13–17, 2005.

[39] A. Vahdat and D. Becker, "Epidemic Routing for Partially Connected Ad Hoc Networks," Technical Report, Duke University, Apr. 2000.

[40] D. J. Daley and J. Gani, *Epidemic Modelling*, Cambridge University Press, 1999.

[41] M. E. J. Newman, "Spread of Epidemic Disease on Networks," *Physical Review E* 66, 016128, 2002.

[42] L. A. Meyers, "Contact Network Epidemiology: Bond Percolation Applied to Infectious Disease Prediction and Control," *Bulletin of the American Mathematical Society*, Vol. 44, Jan. 2007.

[43] M. E. J. Newman, I. Jensen, and R. M. Ziff, "Percolation and epidemics in a two-dimensional small world," *Physical Review E* 65 021904, 2002.

[44] A. Jindal and K. Psounis, "Performance Analysis of Epidemic Routing under Contention," IEEE Workshop on Delay Tolerant Mobile Networks, Vancouver, Canada, July 2006.

[45] D. Nain, N. Petigara, and H. Balakrishnan, "Integrated Routing and Storage for Messaging Applications in Mobile Ad Hoc Networks," Proceeding of WiOpt, Autiplis, France, Mar. 2003.

[46] J. Su, A. Chin, A. Popivanova, A. Goel, and E. de Lara, "User Mobility for Opportunistic Ad-Hoc Networking," 6th IEEE Workshop on Mobile Computing Systems and Applications (WMCSA), English Lake District, UK, Dec. 2–3, 2004.

[47] T. Small and Z. J. Haas, "Resource and Performance Tradeoffs," ACM SIGCOMM, Philadelphia, Pennsylvania, Aug. 22–26, 2005.

[48] Z. J. Haas and T. Small, "A New Networking Model for Biological Applications of Ad Hoc Sensor Networks," IEEE/ACM Transactions on Networking, Vol. 14, No. 1, Feb. 2006.

[49] S. K. Yoon and Z. Haas, "Efficient Tradeoff of Restricted Epidemic Routing in Mobile Ad Hoc Networks," Milcom, Orlando, FL, Oct. 2007.

[50] T. Spyropoulos, K. Psounis, and C. S. Raghavendra, "Spray and Wait: An Efficient Routing Scheme for Intermittently Connected Mobile Networks," ACM SIGCOMM Workshop on Delay Tolerant Networking, Philadelphia, Pennsylvania, Aug. 22–26, 2005.

[51] T. Spyropoulos, K. Psounis, and C. S. Raghavendra, "Spray and Focus: Efficient Mobility-Assisted Routing for Heterogeneous and Correlated Mobility," PerCom Workshop on Intermittently Connected Mobile Ad Hoc Networks (ICMAN), White Plains, New York, Mar. 2007.

[52] K. A. Harras, K. C. Almeroth, and E. M. Belding-Royer, "Delay Tolerant Mobile Networks (DTMNs): Controlled Flooding in Sparse Mobile Networks," Proceedings of IFIP-TC6 Networking Converence, Vol. 3462, pages 1180–1192, Waterloo, Canada, May 2–6, 2005.

[53] Z. J. Haas, J. Y. Halpern, and L. Li, "Gossip-Based Ad Hoc Routing," IEEE INFOCOM, New York, June 23, 2002.

[54] S. K. Yoon and Z. Haas, "Tradeoff between Energy Consumption and Lifetime in Delay Tolerant Mobile Network," Milcom, San Diego, California, Nov. 2008.

[55] J. Chang and L. Tassiulas, "Energy Conserving Routing in Wireless Ad-Hoc Networks," INFOCOM, Conference on Computer Communications, Tel Aviv, Israel, Mar. 2000.

[56] R. C. Shah and J. M. Rabaey, "Energy Aware Routing for Low Energy Ad Hoc Sensor Networks," IEEE Wireless Communications and Networking Conference, Orlando, Florida, Mar. 17-21, 2002.

[57] M. Maleki, K. Dantu, and M. Pedram, "Power-Aware Source Routing Protocol for Mobile Ad Hoc Networks," International Symposium on Low Power Electronics and Design, Monterey, California, Aug. 12–14, 2002.

[58] S. Jain, M. Demmer, R. Patra, and K. Fall, "Using Redundancy to Cope with Failures in a Delay Tolerant Network," ACM SIGCOMM, Philadelphia, Pennsylvania, Aug. 22–26, 2005.

[59] Y. Wang, S. Jain, M. Martonosi, and K. Fall, "Erasure-Coding Based for Opportunistic Networks," Proc. ACM SIGCOMM, Workshop on Delay Tolerant Networking and Related Topics (WDTN-05), Aug. 2005.

[60] Y. Lin, B. Liang, and B. Li, "Performance Modeling of Network Coding in Epidemic Routing," Proc. 1st International MobiSys Workshop on Mobile Opportunistic Networking, San Juan, Puerto Rico, June 11, 2007.

Chapter 4

A Routing-Compatible Credit-Based Incentive Scheme for DTNs

Haojin Zhu, Xiaodong Lin, Rongxing Lu, Yanfei Fan, and Xuemin (Sherman) Shen

4.1 Introduction 102
4.2 Related Work 105
4.3 System Model and Design Goals 106
    4.3.1 Network Model 106
    4.3.2 Data Forwarding Strategy 106
    4.3.3 Rewarding Model 107
    4.3.4 Attack Model 107
    4.3.5 Design Goals 108
4.4 The Proposed SMART Scheme 108
    4.4.1 Pairing Technique 109
    4.4.2 The Overview of SMART 109
        4.4.2.1 Preventing Layer Injection or Nodular Tontine Attack 109
        4.4.2.2 Motivating Nodes to Submit Coins 111
    4.4.3 The SMART Scheme 112
        4.4.3.1 System Initialization 112
        4.4.3.2 Bundle Generation 113
        4.4.3.3 Bundle Forwarding 113
        4.4.3.4 Charging and Rewarding 114
    4.4.4 Efficiency Enhancement 114

    4.4.4.1   Reducing the Transmission and Computation
               Overhead With Aggregate Signature 115
    4.4.4.2   Efficient Fragmentation Authentication with
               Merkle Hash Tree 115
4.5   Performance Evaluation 117
    4.5.1   Cryptographic Overhead Evaluation 117
        4.5.1.1   Communication Overhead 117
        4.5.1.2   Computation Cost 118
    4.5.2   Simulation 118
        4.5.2.1   Simulation Setup 118
        4.5.2.2   Incentive Effectiveness 119
        4.5.2.3   Scenario I: Impact of Traffic Load 120
        4.5.2.4   Scenario II: Impact of Forwarding Copy
                   Number 121
4.6   Open Issues 123
    4.6.1   Public Key Revocation in DTNs 123
    4.6.2   Reputation-Based Incentive Scheme 123
4.7   Conclusion 124

4.1   Introduction

MOST popular Internet applications rely on the existence of a contemporaneous end-to-end link between source and destination, with moderate round-trip times and small packet loss probabilities. This fundamental assumption is not expected in some challenged networks, which are often referred to as Delay/disruption-Tolerant Networks (DTNs). Applications of this emergent communication paradigm are wide ranging and include low-cost Internet service provision in remote or developing localities[1], vehicular DTNs for dissemination of location-independent information (e.g. local ads, traffic reports, parking information, etc.) [2, 3] or for providing multi-hop Internet access [4], social-based networks to allow humans to communicate without network infrastructure [5, 6], pocket switched networks [7], underwater networks [8], etc. Take Figure 4.1(a) as an example. In DTNs, the in-transit messages, also named bundles, can be sent over an existing link and buffered at the next hop (Node B) until the next link (Node C) in the path appears. This message propagation process is usually referred to as the "store-carry-and-forward" strategy, and the routing is made in an "opportunistic" fashion.

Previously reported studies have focused on opportunistic data propagation in DTNs [10, 9, 11, 12, 13], which depends on the hypothesis that each individual node is ready to forward packets for others. This hypothesis, however, might be easily violated in the presence of selfish nodes or even malicious ones, which may choose to save their precious wireless resources by refusing to serve as bundle relays [14]. This can be illustrated by Figure 4.1(b), in which

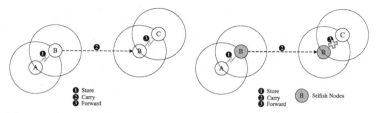

(a) DTN data forwarding without selfish nodes

(b) DTN data forwarding with selfish nodes

Figure 4.1: DTN data forwarding with or without selfish nodes.

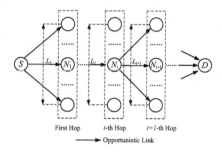

Figure 4.2: A generalized data forwarding strategy.

a selfish node B receives packets from A while it does not forward the packets to the next hop receiver C. Such selfish actions may be more challenging for researchers in certain applications of DTNs such as vehicular DTNs and social networks, which are decentralized and distributed over a multitude of devices that are controlled and operated by individuals. In these applications, it is highly possible that there exist some selfish users who may not want to forward such bundles without compensation. Further, even from the security point of view, naive packet forwarding may open a new door for malicious users, who may intentionally try to launch Denial-of-Service (DoS) attacks on the network by flooding the network with dummy messages. Thus, to deploy an applicable delay tolerant network in real-world scenarios, proper incentives and security mechanisms should be in place.

One of the most promising ways to address the selfishness issue and stimulate cooperation among selfish nodes in DTNs is using *incentive schemes*, which basically fall into two categories: reputation-based or credit-based schemes. Reputation-based schemes rely on individual nodes to monitor neighboring nodes' traffic and keep track of each others' reputation so that uncooperative nodes are eventually detected and excluded from the networks [15, 17, 14, 16], while credit-based schemes introduce some form of virtual currency to regulate the packet-forwarding relationships among different nodes [18, 19, 20]. The previously reported incentive schemes, which were proposed for conventional mobile ad hoc networks, may not be suitable for DTNs, for the following two reasons. Firstly, a common assumption adopted in existing incentive schemes is that a full end-to-end path between source and destination can be determined before data forwarding occurs. This assumption does not hold in DTNs due to its intrinsic opportunistic forwarding nature. Secondly, the reported schemes are designed mainly for single copy forwarding. However, multi-copy forwarding or even flooding is often adopted to enhance the reliability of DTN communication [9], which makes most existing incentive schemes incompatible with diverse DTN routing.

To address the above research issues, we propose a Secure Multi-Layer Credit-based Incentive (SMART) scheme for DTNs afflicted with selfish nodes. Similar to other credit-based incentive schemes, SMART uses credits to provide incentive to selfish nodes. One of its novel and distinguishing features is that SMART allows the credit to be transferred/distributed by the current intermediate node without the involvement of the sender. Such a design is well suited for DTNs since, in DTNs, the bundle sender cannot predict the bundle forwarding path and intermediate nodes may also suffer from delayed or frequent loss of connectivity to the bundle sender.

To be specific, SMART is based on the notion of a *layered coin* that provides virtual electronic credits to charge for and reward the provision of data forwarding in DTNs. Such a coin is comprised of multiple layers, each of which is generated by the source/destination or an intermediate node. The first layer, also named the *base layer*, is generated by the source to indicate payment rate (credit value), remuneration conditions, class of service (CoS) requirement, and other reward policies. During the subsequent bundle propagation process, each intermediate node will generate a new layer based on the previous layers by appending a non-forgeable digital signature. This new layer is also called the *endorsed layer*, which implies that the forwarding node agrees to provide forwarding service under the predefined CoS requirement and will be rewarded according to the reward policy in the future. With endorsed layers, it is easy to track the propagation path and determine each intermediate node by checking the signature of each endorsed layer. In the rewarding and charging phase, if the provided forwarding service satisfies remuneration conditions defined in the predefined reward policy, each forwarding node along one or multiple path(s) will share the credit defined in this coin depending on different data forwarding algorithms (single-copy/multi-copying forwarding) and the actual forwarding results (bundle delivered along one or multiple paths).

However, the main challenge in designing the SMART is to ensure that the security properties of the scheme are not compromised. Since all security related to a coin, especially during the store-carry-and-forward process, is managed by the intermediate nodes, a selfish node (or even a group of colluding nodes) may attempt to cheat the system to maximize its expected welfare. As an example, a selfish node may arbitrarily inject a fake layer into the current coin or remove several valid layers from it, if such actions can maximize its welfare. This is the security perspective of SMART. Secondly, any security functionality will incur extra computation and transmission overhead. A secure credit-based incentive scheme should be efficient enough to not significantly compromise the system performance. This is the performance perspective of our system.

In this chapter, firstly, we propose a secure multi-layer credit-based incentive scheme to stimulate cooperation among selfish nodes in DTNs. The proposed scheme can be made compatible with diverse data-forwarding algorithms in DTNs. Secondly, SMART can withstand a wide range of cheating actions because of its novel *layer concatenation* technique. Thirdly, we propose two performance-optimization techniques to minimize the computation and transmission overhead. Further, SMART is a one-way, non-interactive protocol, which is particularly suitable for DTNs, where interactive communication suffers from long round-trip delays and frequent disconnection [1]. Lastly, extensive simulations are conducted to demonstrate the efficiency and effectiveness of the proposed SMART scheme.

4.2   Related Work

There is a large amount of literature on incentive mechanisms for different kinds of networks. These reported mechanisms basically fall into two categories: reputation-based schemes and credit-based schemes.

Reputation-based schemes rely on the individual nodes to monitor neighboring nodes' traffic and keep track of each others' reputation so that uncooperative nodes can eventually be detected and excluded from the networks [15, 17, 14, 16]. On the other hand, credit-based incentive schemes introduce some form of virtual currency to regulate the packet-forwarding relationships among different nodes. There are two different ways to realize such kinds of credits: game theory-based schemes and security protocol-based schemes. The first approach tries to investigate such non-cooperative communication scenarios within a game theory framework [24, 25, 21] while the second approach focuses on ensuring the security of the credits by using various cryptographic tools [18, 20]. Most of these schemes always assume that an end-to-end path exists and is determined before the data forwarding process. However, this assumption obviously does not hold in DTNs, which makes them not suitable in DTNs. In [19], a virtual cash-based incentive scheme is proposed to stimulate commercial advertisement dissemination in vehicular networks. In [6], it is suggested to use a multilevel coupon-based scheme to stimulate exchanging information about places of interest or local restaurants. However, in both schemes, the focus is on how to stimulate the advertisement dissemination and transmission is based on simple broadcasting while the DTN routing is not taken into consideration.

We incorporate a secure credit-based incentive scheme into the DTN data routing/forwarding, which distinguishes SMART from previous work. The existing routing or data forwarding schemes in DTN can be categorized into single copy schemes and multi-copy schemes. Some protocols (e.g., First Contact [11] and Direct Transmission/Delivery [10]) generate just a single copy, others enable the source to limit the forwarding copies to a fixed number [9], while epidemic [12] and probabilistic routing [13] potentially create an "infinite" number of messages. A recent study shows that even though single-copy schemes can considerably reduce resource waste, they are often orders of magnitude slower than multi-copy algorithms and are inherently less reliable [9]. Therefore, in this study, we consider a generalized multi-copy data forwarding scheme as the foundation, and therefore our SMART can be made compatible with diverse multi-copy data forwarding schemes.

4.3 System Model and Design Goals

This section describes our system model and design goals.

4.3.1 Network Model

We consider a general delay tolerant network formed by a set of mobile devices owned by individual users. Each node i is assumed to have a unique non-zero identifier $\mathcal{N}_i$, which is bound to a specific public key certificate. We use node i or $\mathcal{N}_i$ interchangeably hereafter. We also assume that each node has limited transmission and reception capabilities so that two nodes outside the transmission range of each other can communicate only via a sequence of intermediate nodes in a multi-hop manner. End-to-end connections are not always guaranteed, and routing, therefore, is made in an "opportunistic" way. Similar to other credit-based schemes such as [24, 19], we assume that there exists in our scheme an *Offline Security Manager (OSM)*, which is responsible for key distribution, and a *virtual bank* (VB), which takes charge of credit clearance. In many DTN application scenarios, there exist some special network components which can serve as the VB, such as roadside unit (RSU) in vehicular DTNs [19] and information publisher in social networks [6]. The DTN nodes can exploit opportunistic links to these network components to submit collected coins to the VB. Before joining the DTN network, every DTN node should be registered with the OSM and obtain its public key certificate. At the clearance phase, the DTN nodes submit the collected layered coins to the VB for receiving their rewards.

4.3.2 Data Forwarding Strategy

We consider a generalized multi-copy data forwarding architecture: as shown in Figure 4.2, for every bundle, B, originating from the source node, $\mathcal{S}$, L_1 copies of B are initially spread by the source and, then, at every subsequent forwarding node $\mathcal{N}_i$, L_i message copies will be opportunistically propagated to the next hops. It is worth pointing out that existing DTN routing schemes can be treated as special cases of this routing model. For a single copy-based forwarding scheme [11, 10], we can choose $\{L_i = 1 | i = 1, 2, \ldots, m\}$, where m is the total hop number of this forwarding path. For epidemic and probabilistic routing [12, 13], $\{L_i | i = 1, 2, \ldots, m\}$ a specific large number can be chosen. On the other hand, if a spray and wait routing scheme is chosen as the basic data forwarding scheme [9], we can assume $\{L_1 = L, L_i = 1 | i = 2, \ldots, m\}$, where L is the the chosen forwarding copy number.

4.3.3   Rewarding Model

There are several available rewarding models which can be adopted in SMART. For example, a popular charging method in [18] is paying per packet, which means for each successfully transmitted unit-sized packet, each of N intermediate nodes should receive λ credits, while the source needs to pay $\lambda * N$ in total. However, we argue that this method is not suitable for opportunistic data forwarding in that it is difficult for the source to predict how many copies or hops are needed to successfully deliver a message to the destination. Therefore, in this study, we consider a profit sharing model, which means that the intermediate nodes involved in a successfully bundle delivery will be paid with a dividend of the total credit provided by the source node. The source node can also specify a diverse, case-by-case basis rewarding requirements in the base layer of a layered coin, which can be regarded as a part of DTN routing policy [26]. For example, a bundle should be successfully delivered within a particular TTL (Time-to-live) period or only the intermediate nodes along the first successfully delivery path can be remunerated. The study on rewarding policy is still an open problem and therefore deserves more investigation in future incentive related research.

4.3.4   Attack Model

Due to their selfish nature, mobile nodes will try to cheat the system to maximize their welfare. In particular, a selfish node can exhibit one of the three following selfish actions:

1. **Credit Forgery Attack (or Layer Injection Attack:)** A selfish node may attempt to forge a valid credit (e.g. collude with other nodes to inject non-existent layers into a valid layered coin) to reward itself for work it did not do or for more than it has done.

2. **Nodular Tontine Attack (or Layer Removal Attack:)** Unlike in a layer injection attack, when receiving a multi-level credit, a selfish node may try to remove one or several existing layers which have been generated by the previous forwarding nodes. This attack is particularly effective in profit-sharing systems, where the profits of the removed nodes will be shared by the remaining nodes. In this sense, it is similar to a

tontine system,[1] in which participants share a common fund and have been known to try to kill each other off, thereby increasing their shares. Therefore, we denote this kind of attack as a nodular tontine attack.

3. **Submission Refusal Attack:** In DTNs, due to the lack of end-to-end connection, a source node as well as other intermediate nodes may not have a clear idea about the forwarding progress and thus it relies on the last forwarding node to submit the generated layered coins to a VB for clearance. However, if colluding with the source node, the last intermediate node may refuse to submit the received credits and receive behind-the-scenes compensation from the source node.

Note that any of the selfish actions above can be further complicated by the collusion of two or more nodes. However, we only consider each selfish node with a unique identity as well as a corresponding public key certificate. Similar to [25], in this work, we consider a general DTN network and thus assume that no "extra communications" exist among the DTN nodes.

4.3.5   Design Goals

The design goals include

- *Effectiveness*: The proposed scheme should be effective in stimulating cooperation among the selfish nodes.

- *Security*: It should be secure and robust from various attacks.

- *Efficiency*: It should work efficiently without introducing much extra communication and transmission overhead.

- *Generality*: It should be compatible with the most popular DTN routing schemes.

4.4   The Proposed SMART Scheme

In this section, we first provide some preliminary background, which is the design foundation of SMART. Then we give an overview of the SMART scheme, followed by a detailed presentation of SMART. Finally, we introduce two efficient performance enhancement methods.

---

[1]The tontine, named after Lorenzo Tonti, a Neapolitan banker who started such a scheme in 1653, refers to a system in which each investor pays a sum into the fund, and in return receives dividends from the capital invested; as each person dies his share is divided among all the others until only one is left, reaping all the benefits.

4.4.1   Pairing Technique

SMART is based on bilinear pairing, which is briefly introduced below. Let $\mathbb{G}$ be a cyclic additive group and $\mathbb{G}_T$ be a cyclic multiplicative group of the same order, q, i.e., $|\mathbb{G}| = |\mathbb{G}_T| = q$. Let P be a generator of $\mathbb{G}$. We further assume that $\hat{e} : \mathbb{G} \times \mathbb{G} \rightarrow \mathbb{G}_T$ is an efficient admissible bilinear map with the following properties:

- Bilinear: for $a, b \in \mathbb{Z}_q^*$, $\hat{e}(aP, bP) = \hat{e}(P, P)^{ab}$.

- Non-degenerate: $\hat{e}(P, P) \neq 1_{\mathbb{G}_T}$.

- Computable: there is an efficient algorithm to compute $\hat{e}(P_1, Q_1)$ for any $P_1, Q_1 \in \mathbb{G}$.

According to [27], such an admissible bilinear map $\hat{e}$ can be constructed by Weil or Tate pairings on the elliptic curves.

4.4.2   The Overview of SMART

Before presenting our SMART scheme, we first introduce a naive multi-layer coin scheme. In such a naive scheme, the data forwarding process can also be regarded as a layered coin generation process. When a node sends its own messages, the node will lose credit (or virtual money) to the network because other nodes incur a cost to forward the messages. The bundle sender first generates the base layer of a layered coin and then sends it together with the original bundles to a certain number of downlink nodes. At each subsequent hop, each intermediate node generates a new endorsed layer based on the previous layered coin. It is obvious that with layered coins each hop of a successful data-forwarding process can be easily tracked. After that, each intermediate node periodically submits its collected layered coins to the VB, which can calculate credits for each intermediate node and also make a charge on the bundle senders. Note that, since only the nodes on the successful delivery path are rewarded, each intermediate node can launch different kinds of attacks on this naive system. In the following sections, we progressively determine what SMART needs in order to prevent the various attacks.

4.4.2.1   Preventing Layer Injection or Nodular Tontine Attack

Layer injection attack and nodular tontine attack are two ways to cheat the SMART. In a layer injection attack, several nodes may collude with each other to cheat extra credits. We assume that the total number of nodes along the successful delivery paths is m, and the source node is going to reward these m nodes with α credits. Each node is to receive α/m credits. However, if a malicious node colludes with other n nodes to launch an layer injection attack, the colluding group will receive $\alpha * ((n + 1)/(n + m) - 1/m)$ extra credits. On the other hand, in a nodular tontine attack, an intermediate node tries to

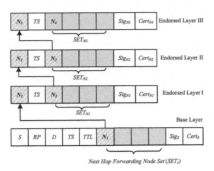

Figure 4.3: An example of layered coin for a single forwarding path.

obtain extra credits by removing the endorsed layers generated by previous intermediate nodes. When a misbehaving node removes n layers from the original layered coin, it can make an extra profit of $\alpha * (1/(m-n) - 1/m)$.

The main reason behind Layer Injection and Nodular Tontine Attacks is that the naive multi-layer incentive scheme lacks any integrity protection mechanism to prevent the misbehaving nodes from arbitrarily injecting or removing layers. To thwart these attacks and ensure the security of layered coins, we introduce a *layer concatenation* technique, which tries to concatenate different layers with each other by injecting the generator information of the next layer into the pervious layer. The basic idea of layer concatenation can be seen in Figure 4.3. Starting from the source node, each node stores identification information about the next forwarding node set (SET), which includes all the next-hop forwarders, in its layer. For example, in Figure 4.3, the identity of first intermediate node $\mathcal{N}_1$ is embedded in the base layer. This design disallows any subsequent forwarding nodes from removing endorsed layer I and its generator $\mathcal{N}_1$ from the layered coin since any attacker has to forge a new, non-$\mathcal{N}_1$-included base layer to replace the current one though this cannot be achieved without the private key of the bundle sender. Similarly, the second intermediate node $\mathcal{N}_2$ is also defined in the endorsed layer generated by $\mathcal{N}_1$. Such a process will continue until the last endorsed layer generated by the destination. It is obvious that, with this layer concatenation technique, the different layers can form a linkable layer chain. Each following node can easily detect the layer injection or nodular tontine attacks by checking the linkability of this layer chain.

From Figure 4.3, we can further describe the components of a layered coin. A layered coin is composed of a *base layer* and multiple *endorsed layers*. A base layer is comprised of S and $Cert_S$, which are the identity and public key certificate of the source node, respectively; RP, which refers to the CoS requirements and rewarding policy proposed by the source, D, which is the identity of the destination node; TS and TTL, which refer to the bundle creation timestamp and time-to-live information, respectively; the *forwarding node set (SET)*, which includes all the possible forwarding nodes in the next hop and *Sig*, which is the signature generated by the source node to protect the authenticity and integrity of the above mentioned information. Similar to the base layer, an endorsed layer includes node identity, TS, SET, and a supporting signature.

4.4.2.2 Motivating Nodes to Submit Coins

We consider a countermeasure to the third type of selfish actions. As we discussed before, due to lack of end-to-end connections in DTNs, SMART requires that the intermediate nodes opportunistically submit layered coins for clearance. However, the last intermediate node, the one that determines if a full linkable layer chain can be established, may collude with the sender to attack this system. In particular, if the last intermediate node does not submit the layered coin to the VB and loses the α/m credit, the sender can save α credit. In particular, if the sender gives the last intermediate node a behind-the-scenes compensation of $\alpha/m + \epsilon$, where $\epsilon > 0$, the last node will be better off while the sender still enjoys a net gain of $\alpha * (1 - 1/m) - \epsilon$. However, the other nodes except the sender and the last intermediate node will receive nothing, which may lead to a serious fairness issue.

In order to prevent this cheating action, we propose two strategies to discourage the bundle sender from colluding with the last forwarding node. The first is a *charge-model*-based solution [24]. For every forwarding request, SMART requires that the VB charges the sender an extra amount of credit, α, even if the last intermediate node does not submit the layered coins for clearance. This extra charge is reasonable since, even though it seems no successful delivery path exists, this data forwarding still incurs forwarding costs to all the forwarding nodes involved. This extra charge goes to the VB, which either keeps it or returns the credit back to the involved forwarding nodes uniformly. Given such extra charges, even a colluding group cannot benefit from this cheating action.

SMART can also reduce the risk of the submission refusal attack by using multiple copy forwarding. We assume that the source node colludes with n_a forwarding nodes to launch a submission refusal attack. Let n_c denote the number of copies transmitted for each message, and d refer to the average number of one-hop neighbors of a DTN node. To maximize the attacking effect, we consider that all of the colluding nodes are located in the destination's transmission range. Given this setting, the probability of successfully launching a submission refusal attack can be defined as the probability that every successful delivery path is controlled by the colluding nodes. In other words, the probability of successfully defending a submission refusal attack (or SR rate) is the probability that at least one forwarding path does not involve any

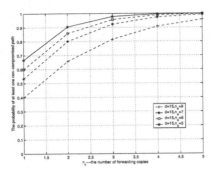

Figure 4.4: The probability that there exists at least one non-compromised path under different n_c.

colluding nodes. SR can be computed with the following equation:

$$FI = \begin{cases} 1 - \prod_{i=1}^{n_c} \frac{n_a-i+1}{d-i+1}, & \text{if } n_c \leq n_a \\ 1, & \text{if } n_c > n_a \end{cases} \qquad (4.1)$$

Figure 4.4 shows the SR under different n_c, d, and n_a values. We notice that SR grows very quickly when n_c increases. For example, when $d = 15$, $n_a = 9$, and $n_c = 5$, SR is approximately 95.8%. Therefore, depending on the level of security required and the potential number of colluding nodes in the network, OSM can find an optimal n_c that achieves a good balance between security and efficiency.

4.4.3   The SMART Scheme

In this subsection, we present the details of the SMART scheme, which includes "System Initialization," "Bundle Generation," "Bundle Forwarding," and "Charging and Rewarding" steps.

4.4.3.1   System Initialization

OSM adopts bilinear pairing system parameters $(q, \mathbb{G}, \mathbb{G}_T, \hat{e}, P)$ as the system parameters. In addition, two hash functions are formed: $H : \{0,1\}^* \rightarrow \{0,1\}^*$ and $H_2 : \{0,1\}^* \rightarrow \mathbb{G}$. The system parameters $(q, \mathbb{G}, \mathbb{G}_T, \hat{e}, P, H, H_2)$ will be preloaded in every DTN node. Each node, $\mathcal{N}$, randomly chooses $sk_\mathcal{N} \in \mathbb{Z}_q^*$ as its private key, which corresponds to the public key expressed as $PK_\mathcal{N} = sk_\mathcal{N} P$. Then it contacts the OSM to obtain its corresponding public key certificate.

4.4.3.2   Bundle Generation

When a bundle sender, $\mathcal{S}$, is going to send a bundle B to the destination $\mathcal{D}$, after determining the next hop forwarding node set SET_S, $\mathcal{S}$ signs on the bundles with its private keys sk_S by computing $Sig_S \leftarrow sk_S H_2(B||S||RP||D||TS||TTL||SET_S)$. Here, we use the Boneh, Lynn, and Shacham (BLS) signature [28] as the underlying building block to generate the supporting signature. Thus, $\mathcal{S}$ obtains the base layer as $B\_layer = (\mathcal{S}, RP, D, TS, TTL, SET_s, Sig_S, Cert_S)$. Then $\mathcal{S}$ forwards the bundle as well as the base layer to the next forwarding nodes as follows:

$$\mathcal{S} \rightarrow SET_S : B, B\_Layer$$

Note that, in a multi-copy opportunistic data forwarding algorithm, a bundle may be forwarded along with multiple paths. Each forwarding path may form its layered coin even though the generated coins share the same base layer. Without loss of generality, in the following section, we take a single forwarding path $\mathcal{S} \rightarrow \mathcal{N}_1 \rightarrow \mathcal{N}_2 \rightarrow \ldots \mathcal{N}_i \ldots \rightarrow \mathcal{N}_m \rightarrow \mathcal{D}$ as an example to show the details of the basic SMART scheme, where $\mathcal{N}_m$ represents the last intermediate node.

4.4.3.3   Bundle Forwarding

When an intermediate node, $\mathcal{N}_i$, receives the bundle as well as the layered coin which includes a base layer and multiple endorsed layers, it performs the following steps to authenticate the layered coin:

1. Check if the bundle is in their lifetime.

2. Check the linkability of the layer chains.

3. Verify the sender's certificate and check the supporting signature of the base layer by verifying if $\hat{e}(P, sig_s) = \hat{e}(PK_S, H_2(B||S||RP||D||TS|| TTL||SET_S))$ holds.

4. Verify the intermediate nodes' certificates and check the endorsed layers one by one.

After performing the above verifications and determining the next hop forwarding node set, $SET_{\mathcal{N}_i}$, $\mathcal{N}_i$ creates an additional endorsed layer by computing $Sig_{\mathcal{N}_i} \leftarrow sk_{\mathcal{N}_i} H_2(B||B\_Layer||\mathcal{N}_i||TS||SET_{\mathcal{N}_i})$ and thus obtains the i-th endorsed layer, $E\_Layer_i = (\mathcal{N}_i, TS, SET_{\mathcal{N}_i}, Sig_{\mathcal{N}_i}, Cert_{\mathcal{N}_i})$. Then $\mathcal{N}_i$ forwards the bundle as well as the layered coin to the next forwarding node set as follows:

$$\mathcal{N}_i \rightarrow SET_{\mathcal{N}_i} : B, B\_Layer, E\_Layer_1, \ldots, E\_Layer_i$$

The verification of the supporting signature of i-th endorsed layer is performed by computing if $\hat{e}(P, Sig_{\mathcal{N}_i}) = \hat{e}(\mathcal{N}_i, H_2(B||B\_Layer||\mathcal{N}_i||TS||SET_{\mathcal{N}_i})$ holds.

Similar steps are also taken by each intermediate node before the bundles reach the destination, $\mathcal{D}$. When the destination receives the bundles, it may also check the bundles' lifetime, senders and forwarders' certificates, and the layered coins one by one. If the verification passes, it may generate a special endorsed layer as the receipt: $Sig_{\mathcal{D}} \leftarrow sk_{\mathcal{D}}H_2(B||B\_Layer||\mathcal{D}||TS)$. Thus, it obtains the endorsed layer $E\_Layer_{\mathcal{D}} = (\mathcal{D}, TS, Sig_{\mathcal{D}})$. Then $\mathcal{D}$ sends it to $\mathcal{N}_m$ as follows:

$$\mathcal{D} \rightarrow \mathcal{N}_m : B, E\_Layer_{\mathcal{D}}$$

Thus, the last intermediate node obtains a complete layered coin $B, B\_Layer, E\_Layer_1, \ldots, E\_Layer_i, \ldots, E\_Layer_m, E\_Layer_{\mathcal{D}}$, which will be submitted to the virtue bank for clearance in the future.

4.4.3.4 Charging and Rewarding

After a batch of a given size of layered coins is gathered, the last intermediate node may connect to the VB and submit the collected layered coins for clearance. After receiving the submitted layered coins, the VB first checks the certificates of each node in the forwarding path and then verifies the legitimacy of the layered coins. The VB also checks whether these layered coins have been deposited before by inquiring as to the sender's previous record. If all verifications pass, a predefined amount of the credit will be shared by all of the forwarders under a particular predefined rewarding policy.

The credit calculation should take bundle fragmentation into consideration. In DTNs, when a message is large, it may not be possible to send the entire message at once. One possible solution is to split the message into smaller pieces and let each become its own bundle, or "fragment bundle," and send some pieces of a large message through the current link and the rest of the message through another link later to make the best use of limited resources. Bundle fragmentation is regarded as a unique characteristic of DTN forwarding [29] and a recent study shows that the fragment size may follow a certain distribution in practice [30]. As the general discussion on credit calculation, we assume that there are n intermediate nodes participating in a successful bundle forwarding and each node $\mathcal{N}_i | 1 \leq i \leq n$ forwards δ_i percentage of fragments, where $0 < \delta_i \leq 1$. Then, the node $\mathcal{N}_i$ will receive $Cred_{\mathcal{N}_i} = \alpha * \delta_i / \Sigma_{j=1}^n \delta_j$ credits, where α is the total credits provided by the bundle sender.

4.4.4 Efficiency Enhancement

In this subsection, we propose two methods to further improve the computation and transmission efficiency of the SMART scheme.

4.4.4.1 Reducing the Transmission and Computation Overhead With Aggregate Signature

Signature transmission and verification contribute to most of the transmission and computation overhead incurred by SMART transmission and verification. Therefore, reducing the signature size and increasing the verification efficiency is a major concern in the practical deployment of the SMART scheme. Here, we take advantage of the aggregated signature to reduce the transmission and verification cost.

An aggregate signature is a digital signature that supports aggregation of n distinct signatures issued by n distinct signers to a single short signature [28]. This single signature (and the n original messages) will convince the verifier that the n signers indeed sign the n original messages. With aggregate signature, it is possible for the intermediate nodes to aggregate the received layered coins into a short one.

Step 1: Layered Coin Aggregation. Let an intermediate node $\mathcal{N}_m$ receive a layered coin which is constituted with a base layer $B\_layer = (S, RP, D, TS, TTL, SET_s, Sig_S, Cert_S)$ and multiple endorsed layer $E\_Layer_i = (\mathcal{N}_i, TS, SET_{\mathcal{N}_i}, Sig_{\mathcal{N}_i}, Cert_{\mathcal{N}_i})|1 \le i \le m-1$, where $S \to \mathcal{N}_1 \ldots \to \mathcal{N}_i \ldots \to \mathcal{N}_m$ is the current forwarding path. For the simplicity of presentation, we assume that $M_0 = B||S||RP||D||TS||TTL||SET_S$ and $M_i = B||B\_Layer||\mathcal{N}_i||TS||SET_{\mathcal{N}_i}$, where $1 \le i \le m-1$. Thus, the layered coin signatures can be represented as $Sig_S \leftarrow sk_S H_2(M_0)$ and $\{Sig_{\mathcal{N}_i} \leftarrow sk_{\mathcal{N}_i} H_2(M_i)|1 \le i \le m-1\}$. To aggregate the layered coin, node $\mathcal{N}_m$ can compute and obtain the aggregate signature: $Sig_{agg} \leftarrow Sig_S \prod_{i=1}^{m-1} Sig_{\mathcal{N}_i}$. In the subsequent bundle forwarding process, node $\mathcal{N}_m$ could transmit aggregate signature Sig_{agg} rather than transmit the signatures one by one. Therefore, the transmission overhead can be reduced.

Step 2: Layered Coin Batch Verification. Given the aggregate signature Sig_{agg}, the message M_0 and $\{M_i|1 \le i \le m-1\}$ on which it is based, and public keys PK_S and $\{PK_{\mathcal{N}_i}|1 \le i \le m-1\}$, node $\mathcal{N}_m$ can verify the aggregate signature by checking if $\hat{e}(Sig_{agg}, P) = \hat{e}(PK_S, H_2(M_0)) \prod_{i=1}^{m-1} \hat{e}(PK_{\mathcal{N}_i}, H_2(M_i))$.

It is observed that the computation cost that the intermediate node spends on verifying m signatures is reduced from $2m$ pairing operations to $m+1$ pairing operation, where pairing operation is the most computationally expensive operation in the SMART scheme. Thus, this batch verification can dramatically reduce the verification delay, particularly when verifying a large number of layered coins.

4.4.4.2 Efficient Fragmentation Authentication with Merkle Hash Tree

To support layered coin-based fragment authentication in SMART, one possible way is to make each fragment self-authenticating by attaching a layered coin to the end of each fragment separately. However, this approach may lead to a more serious performance issue since the intermediate nodes have to spend more computational efforts on verifying a growing number of signatures.

The Merkle tree [31] (also called binary hash tree) is a complete binary tree equipped with a function hash and an assignment Ω, which maps a set of nodes to a set of fixed-size strings. In a Merkle tree, the leaves of the tree contain the data, and the value of an internal tree node is the hash value of the concatenation of the values of its two children. Merkle tress have been applied in DTNs to realize efficient bundle authentication [33]. Here, we extend it to support efficient implementation of a credit-based incentive scheme, or a Merkle Hash Tree-based SMART scheme (MHK-SMART).

Building Merkle Tree: To build a Merkle tree for our problem, the sender constructs N leaves $\{\Omega_i = H(F_i)|i = 1,\ldots,m\}$ with each leaf corresponding to a fragment bundle, where $\{F_i, |i = 1,\ldots,m\}$ refer to m fragments. The bundle sender then builds a complete Merkle tree with these leaves. The Ω value of each node is defined as the following:

$$\Omega(V) = H(\Omega(V_{left})||\Omega(V_{right})))$$

where we use V to denote an internal tree node, and V_{left} and V_{right} to denote V's two children. Figure 4.5 shows an example of constructing such a Merkle tree. To add a credit-based incentive scheme to these bundles, the bundle sender only needs to generate a layered coin based on the root of the Merkle tree, which replaces the original bundle as the signed message.

Fragment Authentication with Merkle Tree-Based Incentive Scheme: To authenticate a particular fragment such as F_1, the intermediate node needs the set of hash value $\Omega_2, \Omega(B), \Omega(D)$ and the base layer which is a signature on the root $\Omega(E)$. The verifier can calculate each hash in the path from F_1 leaf node to the root node, and finally check the validity of the layered coin. Note that to verify m fragments, it only performs one signature verification operation instead of verifying m signatures in total.

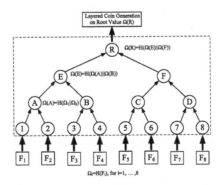

Figure 4.5: An example of Merkle tree building.

4.5   Performance Evaluation

In this section, we evaluate the performance of SMART from several aspects. Our evaluation starts with the cryptographic cost evaluation, which summarizes the computation and transmission cost incurred by the cryptographic operations in a SMART scheme. Then, by considering cryptographic cost as the system parameter, we further demonstrate the effectiveness and efficiency of SMART in stimulating selfish nodes with extensive simulations. The evaluated schemes include the basic SMART, Agg-SMART and MKH-SMART. Note that Agg-SMART and MKH-SMART can be jointly considered in simulation as Optimized SMART.

4.5.1   Cryptographic Overhead Evaluation

4.5.1.1   Communication Overhead

One of the major advantages of SMART is the reduction of transmission cost. It is observed that the communication cost of layered coin is dominated by the size of supporting signatures generated by the intermediate nodes. To ensure the security of the protocol, the elements in $\mathbb{G}$ could be up to 160 bits. We summarize the approximated length of components of a layered coin in SMART as shown in Table 4.1. Note that L refers to the number of copies adopted in the bundle forwarding scheme. In the following performance analysis section, we take $L = 4$ as an example.

Table 4.1: The Size of Each Component of Layered Coin (Bytes)

Base	Comp	$\mathcal{S}$	RP	D	TS		Total Size	
	Size	4	10	4	4			
Layer	Comp	TTL	SET	Sig	Cert		66+4L	
	Size	4	4L	20	20			
Endorsed	Comp	$\mathcal{N}_i$	TS	SET	Sig	Cert	Total Size	
Layer	Size	4	4	4L	20	20	48+4L	

For m layered coins corresponding to m bundle fragments, each of which is accompanied with n endorsed coins, the total size of the layered coins (including both of the base layers and endorsed layers) without aggregation should be $82m + 62mn$. However, in our Agg-SMART scheme, the total size can be reduced to $82m + (42n + 20)m$ by taking advantage of aggregation signature. Under the same parameter, if every k fragments can be rebuilt with a Merkle hash tree, the total size of MKH-SMART can be further reduced to $82m/k + (42n + 20)m/k$. In other words, after adopting two optimization methods the transmission overhead of a basic SMART will be reduced from $82m + 62mn$ to $82m/k + (42n + 20)m/k$.

4.5.1.2   Computation Cost

The computation costs are measured by the most expensive pairing (Pair) and point multiplication (Pmul) operation. In the basic SMART scheme, a Pmul operation is involved for each base layer or endorsed layer generation while two pairing operations are necessary for verification. To investigate the performance of the proposed SMART scheme, we first study the time for (Pmul) operation and Pair operation. We evaluate the delay of cryptographic operations on an Intel Pentium 4 3.0 GHz machine with 1 GB RAM running Fedora Core 4 based on cryptographic library MIRACL [32], which is shown in Table 4.2.

Table 4.2: Cryptographic Operations Execution Time

	Descriptions	Execution Time
T_{pmul}:	The time for one point multiplication in G	0.86 ms
T_{pair}:	The time for a pairing operation	4.14 ms

Here, we focus on the cost of verifying operation in SMART since the verification operation will be operated at each hop. Based on the execution time results, we have the verification cost for the nth intermediate node in the basic SMART as $T_{\mathrm{SCI}} = 2 * mn * T_{\mathrm{pair}}$, where m and n refer to the number of fragments. In the Agg-SMART scheme, by using aggregate signature and batch verification technique, the verification cost can be reduced to $T_{\mathrm{agg-SCI}} = m * (n + 1)(T_{\mathrm{pair}} + T_{\mathrm{pmul}})$. The verification cost can be further reduced in the MKH-SMART scheme. Given every k fragment can be rebuilt with a Merkle hash tree, the total verification cost of MKH-SMART can be further reduced to $T_{\mathrm{MKH-SCI}} = m/k * (n + 1)(T_{\mathrm{pair}} + T_{\mathrm{pmul}})$.

After determining the cryptographic overhead, in the following sections, we will evaluate the performance of SMART by implementing SMART and optimized SMART on a specific DTN routing protocol.

4.5.2   Simulation

In this section, we evaluate the performance of SMART by simulations.

4.5.2.1   Simulation Setup

We implement our SMART scheme on a publicly available DTN simulator *Opportunistic Networking Environment (ONE) simulator* [34] and evaluate its performance under a practical application scenario: vehicular DTNs. We run our simulation with 250 vehicles uniformly deployed in an area of 4000 by 4000 meters.The average speed of vehicles varies from 10 km/h $\sim$ 50 km/h (or 2.7 m/s $\sim$ 13.9 m/s) and the transmission coverage of cars is 300 m. The map adopted in the study is extracted from a real city map, which makes the model realistic. Each vehicle is first randomly scattered on one position of the roads and moves towards another randomly selected position along the paths in the map. The details of our simulation parameters are summarized in Table 4.3.

Table 4.3: Simulation Parameters

Parameter	Value Range
Duration	12 hrs
Number of nodes	250 nodes
Speed of nodes	10 km/h ~ 50 km/h
Transmission coverage	300m
Mobility Model	Map-based mobility model
Message size	5 m
Fragmentation size	10k ~ 100k
Message generation interval	5s ~ 45s
Routing Protocol	Spray and Wait routing protocol
Number of forwarding copies	1 ~ 32 copies

Based on these parameters, we implement our SMART on top of a typical multi-copy DTN routing protocol, Spray and Wait routing (SW) protocol, the effectiveness and efficiency of which has been demonstrated in [9]. Generally speaking, spray and wait is available in the normal (non-binary) and the binary variants. In this simulation, we choose binary spray and wait (SWB) as a basic routing protocol. However, it is important to point out that a SMART scheme can also be applied to other routing schemes if we choose a corresponding forwarding copy number for each forwarding hop.

4.5.2.2 Incentive Effectiveness

We start our evaluation by observing the incentive effectiveness of the SMART scheme. We define two kinds of selfish scenarios: individual selfishness and mass selfishness. In an individual selfishness case, only a small number of selfish nodes may not be willing to forward packets for others, even though they still expect others to forward packets on their behalf. On the other hand, mass selfishness can be defined such that every node has an "intrinsic" selfish nature so that it may probabilistically drop a certain percentage of messages instead of forwarding them. The incentive effectiveness can be measured by the message delivery probability which is shown in Figure 4.6. For the mass self-ishness, packet dropping probability of each network node increases from 10 to 40 percent, while the average successful delivery rate will drop from 56.39% to

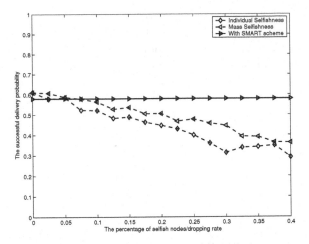

Figure 4.6: Effect of incentive scheme.

36.31%. On the other hand, as for the individual selfishness, if 10 to 50 percent of network nodes are selfish nodes, the average successful delivery rate will dramatically decrease from 52.21% to 29.08%. This result demonstrates that the average network throughput could degrade significantly when the selfish nodes or selfish behaviors exist. However, with SMART in place, nodes are naturally motivated to participate in bundle forwarding to earn as many credits as possible. Though the successful delivery rate of SMART is slightly lower in the beginning due to the extra security overheads, the network throughput would keep relatively stable since SMART can successfully stimulate selfish nodes in packet forwarding. This demonstrates the incentive effectiveness of SMART. In the following section, we will discuss the other important metrics related to SMART-based DTN routing: delivery ratio, overhead ratio, average latency, and number of forwarding copies.

4.5.2.3 Scenario I: Impact of Traffic Load

To evaluate the practicality of a SMART scheme, we first examine the system performance under different sending frequencies by adjusting the message generation interval, which is initialized to 35s, and then gradually decreased to 5s. Figure 4.7 shows system performance of original SWB routing protocol without any incentive scheme, SWB with SMART scheme, and SWB with optimized SMART scheme. The network performance can be measured in terms of three metrics: successful delivery rate, overhead ratios, and average latency. Figure 4.7(a) shows the relationship between the successful delivery rate and the message sending frequency. It is clear that a higher message sending frequency would result in a lower delivery rate in different SWB scenarios due to the increased number of forwarding messages. However, we can

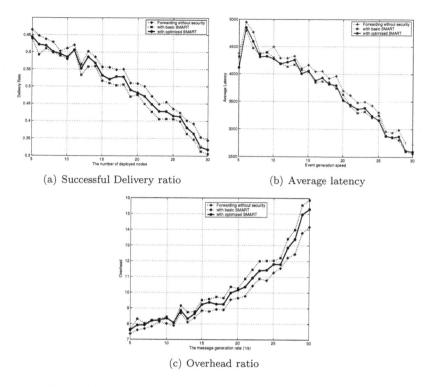

(a) Successful Delivery ratio

(b) Average latency

(c) Overhead ratio

Figure 4.7: Impact of network load on system performance.

also see that the performance of SWB with SMART scheme, and, optimized SMART, is very close to that of SWB without any security add-ons. For example, when a high message forwarding frequency is in place (e.g., message generation interval is set to 5s), a SMART scheme incurs a 13.3% decrease of successful delivery rate while the optimized SMART scheme only incurs an 8.3% decrease. Figure 4.7(b) shows the average latency of different scenarios. It is observed that after a small growth period, the average latency will decrease quickly and optimized SMART has a comparable performance with the SMART scheme, both of which are less than the no-security system. This is mainly caused by the dramatically decreased delivery rate. Figure 4.7(c) demonstrates that SMART and optimized SMART only have a slightly larger overhead than no-incentive SWB. However, the increased overhead is not so significant and thus they have a similar overall performance.

4.5.2.4   Scenario II: Impact of Forwarding Copy Number

Multi-copy data forwarding is a major characteristic of DTN data forwarding. In this section, we investigate the impact of number of copies on the system performance and also study how to find an optimal forwarding copy number

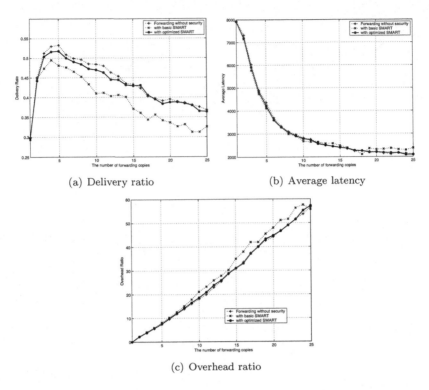

(a) Delivery ratio   (b) Average latency

(c) Overhead ratio

Figure 4.8: Impact of forwarding copy number on system performance.

with/out incentive mechanism. In Figure 4.8, the number of forwarding copies is initially set to 1 and then increased one by one. It is obvious that the delivery rate will grow very fast in the beginning and then decrease after a specific threshold (for example, 5 in Figure 4.8(a)). This shows that an optimal copy number exists to achieve a highest successful delivery rate. On the other hand, from Figure 4.8(b) and Figure 4.8(c), it is observed that the average latency will decrease with the increased copies while overhead will increase significantly along the forwarding copies. By jointly considering the system performance and ensuring a certain security level, the OSM can choose an optimal forwarding copy number to find a balance between the security and system performance.

In summary, the above simulation results demonstrate that SMART is indeed a viable, lightweight solution for stimulating bundle forwarding in a DTN environment.

4.6 Open Issues

In previous sections, we have introduced the SMART scheme in detail, which can be used to stimulate routing and data forwarding in DTNs. In this section, we further discuss other challenges related to secure incentive design in DTNs.

4.6.1 Public Key Revocation in DTNs

Public key management is the foundation of any security protocols. In a secure incentive scheme, any misbehaving or malicious nodes will pay the penalty of having their public key certificates revoked. Even for those selfish nodes which run out of their credits, one possible punishment action is also revoking their certificates or reducing their CoS right by revising their certificates. However, public key revocation still represents a great challenge in delay tolerant networks. In a traditional Public Key Infrastructure (PKI), the most commonly adopted certificate revocation scheme is through a Certificate Revocation List (CRL), which is a list of revoked certificates stored in central repositories prepared by the Certificate Authorities (CAs). However, in DTNs, the nodes may suffer from delayed or frequent loss of connectivity to CRL servers [22]. In [23], the use of periodical public key updating is suggested to replace the traditional public key revocation, although, in the real world, public key distribution is also a challenging problem and may lead to a lot of extra management costs. Another possible way to address public key revocation in DTNs is by using cooperative CRL distribution [22], which needs further investigation to find an improved method.

4.6.2 Reputation-Based Incentive Scheme

In this chapter, we have introduced a credit-based incentive scheme. However, the reputation-based incentive scheme in DTNs is still an open problem. There are two major challenges for designing a reputation-based scheme in DTNs. The first challenge comes from the reputation measurement. Existing reputation-based incentive schemes rely on the neighboring nodes to detect if a packet has been successfully forwarded. This method may not work well in DTNs since neighboring nodes do not have a clear way of distinguishing if the packets have been forwarded or not and thus the forwarding actions may happen out of the sensing range of the neighboring nodes due to the unique "store-carry-and-forward" DTN transmission. The second challenge is how to efficiently and effectively propagate the reputation. Due to these two challenges, designing a reputation-based incentive scheme is still an open problem.

4.7   Conclusion

In this chapter, we have proposed a secure multi-layer credit-based incentive (SMART) scheme to stimulate cooperation in packet forwarding for delay tolerant networks. We have also proposed two efficiency optimization methods to reduce the transmission and computation overhead. The SMART is compatible with diverse existing routing schemes and is expected to improve the system performance of DTNs that suffer from selfishness. We also demonstrated the efficiency and effectiveness of SMART through extensive simulations under different system parameters. Our future work includes a reputation-based incentive scheme or a secure incentive compatible routing scheme for DTNs.

Bibliography

[1] A. Kate, G. Zaverucha, and Urs Hengartner, "Anonymity and security in delay tolerant networks," Proc. of *SecureComm 2007*, Sept. 2007.

[2] H. Zhu, R. Lu, X. Lin, and X. Shen, "Security in service-oriented vehicular networks," to appear in *IEEE Wireless Communication, Special Issue on Service-Oriented Broadband Wireless Network Architecture*, Vol. 16, No. 4, Aug. 2009.

[3] R. Lu, X. Lin, H. Zhu, P.H. Ho, and X. Shen, "ECPP: Efficient Conditional Privacy Preservation Protocol for Secure Vehicular Communications," Proc. of *IEEE Infocom'08*, Phoenix, Arizona, USA, Apr. 14–18, 2008.

[4] J. Zhao and G. Cao, "VADD: Vehicle-Assisted Data Delivery in Vehicular Ad Hoc Networks," Proc. of *Infocom'06*, 2006.

[5] P. Hui, J. Crowcroft, and E. Yoneki, "Bubble Rap: Social-based Forwarding in Delay Tolerant Networks," Proc. of *MobiHoc'08*, 2008.

[6] A. Garyfalos and K. C. Almeroth, "Coupons: A Multilevel Incentive Scheme for Information Dissemination in Mobile Networks," *IEEE Trans. on Mobile Computing*, Vol. 7, No. 6, June 2008.

[7] A. Chaintreau, P. Hui, J. Crowcroft, C. Diot, R. Gass, and J. Scott, "Impact of Human Mobility on the Design of Opportunistic Forwarding Algorithms," Proc. of *IEEE Infocom'06*, 2006.

[8] J. H. Cui, J. Kong, M. Gerla, and S. Zhou, "Challenges: Building Scalable Mobile Underwater Wireless Sensor Networks for aquatic applications," *IEEE Network, Special Issue on Wireless Sensor Networking*, May 2006.

[9] T. Spyropoulos, K. Psounis, and C. S. Raghavendra, "Efficient routing in intermittently connected mobile networks: the multiple-copy cast," *IEEE/ACM Trans. on Networking*, Vol. 16, No. 1, Feb. 2008.

[10] T. Spyropoulos, K. Psounis, and C. S. Raghavendra, "Efficient routing in intermittently connected mobile networks: the single-copy cast," *IEEE/ACM Trans. on Networking*, Vol. 16, No. 1, Feb. 2008.

[11] S. Jain, K. Fall, and Rabin Patra, "Routing in a Delay Tolerant Network," Proc. of the *ACM SIGCOMM*, 2004.

[12] A. Vahdat and D. Becker, "Epidemic routing for partially connected ad hoc networks," Technical Report CS-200006, Duke University, Apr. 2000

[13] A. Lindgren, A. Doria, and O. Schelen, "Probabilistic Routing in Intermittently Connected Networks," Proc. of *SAPIR*, 2004.

[14] S. Marti, T. Giuli, K. Lai, and M. Baker, "Mitigating Routing Misbehavior in Mobile Ad Hoc Networks," Proc. of *ACM Mobicom*, Boston, Massachusetts, Aug. 2000.

[15] Q. He, D. Wu, and P. Khosla, "SORI: A Secure and Objective Reputation-Based Incentive Scheme for Ad Hoc Networks," Proc. of *WCNC 2004*, Atlanta, Georgia, Mar. 2004.

[16] Y. Zhang and Y. Fang, "A Fine-grained Reputation System for Reliable Service Selection in Peer-to-peer Networks," *IEEE Trans. on Parallel and Distributed Systems*, Vol. 18, No. 8, pages 1134–1145, Aug. 2007.

[17] S. Buchegger and J. Le Boudec, "Performance Analysis of the CONFIDANT Protocol: Cooperation of Nodes — Fairness in Distributed Ad-Hoc Networks," Proc. of *IEEE/ACM Workshop on Mobile Ad Hoc Networking and Computing (MobiHOC)*, Lausanne, Switzerland, June 2002.

[18] Y. Zhang, W. Lou, W. Liu, and Y. Fang, "A Secure Incentive Protocol for Mobile Ad Hoc Networks," *ACM Wireless Networks*, Vol. 13, No. 5, pages 569–582, Oct. 2007.

[19] S. B. Lee, G. Pan, J-S Park, M. Gerla, and Songwu Lu, "Secure Incentives for Commercial ad Dissemination in Vehicular Networks," Proc. of *MobiHoc'07*, Sept. 2007.

[20] R. Lu, X. Lin, H. Zhu, C. Zhang, P.H. Ho, and X. Shen, "A Novel Fair Incentive Protocol for Mobile Ad Hoc Networks," Proc. of *IEEE WCNC'08*, Las Vegas, Nevada, USA, Mar. 31 – Apr. 3, 2008.

[21] M. Felegyhazi, J.-P. Hubaux, and L. Buttyan, "Nash Equilibria of Packet Forwarding Strategies in Wireless Ad Hoc Networks," IEEE Trans. on Mobile Computing, 2006.

[22] X. Lin, R. Lu, C. Zhang, H. Zhu, P.H. Ho, and X. Shen, "Security in Vehicular Ad Hoc Networks," *IEEE Communications Magazine*, Vol. 46, No. 4, 88–95, 2008.

[23] A. Seth, U. Hengartner, and S. Keshav, "Practical Security for Disconnected Nodes," in Proc. of *NPSec'05*, Nov. 2005.

[24] S. Zhong, J. Chen, and Y. Yang, "Sprite: A Simple, Cheat-proof, Credit-based System for Mobile Ad-hoc Networks," Proc. of *IEEE Infocom'03*, 2003.

[25] S. Zhong, L. Li, Y. Liu and Y. Yang, "On Designing Incentive-compatible Routing and Forwarding Protocols in Wireless Ad-hoc Networks" *Wireless Networks*, Vol. 13, pages 799–816, 2007.

[26] S. Symington, S. Farrell, H. Weiss, and P. Lovell, "Bundle security protocol specification," draft-irtf-dtnrg-bundle-security-05.txt, work-in-progress, Feb. 2008.

[27] D. Boneh and M. Franklin, "Identity Based Encryption from the Weil Pairing," Proc. of Crypto'01, LNCS, Vol. 2139, pages 213–229, Springer-Verlag, 2001.

[28] D. Boneh, B. Lynn, and H. Shacham, "Short Signatures from the Weil Pairing," *Journal of Cryptology*, Vol. 17, No. 4, pages 297–319, 2004.

[29] S. Farrell and V. Cahill, "DTN: An architectural retrospective," *IEEE J. Sel. Areas Commun.*, Vol. 26, No. 5, pages 828–836, June 2008.

[30] M. Pitkanen, M. Keranen, and J. Ott, "Message Fragmentation in Opportunistic DTNs," Proc. of *WoWMoM 2008*, 2008.

[31] R. Merkle, "Protocols for Public Key Cryptosystems," Proc. of *IEEE S&P*, pages 122–133, 1980.

[32] Multiprecision Integer and Rational Arithmetic C/C++ Library (MIRACL).

[33] N. Asokan, K. Kostiainen, P. Ginzboorg, J. Ott, and C. Luo, "Applicability of Identity-based Cryptography for Disruption-tolerant Networking," Proc. of *the First International MobiSys Workshop on Mobile Opportunistic Networking (MobiOpp)*, June 2007.

[34] The One Simulator: http://www.netlab.tkk.fi/tutkimus/dtn/theone/.

[35] J. C. Cha and J. H. Cheon, "An Identity-based Signature from Gap Diffie-Hellman Groups," Proc. of *PKC'03*, Vol. 2567, pages 18–30, 2003.

Chapter 5

R-P2P: a Data-Centric Middleware for Delay Tolerant Applications

Corrado Moiso, Antonio Manzalini, Francesco De Pellegrini, Iacopo Carreras, Daniele Miorandi, and Athanasios Vasilakos

5.1	Application Scenario	129
5.2	Related Works	131
	5.2.1 Novelty of the Proposed Architecture	132
5.3	Sketch of the System	132
	5.3.1 Query Diffusion and Data Retrieval	134
	5.3.2 Diffusion of Advertisements	136
5.4	Functional Blocks	138
	5.4.1 User Nodes	139
	5.4.1.1 Service Container	139
	5.4.1.2 Interaction Controller	140
	5.4.1.3 Content Manager and Data	143
	5.4.1.4 Network Interface	143
	5.4.2 Throwbox Nodes	144
5.5	System Dimensioning	148
	5.5.1 RP2P Design Space	149
	5.5.2 Application Requirements	150
	5.5.3 System Dimensioning	150
5.6	Preliminary Implementation	151
	5.6.1 U-Hopper: The Mobile Part of R-P2P	151
	5.6.2 Throwboxes	153

5.7 Conclusions 154
5.8 Acknowledgments 154

The diffusion of electronic devices equipped with computing and wireless communications capabilities has changed the notion of interaction with the technology in everyday life. The evolution of information technologies is not always linear, though. One instructive example is the success of guidance systems, traditionally based on the GPS technology. GPS technology, in fact, was designed intentionally for positioning systems, and requires an extremely expensive satellite infrastructure. But, surprisingly, such technology is facing unexpected competitors. Cellular phones, for example, can provide positioning functionalities based on base stations triangulation, and projecting detected positions on a map. The resolution of this technique is on the order of a kilometer at present, and of course it is not suitable for vehicular road guidance systems. Nevertheless, it is applicable to a range of less demanding location-based applications: in general, consensus exists on the fact that the massive deployment of communicating and computing devices has the potential to be exploited in novel and unplanned manners.

This relates to the need for coupling environmental or context data to one or more services running on user devices. In practice, data of interest for an application can be retrieved via available sensor measurements or directly from devices in the surroundings. For a wide class of applications, in particular, a key feature of the available data, e.g., the customary example of a temperature measurement or the position of the closest Starbucks coffee, is the locality of such information with respect to the user position. A self-explanatory example is represented by the fast growing field of Bluetooth advertisement (see for example `http://www.kombok.it` and `http://www.bluetooth-advertising.co.uk`). To this aim we use the generic term *source* to mean a wireless device diffusing information pertinent to a given location.

Translated into networking requirements, location-based data distribution would require a fine-grained deployment, bridging possibly different technologies in order to reach the needed wireless coverage. But, at present, such a fine-grained wireless coverage has a unique candidate, i.e., a cellular network. The idea of injecting and distributing local data into the operators' networks poses several technical concerns, though, especially in terms of scalability.

Nevertheless, alternative architectures are possible. Recently, several legacy devices, i.e., standard cellular phones, have been enabled with proximity wireless communications. In particular, Bluetooth or IEEE802.11 interfaces would potentially lead to systems composed of several handheld devices coming now and then into radio range, forming a mobile ad hoc network based on one-hop communications. These systems are referred to as opportunist networks. Their connectivity pattern is typically intermittent due to sudden and repeated changes of environmental conditions: nodes moving out of range, drop of link capacity, or short end-node duty cycles [7, 15]. In this chapter we consider a scenario where several wireless mobile devices cooperate in order to feed local data to applications leveraging opportunistic communications. The opportunities for pairs of nodes to communicate are rather short and, with respect to this, communication among nodes is conveniently described by the set of nodes' *contacts* [6, 16].

The use of such systems appears promising for two main reasons. First, there is no need for central data management. Second, it is inherently localized, because only communications with nodes in radio range are possible. Of course, even though they natively support data diffusion, opportunistic communications have several limitations. In terms of delay, for example, no guarantee can in principle be provided so that applications running on top of such networks are required a certain degree of *delay-tolerance*. Furthermore, the amount of data that can be exchanged per contact is dictated by the finite duration of contacts, the need to trade off device discovery duty cycles for energy savings, and, last but not least, by the limited bandwidth of current technologies, e.g. Bluetooth. As it will be clear in the following, for the above reasons we enforced the opportunistic network with an interconnected network of throwboxes.

The main contribution of this chapter is the description of a system, namely R-P2P (Reverse Peer to Peer) and the related middleware interfacing mobile devices and throwboxes. The baseline functionality provided by R-P2P is the retrieval of data generated by sources deployed in the area of interest to the user. R-P2P is novel since it decouples the query diffusion and the data retrieval process using two separate network extensions: an opportunistic network supports the diffusion of queries and advertisements, whereas a network of interconnected throwboxes supports the retrieval of data injected into the system.

In the following description of the architecture, cooperation and security issues are out of the scope of the chapter; nevertheless, enforcing users' cooperation and ensuring security are a clear requirement for the R-P2P system.

5.1 Application Scenario

As a reference scenario we represent a simple use case. Our virtual user Sonia is visiting Trento, Italy, during the end of May. She came to town in order to attend the Festival of the Economy, a known event involving Italian institutions, the local municipality, and several Italian and foreign companies who organize events for the whole duration of the 3-day festival (see www.festivaleconomia.it). She knows that, apart from the scheduled institutional events, there are several satellite social events occurring in parallel, some of them with short or almost no advance notice.

This scenario fits the typical requirements for the usage of the novel system described in the following:

- during a limited period, a high density of outside people gather in town; they are typically interested in information relative to the specific location and in events occurring in the area — this requires an *easy to deploy and light architecture*;

- visitors have a broad range of interests, covering events related to the festival, but also satellite events occurring in ad-hoc fashion, e.g., restaurants, special offers, wine bars, or parties — this requires a *general purpose data sharing* system;

- each user will use local information to decide on the eligibility of some event for his/her schedule, but *most relevant contents will be generated by users on the fly, including opinions, tastes and ratings* — this excludes Web-based only publishing solutions;

R-P2P is designed to meet the above requirements. While Sonia walks in Trento, her handheld device running R-P2P will forward queries for certain information, which might become available at some of the other persons' devices running R-P2P; R-P2P will also allow transparent exchange of data during contacts. Sonia will verify from time to time the display of her handheld device: an application will integrate fresh information on ongoing events and information gathered by other users on a map, as depicted in Figure 5.1.

In the example, Sonia will be interested in a concert and a talk: based on some user's feedback, she will opt for the concert. She will also obtain some

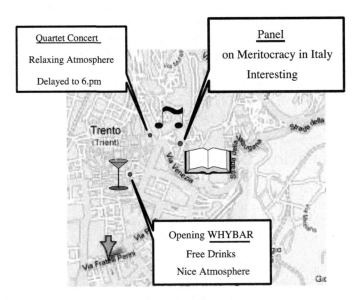

Figure 5.1: A graphical representation of the use case, enriching traditional mapping systems with notifications based on user-generated contents.

multimedia material on the place the concert will be hosted in, a nice palace in downtown Trento. At the concert, she will confirm the opinion of other visitors, just by filling a short rating form on her smart-phone. Later on that evening, she also will be able to plan an unexpected stop at wine bar, whose owner is promoting the opening with free drinks; the owner just advertised the happening using a commercial Bluetooth advertisement device over the R-P2P network, diffusing some adds and storing some related multimedia material over R-P2P.

5.2   Related Works

The possibility of running applications on top of opportunistic networks is recently gathering some attention. In [22] the operation of application protocols such as HTTP or e-mail in delay tolerant networks (DTNs) is addressed: the issue there is how to adapt them in order to run under a regime of intermittent connectivity. A dedicated application-aware module pilots storing and forwarding operations of DTN modules using a suitable "application-hint" header protocol. [21] describes general principles for such adaptation.

We notice that the system considered in this chapter is not end-to-end, but data-centric, so that traditional end-to-end DTN techniques do not apply.

Other authors have explored the technical feasibility of P2P systems in the presence of intermittent connectivity. The work in [23] proposes a system, namely $7DS$, which runs as an application over mobile hosts. The aim of $7DS$ is to provide enhanced Internet connectivity sustaining cooperative data exchange with mobile peers and a mechanism for data dissemination. $7DS$ implements a passive/active scheme and a mobile host that relies on local peer connections to retrieve data from local servers. In [17], the authors introduce passive distributed indexing (PDI), a P2P overlay designed for mobile applications. PDI provides a query diffusion protocol on a restricted overlay of mobiles covering a few hops. The work in [29] proposes a delay-tolerant overlay for a mobile ad-hoc network, namely ASOS, with the aim of accomplishing disruption tolerant operations. In ASOS queries and data are disseminated on a MANET: several issues are explored on data diffusion and optimization of data retrieval.

Several related works, in particular, proposed solutions to cope with frequent disruptions using cooperative caching techniques [24, 30, 18, 12]. The main effort is to sustain data flows even in the presence of high route failure rates, inherently due to mobility [12]. The authors of [24] proposed a hybrid technique using both peer-to-peer communications and APs in radio range. Also, in [30], the authors describe existing trade-offs depending on the mobility and cache size trading off local caching and queries issued to central-servers. Techniques for probabilistic quorum construction are applied to ad-hoc networks in [18], based on the seminal work [20].

5.2.1 Novelty of the Proposed Architecture

The R-P2P system works based on diffusion of queries and avertisements and on data retrieval. Most related works mentioned before share the common view that *both* queries diffusion *and* data retrieval occur on top of the same MANET. We take a different approach for two main reasons. First, at the current status of the technology, measurements performed on the field show that the duration of contacts is not always such as to guarantee complete data exchange for large data files [14]. Also, a further problem holds with respect to data retrieval: at the time when a query reaches a node able to retrieve the queried data, the potential return path through which the query is forwarded likely disappeared. For such a reason, all the above mentioned approaches typically foresee a large overhead in terms of data dissemination: basically they try to reach back to the source of the query via controlled flooding.

In sight of the above technical concerns, the major novelty of the proposed system architecture is that it couples a Distributed Storage System (DSS), e.g., a Distributed Hash Table (DHT), deployed over an interconnected network, to a DTN network of mobile nodes. The middleware described in the following, in particular, is meant to confine the forwarding of queries and advertisements to the opportunistic part of the network, whereas the data retrieval is operated through a set of interconnected throwboxes. The net effect is that *we split the forward and the return path used in the query/retrieve data process*. Also, the proposed technique leverages the opportunistic part of the network for queries and advertisements, which are typically very short messages, whereas data retrieval is operated via the interconnected network. Another feature of the proposed solution is that the system architecture is designed to retrieve natively data from local sources in radio range, and to maintain the local scope of the retrieved data.

We called this middleware Reverse Peer-to-Peer (R-P2P), because the system resembles a P2P system, but as it is opposite to a regular p2p system, data are injected into the storage system after the query is produced, whereas in a regular p2p system, queries match data *already* stored in the system.

5.3   Sketch of the System

The architectural skeleton of the proposed architecture is depicted in Figure 5.2. We consider three types of communication devices: *throwboxes, user nodes*, and *sources*. User nodes are mobile handheld devices, such as a mobile phone, a PDA, or a similar electronic device with communication capabilities, carried around by users. Typically, such devices are equipped with a primary wireless interface connecting to an operator network, and a secondary wireless interface, e.g., Bluetooth or IEEE802.11. When in radio range of another device, using these secondary interfaces, user nodes can establish communications in a point to point fashion with no need for the fixed infrastructure. This is usually referred to as opportunistic communications (see [16, 15]).

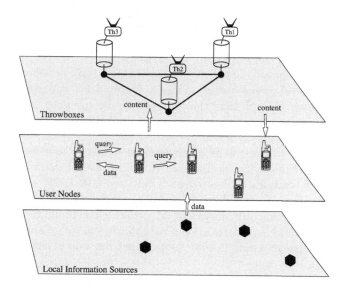

Figure 5.2: Layered representation of the network components: on the upper layer, throwboxes, in the middle layer user nodes, and in the lowest layer localized storage.

Also, user nodes can interact with sources in radio range, namely devices acting as sources of *local* data: these data are consumed by applications running on user nodes. Sources are depicted in Figure 5.2 at the bottom layer: a typical example of such a device can be a laptop equipped with an RF interface acting as a bucket of user-generated content, but we plan to include sensors or RFIDs in the middleware as well.[1] In some other cases, contents might be generated directly onboard of user nodes, e.g., this is the case with a video or a picture taken with a video-phone.

A set of *throwboxes* occupy the third logical layer of the system. Throwboxes can interact within themselves via a conventional interconnected network, e.g., a mesh network. User nodes communicate with a throwbox using the secondary wireless interface.

Notice that the basic architecture of the system described here is similar to what was proposed in [26] for data collection with the use of "mules." In that work, though, data collected by "mules" are finally uploaded to the sink of a sensor network; the use of a query-based mechanism makes R-P2P different and novel, because queries are generated at user nodes and the sink here does not correspond to the origin of the sources. Also, R-P2P is meant to support both **query** and **publish** operations. In particular, issuing a **query**,

---

[1] The issue in such cases is to have a suitable wireless card for the user nodes to interface with those devices, which is beyond the scope of the current work.

a mobile device can retrieve data pertinent to the given location where the user is located. Conversely a user node can `publish` information relevant to the given location in the form of an advertisement.

The pictorial representation of the two mechanisms is reported in Figure 5.3 in the case of `query` operation and in Figure 5.4 in the case of `publish` operation. In order to implement the two basic `query` and `publish` operations, in particular, two main procedures have to be implemented: query/ advertisement diffusion and data retrieval.

The query/advertisement diffusion is carried at the user node layer by leveraging opportunistic forwarding. To this aim several variants are possible, including K-relaying and spray and wait; see [10, 3, 27] and references therein. As we mentioned before, in order to efficiently use the capacity of opportunistic networks, only queries or advertisements should be propagated. In fact, data retrieval is performed only with devices in range and no data are forwarded in a multihopping fashion (notice that one hop exchange of data is possible, though).

5.3.1   Query Diffusion and Data Retrieval

The typical operations for query dissemination follow:

- once a new `query` is issued, a unique query identifier `query-id` is generated, and attached to the `query`; the identifier will later on let the node issuing the query recognize the corresponding data once available; also, optional information released by the source node will provide indication about the intended validity of the query and/or the data and the geographical scope, i.e., the maximum distance within which the query and the potentially retrieved data are considered to be of interest.

- two user nodes coming into reciprocal radio range first check whether they possess data which fulfill the other node pending queries, i.e., queries they generated before; if so they exchange the corresponding data and erase pending queries. As a second step, each user node forwards pending queries and registers new queries received by the peer node (Figure 5.3(a) and (b));

- when in range of a source, a user node queries the source in order to retrieve data of interest according to stored queries (Figure 5.3(c)).[2]

Thus, in R-P2P data retrieval occurs *on-demand*; retrieved data are thereafter made available over a distributed storage system:

- when in radio range of a throwbox, a user node uploads source data retrieved, and erases them from its local storage (Figure 5.3(d)) — uploaded contents are associated to the `query-id` that triggered their retrieval;

---

[2]We assume that suitable mechanisms are used in order to enforce cooperation.

(a) U1 issues a query for data A to node U2

(b) U2 forwards the query to node U3

(c) Data A retrieved by used node U3 from a source

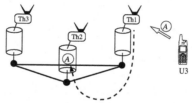

(d) Data A stored to the DSS via Throwbox Th1 by U3

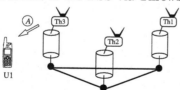

(e) Data A retrieved at Throwbox Th3 by user node U1

Figure 5.3: Sketch of the system operations in the case of query diffusion. The example reports the set of interactions from the query of a given data to the retrieval at a throwbox some time after the query is issued.

- at throwboxes layer, a complete overlay permits sharing data uploaded into the system;

- after disseminating queries, user nodes coming into contact with a throwbox check their own pending `query-ids` for matches with the data retrieved by the system and download the corresponding data (Figure 5.3(e)).

5.3.2　Diffusion of Advertisements

A dual functionality is the dissemination of advertisements: a user node can upload data into a throwbox, and advertise its presence in the throwbox storage system diffusing an advertisement message. Advertisements are tagged with a topic description and an identifier to retrieve (additional) content on the throwboxes. The typical system operations for advertisement dissemination follow:

- when a user node is aiming at propagating an advertisement on a specific topic, it uploads the associated content into the DSS, by interacting with a throwbox in radio range (Figure 5.4)(a). The content is associated to a unique identifier adv-id generated by the user node.

- the user node propagates the information to the devices which are in radio range that a new advertisement is available (Figure 5.4)(b). The message includes the description of the advertisement topic and the adv-id, to be used by devices interested in the advertisement topic to retrieve the associated content from the DSS. In order to avoid flooding, the device could limit the propagation of the advertisement (e.g., in terms of numbers of propagations or time period during which the propagation is performed). Moreover, as for query propagation, optional information released by the source device provides indication about the intended geographical or time validity of the advertisement and constraints on propagation (e.g., maximum number of propagations performed by a single device or the maximum lengths of the propagation chain across different devices).

- when a device receives a message related to the availability of an advertisement, it starts propagating it, according to the propagation rules (if any) included in the message (Figure 5.4)(c). Moreover, it checks whether the associated topic is relevant for it (i.e., for the applications running on it).

- if the device is interested in the topic, it tries to contact a throwbox to retrieve the content associated to the advertisement. When in radio range of a throwbox, it performs a request to download the data associated to the adv-id included in the received message (Figure 5.4)(d).

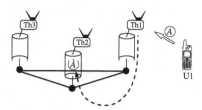

(a) Advertisement A stored at Throwbox Th1 by user node U1

(b) User node U1 forwards the content of the advertisement to U2

(c) User node U2 forwards the content of the advertisement to U3

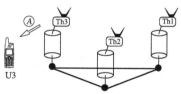

(d) U3 retrieves the advertisement A at Throwbox Th3

Figure 5.4: Sketch of the system operations in the case of advertisement diffusion. The example reports the set of interactions from the placement of an advertisement in a certain throwbox to the retrieval of the complete advertisement of an interested node.

5.4   Functional Blocks

The identified architecture has been in part implemented by a set of functional blocks as depicted in Figure 5.5. As concerns user nodes, 4 main modules exist:

- Service Container: it executes the service logic, which includes services deployment, execution deprecation, and update. Furthermore, this component issues queries, notifies events, and waits for replies to queries;

- Interaction Controller: this module regulates the interactions with throwboxes and user nodes in order to retrieve data; it also dispatches to the content manager queries, either issued by the service container or by other nodes. Two sub-modules specialize specific functionalities: i) the sub-module *Interaction with user nodes* regulates the temporary storage to the user node data repository of data retrieved from sources in radio range and processes queries coming from other user nodes, ii) the sub-module *Interaction with throwboxes* is in charge of issuing the storage/retrieval of data to throwboxes;

- Content Manager: it processes queries received from peer nodes and verifies whether the required information has to be retrieved (stored) from (to) the data repository on board user nodes;

- Network Interface: provides the basic network functionalities for one-hop communications.

As concerns throwboxes, we find:

- Interaction with user nodes: this module is in charge of processing the requests coming from the user nodes in radio range of the throwbox to either store or retrieve data; such requests are then forwarded to the throwbox primitive module;

- Throwbox Primitives: it issues the appropriate commands to the DSS system in order to handle data to (from) the distributed storage on the set of throwboxes;

- Network Interface: this module provides basic network functionalities for one-hop wireless communications;

- Distributed Storage System: it is used to store data, where data indexing occurs via identifiers plus a locality-based address of the DSS system.[3]

In the following we describe in detail the primitives implemented by each of the modules described before, together with the main operational features.

---

[3]DSS uses a standard wired or wireless network.

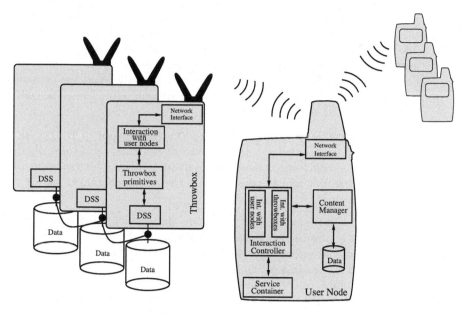

Figure 5.5: Functional blocks representation for the network components: throwboxes and user nodes.

5.4.1 User Nodes

5.4.1.1 Service Container

The service container provides the necessary programming abstractions that allow services to utilize the R-P2P middleware. In particular, it executes services running inside user nodes, and provides facilities for services deployment, execution, deprecation, and update. From the service controller, services are able to (i) query for data, issuing a query that is forwarded to the Interaction Controller; and (ii) generate and advertise contents that can be of interest to other user nodes. These functionalities are enabled through the following primitives:

- search(query) → result: this primitive is invoked by the Service Container towards the Interaction Controller, whenever a service, running on a user node, needs to retrieve some information specified by a certain query. If the requested information is already available in the device's storage, it is returned through the result variable as a contentList. If no local information is matching the specified query, a no-data response is returned through the result variable. In this second case, the underlying Interaction Controller invokes an appropriate callback primitive as soon as the queried content becomes available.

- publish(advertisement): this primitive is invoked by the Service Container when some user/service-generated information is available. Such information takes the form of an advertisement, and is composed by its content and a topic (e.g., food, entertainment, shops, etc.), i.e., a meta-description of such content. As we will see in the following, the content is then stored in the Throwboxes storage system, while the topic only is diffused among user nodes.

- subscribe(topic): this primitive issues the subscription of the Service Manager of a user-node to a specific class of contents; once subscribed, the Interaction controller is in charge of handling the advertisement of interest for the user and to pass them to the Service Manager calling a notify primitive.

Each query and advertisement issued by the Service Container is accompanied by some control information (*query-info* and *advertisement-info*) adding some additional information such as the spatial/temporal validity of the query/ advertisement(e.g., timeouts or a maximum distance for the forwarding of the query/advertisement). This prevents devices from propagating messages exceeding a given timeframe or outside a given spatial range.

5.4.1.2 Interaction Controller

The Interaction Controller "glues together" the various system components and regulates the interactions and flows of information inside the system. In particular, this module is in charge of processing the messages received from a user node or a throwbox: the module recognizes the incoming message as originated by a throwbox or a user node, and it then delegates other system components to process the request. In particular, it communicates with the Content Manager for resolving querys and retrieving/storing contents. Clearly, the Interaction Controller is linked to the Service Container for interfacing with the hosted services, thus invoking appropriate callbacks whenever queried content or advertisements are received. Finally, as detailed in the following, it handles the interaction flow with both user nodes and throwboxes.

The data types used by this module are the queryList, which is a list of query items, and the contentList, which is a set of content items. Table 5.1 summarizes the data types that are used inside the system.

The Interaction Controller can perform a notify towards the Service Container

- notify( ): this primitive is invoked by the Interaction Controller whenever a content matching a query previously issued by a service is found. In this case, the Interaction Controller notifies the Service Container of the retrieved contentList, and the Service Container then notifies this to the various services hosted. It is also possible that the Interaction Controller notifies an advertisement related to a subscribed topic to the Service Container.

Table 5.1: Summary of Data Types Used Inside the System

Name	Description	Modules
query	A query specifies the content that a service is requesting. A query is alway associated with (i) a unique query identifier query-id, which is determined by the node issuing the query (ii) a *query-info*, which contains side information such as the spatial/temporal validity of the query, and the *ID* of the node issuing the query.	Service Container
content	A content represents the information requested by a service through a query, or through the subscription to some interest.	Service Container, Content Manager
advertisement	An advertisement contains some user/service generated information. Any advertisement is accompanied by some control information, and in particular (i) a topic description (e.g., food, entertainment, shops, etc.) (ii) the advertisement body (e.g., a short text message) (iii) an identifier for the retrieval of associated data on the throwboxes.	Service Container, Interaction Controller
queryList	A queryList contains a set of querys. Typically, a queryList includes queries issued by a a node itself, and queries issued by other nodes.	Service Container, Interaction Controller
contentList	A contentList contains a set of content items. These are sent as a response to a specific query issued by a user node.	Service Container, Interaction Controller

Interactions with user nodes

This submodule rules the interactions with user nodes. In particular, it provides the following primitives:

- search(queryList) → result: this primitive triggers the transmission of a search message to user nodes in the communication range. Such a message is composed as <search; queryList>, where search specifies the operation, while queryList is the list of queries for which the user is asking content. In particular, the queryList will contain user node's own queries, for which content is possibly expected, and other users' queries, for which other users are expected to simply relay such querys. In case content is returned, this will be contained in the result variable;

- send(contentList) → result: this primitive issues the transmission of the content to be sent to the peer node to fulfill some of the queries; basically it passes to that module the message <send; contentList>. It receives the result of the operation, e.g. success or communication-failure in case a communication error is reported, the result is then passed to the Content Manager for dispatching data and in case a communication failure occurs, a basic control flow functionality processes retransmissions or drops the data transfer;

- event(advertisement-info): this primitive sends the message <advertisement-info> to all the user nodes in radio range, thus notifying the interested user that a publish() primitive has been invoked by the Service Container, with some content potentially interesting.

Interactions with throwboxes

This sub-module rules the interaction of a user node with a throwbox in radio range in order to store or retrieve data. We assume that identifiers of pending queries are passed to the sub-module by the Interaction Controller: the node's own pending queries are matched with corresponding data on the throwbox; conversely, data corresponding to queries of other nodes are pushed to throwboxes. In particular, three primitives are implemented by the module; in the following we refer to the case of querys, the same primitives handle advertisements:

- store(content,content-id) → result: sends the message <store, content, query-id> to a throwbox in radio range to store content in the DSS, and associates it to the identifier query-id; then it waits for a return value reporting on the result of the operation (e.g., success, alreadypresent when another content is already associated to the same query-id, communicationfailure in case a communication error is reported);

- get(query)→ content-id/exception: it sends the message <get, query-id> to a throwbox in radio range; the effect is to retrieve from the DSS the content associated to the identifier query-id. This primitive waits for a return message with the content or an exception indication (e.g., content not yet/no more available, communicationfailure); it does not remove content from the DSS.

- take(query)→ content-id/exception: same as get but the data corresponding to the identifier query is removed from the DSS.

Notice that the user node exits the range of the throwbox, and the exception communicationfailure is passed to the Content Manager in order to manage the partial upload/download of a content.

5.4.1.3 Content Manager and Data

The content manager is in charge of the storage and deletion of contents from the local Data storage. Periodical refreshes permit the freeing of storage: in particular querys, advertisements, and contents are associated with a limited spatial or temporal lifetime. The content manager compares periodically the information on the position of the node and the internal clock in order to free Data from contents out of scope. The Content Manager is accessible from other modules through the following primitives:

- search(query) → contentList: this query is used by the Interaction Controller for querying data that are possibly stored in the permanent repository.

- store(content): this primitive is utilized for storing a content item in the permanent storage of the device.

5.4.1.4 Network Interface

This module implements some basic network functionalities above the Link Layer: it covers a primitive flow control and a basic node discovery. In particular, it regulates beaconing in order to detect nodes in radio range (the rate of beacon transmission influences on the probability of detecting nearby user nodes/throwboxes).

Based on the beacon information, the Network interface module advertises the node type, in order to discriminate either a throwbox or another user node. Once the beacon triggers a connection, this module is responsible for the binding mechanism of the current communication with the Interaction Controller.

Three main functions are implemented: a send primitive to pass messages to the MAC/LL layer together with a fragment/ reassembly primitive for the split and reconstruction of long chunks of transmitted and received data, and a `notify` primitive which passes either received data or transmission reports to the Interaction Controller.

The Network Interface module implements a basic flow control mechanism. In particular, when a large `contentList` is transmitted, the communication may end due to the expiration of the contact, e.g., nodes exited radio range; in such cases a notification message will be passed to the Interaction controller which will pass the list of `query-ids` successfully transmitted to the Content Manager: the corresponding `content` are erased from the local storage. The transmission of the `contentList` will be restored at the next contact with a throwbox. Also, when several devices are transmitting to a throwbox, a back-pressure messaging from throwboxes can regulate concurrent upload flows.

5.4.2 Throwbox Nodes

Throwbox primitives

This module is the dual module of interaction with throwboxes present on board user nodes and implements two dual procedures, `store` and `get`, plus a `take` procedure. This module is also in charge of interworking with the DSS by leveraging features provided by the DSS modules. In what follows we will refer to the practical case of a DSS implemented in the form of a DHT.

`store(content-id)` → `result`: it is used to store into the DHT the content; the indexing is performed by means of the identifier `content-id` and the result of the operation is returned (e.g., `success`, `alreadypresent` when another content with the same `content-id` exists, `DHT-failure`);

`get(query)` → `content-id/` `exception`: it requests to the DHT module to retrieve from the DHT the content indexed with the identifier `query`; if the content is not available, it returns an exception (e.g., content not yet/no more available);

`take(query)` → `content-id/exception`: same as `get` but the data corresponding to the identifier `query` is removed from the DHT.

Network Interface

The functionalities of this component are similar to those of the peer module on user nodes.

DHT

This module implements the local functions for the DSS, implemented as a DHT. We note that, in the proposed system, performance and scalability issues are more relevant than handling of joining/leaving of nodes since the the throwboxes belong to a static network. Chord [28] is our reference DHT implementation on the throwboxes; the choice follows since in R-P2P performance and scalability are more relevant than handling joining/leaving throwboxes.

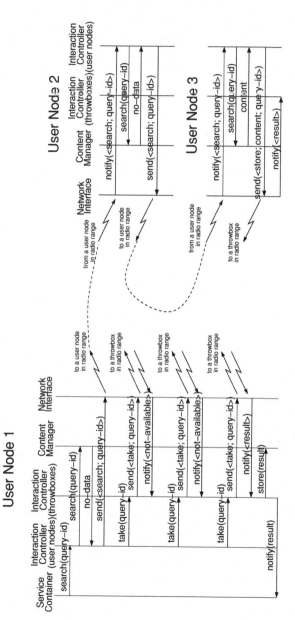

Figure 5.6: Interaction diagram for the sequence of operations involving user nodes.

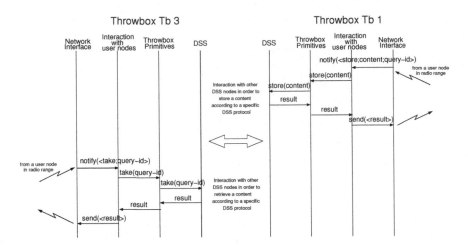

Figure 5.7: Interaction diagram for the sequence of operations involving throw-boxes.

DHT is in charge of storing, retrieving, and removing contents, indexed through identifiers. This module interacts with similar DHT modules deployed in other throwboxes, according to the interconnection topology adopted and implemented by the specific DHT solution.

To improve the performance of the system, contents should be stored by considering their locality: e.g., a content related to a given location L should be stored in a throwbox deployed close to L. In order to fulfill this requirement while maintaining the strengths of DHTs (e.g., balancing node loads and storage usage, fault tolerance and recovery, scalability), the throwboxes are organized as a set of interconnected "localized" DHTs: all the throwboxes located in a given geographical area are part of the same DHT; a "localized" DHT is named through a "location" identifier, such as downtown.trento.it. Such identifiers are configuration parameters of the throwbox, and can be retrieved by user nodes.

The identifiers of stored contents (e.g., the answer to a query) are structured as <location-id, content-id>, where location-id is the location identifier of a "localized" DHT, and content-id is the identifier to be hashed by the DHT.

When a throwbox receives a message from a user node to get/take/store contents, the DHT module has to process it. As a first step, the DHT module checks whether content's location-id is the one of the throwbox; if yes (due to the normal user node mobility, this is the most probable case), the DHT module hashes the content-id and performs the required operations to get/take/store the content. Otherwise, the DHT module accesses a global naming service (e.g., a DNS) for retrieving the address of a throwbox belonging to the remote "localized" DHT, identified by location-id, and forwards the request to it. The procedure to recover the DHT when a throwbox leaves a "localized" DHT (e.g., due to a crash) must be enriched in order to update the entries in the naming service.

In Figure 5.6 and Figure 5.7 we reported the main interaction diagrams for R-P2P.

Beyond DHT for throwboxes organization

In the following we elaborate on alternative solutions for data storage on throwboxes. As concerns R-P2P, in fact, performance and scalability issues are more relevant than handling joining/leaving nodes. This is because throwboxes belong to a static network. Moreover, the search on the throwboxes is done with an exact key. Thus, with regards to performance on scalability and search, the DHT-like structure described above is the best.

However, robustness is also an important factor in the progress of the throwboxes organization and it is indeed needed by the applications lying on top of the R-P2P middleware. From this standpoint, the robustness of DHT-like structures is typically worse than that of unstructured peer-to-peer (P2P) solutions. In particular one customary solution is to use overlay networks: overlay networks create a structured virtual topology above the basic transport protocol level that facilitates deterministic search and guarantees convergence.

One peculiar aspect where R-P2P is novel compared to P2P systems, where queries match data already in the system, is that data are injected into the storage system after the query is produced. However, part of the requested data might be already available on the throwboxes: if this is the case, a new query should not be issued, but data could be simply retrieved from the throwbox system. To address this case, a possible solution is to build an unstructured P2P architecture based on data indexing. Considering the factors including the robustness, the support for multiple keywords, or similar but not identical keywords, and the extension for applications, a challenging target would be to design a semi-unstructured P2P overlay, combining the merits of the structured P2P overlay (i.e. scalability and efficiency) with that of the unstructured P2P overlay (i.e. adaptability and robustness) to build a whole new overlay.

Location criteria

In order to improve the performance of the system, contents should be stored by considering their locality. The design should in turn organize throwboxes based on their geographical position: the distance between two nodes in the corresponding overlay configures then as the Euclidean distance. But, we notice that in distributed environments, performing on-demand network measurements is impractical because it is too costly and time-consuming: distance prediction services appear as a more viable compromise. There are two classes of distance prediction services [32]: *topology-based* using "landmarks" and *coordinate based*. Both these techniques use delay as a distance metric. An enhancement of these two methods combines IP address prefix and delay [13]: in R-P2P, the combination method [13] would permit the definition of the Euclidean distance in the overlay; this would also require defining the distance with corresponding formulas or functions in detail, which are beyond the purpose of the current work.

Throwboxes organization and modeling

Another important issue is the topology of the networked throwboxes. There exist several hierarchical structures suitable to improve the efficiency of the data retrieval and the scalability of the overlay, including the 2-dimensional (i.e., tree-like [25]), the 3-dimensional [19], and N-dimensional [31] topologies. However, there are some faults in normal hierarchical structures because the overhead at individual nodes is determined by the number of levels in the hierarchy. That is, nodes in the upper level will be more important and store more information so that the upper level nodes may be more vulnerable. For such reasons, proper balancing is an important part of the design. A viable solution is to couple a proper load balancing design technique to interconnect the throwboxes with a small-world topology, thus providing good guarantees on the expected data delivery, while offering a certain degree of flexibility in the topology design [11].

5.5   System Dimensioning

Opportunistic communication systems, i.e., delay tolerant networks that are operated in a disconnected regime and use opportunistic forwarding for achieving network-wide communications, are networks where nodes, mobility represents the main "communication medium" for conveying information, and is therefore exploited for achieving system-wide communications. Mobile nodes are typically characterized by limited resources and by a random mobility pattern, thus making message forwarding a very delicate part of the system. In addition, throwboxes or dedicated mobile nodes can be introduced in order to improve performance.

Harmonizing all the requirements deriving from an opportunistic application scenario and mapping them to a specific technological solution represents a demanding task for network and application designers, as it requires accounting for various constraints and factors that will affect the ultimate performance of the system. In the following, we will introduce a design methodology for dimensioning and evaluating an opportunistic communication system, highlighting the key design choices that can impact on the system performance, and pointing out some analytical frameworks that can be used towards this aim.

5.5.1 RP2P Design Space

The first step when designing and dimensioning an opportunistic communication system is to account for all the specific requirements arising from the application scenario at hand, and to understand the space that is left to network designers. In the following, we provide a taxonomy of the various aspects contributing to the operation of an opportunistic communication system:

- **network infrastructure**: the basic infrastructure of *RP2P* system is composed by (i) *relaying nodes*, which are mobile nodes generating/receving data, and relaying and receiving queries. Their number determines the "density" of the network, which is then reflected in the contact opportunities and therefore in the speed of the queries diffusion process; (ii) *throwboxes*, which are dedicated nodes, with greater resources in terms of storage, battery and processing capabilities. Their utilization can greatly increase the performance of the system, especially in the case where relaying nodes are constrained by extremely limited resources. Throwboxes are placed in strategic positions of the served area, and act as gateways towards the fixed infrastructure, thus providing a backbone for the diffusion and retrieval of data;

- **forwarding strategy**: one of the major problems in disconnected mobile networks is to envision efficient mechanisms for disseminating information and letting messages eventually reach their intended destination. The specific forwarding strategy regulates the way messages (queries in the RP2P case) are relayed from one node to the other, and represents one of the key design choices of any opportunistic communication system;

- **mobility pattern of nodes**: as the basis of the message diffusion pattern there are opportunistic communications among nodes, in which a message is relayed when nodes get within mutual communication range. The diffusion of messages is therefore driven by the underlying mobility pattern. Mobility can be constrained in a spatial region, as in the case of a city-wide deployment, can be heterogeneous, as in the case of *pocket switched networks* [8] where nodes, mobility is associated with the social behavior of users, or can be periodic, as in the case of opportunistic communications systems constituted by buses. Each mobility pattern leads to a different performance of the system, and needs to be deeply investigated during the system design phase;

- **communication technology**: the specific communication technology determines (i) when two nodes are within mutual communication range (ii) the amount of data that it is possible to exchange at any contact opportunity. Clearly, this highly impacts the networking performance of these systems, and needs to be carefully evaluated.

5.5.2   Application Requirements

Independently from the specific scenario, the metrics that mostly affect the performance of such systems, and that need to be taken into consideration when evaluating the requirements of any RP2P specific application being developed, can be summarized as follows:

- **delivery delay**: this is the average time that is needed for delivering a message from a source to a destination. This metric is extremely important as it reflects the dynamic behavior of the system, and in particular how efficient a certain technological solution is in letting messages reach their intended destination. Clearly, this metric is affected by many aspects such as, e.g., network infrastructure, mobility pattern of nodes, etc., and is therefore one of the most difficult metrics to evaluate at design phase;

- **utilized resources**: forwarding operations rely on the ability of a node to keep (even for a rather long time) a message in its internal memory. In the RP2P case, only queries are disseminated among nodes. However, in the case of network composed by a very large number of mobile nodes and frequent service data requests, the number of circulating queries (and of utilized resources) can become very large;

- **packet delivery ratio**: due to the intrinsic randomness of the many factors influencing the performance of opportunistic systems, messages can or cannot reach the intended destination. As an example, due to the limited resources available on nodes it can occur that messages are discarded and eventually do not reach any Access Point. The packet delivery ratio provides a measure of how likely this phenomenon is to occur.

5.5.3   System Dimensioning

From this brief introduction emerges the great burden that is posed on network designers, who are currently lacking the necessary tools to (i) first gather the requirements deriving from the application scenario being implemented and then (ii) map such requirements into the most suitable technological solution. It is therefore important to use analytical frameworks which can help network designers in this task, facilitating the process of design and dimensioning of the system.

A first attempt has been presented in [2], where the authors introduced a framework that allows formally classifying mobile wireless networks into some pre-defined categories. Each category is then characterized by a set of routing techniques, which results in the most appropriate for that specific case. The proposed framework results are extremely effective in such classification, but does not provide means to evaluate new forwarding mechanisms, nor to account for important system parameters such as the presence of throwboxes.

In contrast, in [9] authors present a design methodology which is based on three consecutive steps: (i) analysis of requirements, (ii) exploration of the design space, and (iii) performance evaluation. The second phase is supported through an analytical framework which allows easy modeling and evaluation of a wide range of opportunistic communication scenarios, accounting for the network configuration (e.g., number of mobile relays and throwboxes) and for a wide range of relaying schemes. Such framework represents a valuable tool for network designers, as it provides, without running into time-consuming simulations, a preliminary understanding of the performance that it is possible to expect from various technical solutions. Given the number of mobile nodes, a statistical description of their mobility pattern, and the number of throwboxes, it is possible to have an immediate analysis of the metrics defined in Section 5.5.1.

5.6 Preliminary Implementation

We implemented a preliminary version of the R-P2P architecture over off-the-shelf components. In order to comply with the requirements deriving from the specific application scenario considered, we chose to rely mostly on commercially available software/hardware platforms. The architecture is composed of 2 parts: a mobile one, residing on users' personal handheld devices, and a fixed one, acting as the Throwboxes of the R-P2P architecture and interfacing with the fixed infrastructure. In the following, we provide a short overview of both components.

Table 5.2 presents a concise summary of the technologies and devices used in the R-P2P preliminary implementation.

Table 5.2: The R-P2P Architecture Implementation Details in a Nutshell

HW platform	Nokia E65, Nokia N80, Dell600
OS	Symbian OS 9.1, Ubuntu 6.11
SW platform	J2ME (MIDP 2.0), J2SE 1.5
BT version	Bluetooth 1.2

5.6.1 U-Hopper: The Mobile Part of R-P2P

U-Hopper is a User-centric Heterogeneous Opportunistic Middleware [5, 4] residing on mobile nodes. It exploits proximity communication interfaces (i.e., Bluetooth, Wi-Fi, etc.) in order to (i) gather localized information originating from sources embedded in the environment; (ii) opportunistically disseminate queries from other user nodes; and (iii) deliver data to Throwboxes at any occasion. The content being exchanged by user nodes includes data read from T-Nodes or queries generated by user nodes.

In order to embrace the largest number of "potentially available" mobile devices, we selected smartphones as the target platform to develop U-Hopper. In fact, smartphones are nowadays typically carried around by users during their daily activities; currently, also, they have reached a sufficiently large computing and communication power to perform very complex operations.[4] We developed U-Hopper as a Java MIDlet running over J2ME (MIDP profile 2.0), a widely diffused and standardized computing environment currently available on most smartphones shipped today. U-Hopper provides the necessary programming abstractions for implementation and deployment of R-P2P-based services and includes various modules of the R-P2P architecture. The functionalities provided by the U-Hopper middleware include:

- *data gathering*: it is possible to easily program data gathering tasks. The platform then takes care of all operations that are needed in order to first discover such data from the environment, and then issues a callback to all the services interested in the retrieved data. This is implemented as in Figure 5.8: when the two nodes meet, they first exchange their interests, a high-level description of the data they are interested in, and then the data itself. This data is then forwarded to the services.

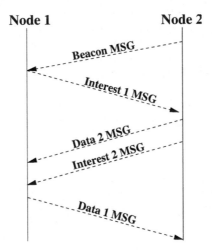

Figure 5.8: U-Hopper data handshake between any two nodes meeting.

- *storage*: the platform allows easy storage and retrieval of data in the permanent repository of the mobile device. The native J2ME storage support is the Record Management Store (RMS), which allows managing data only in a very basic format (byte arrays), making the work of

---

[4] As will be made clear in the following, the U-Hopper middleware can also run on other devices as long as the Java run time environment and some form of proximity communications are supported. Smartphones posses just the *minimum* set of hardware/software requirements to run the middleware platform.

developers cumbersome. We have then extended the J2ME RMS, in order to retrieve data using a declarative SQL-like language. An example of a U-Hopper query is reported in Alg. 1.

Algorithm 1 Example of a U-Hopper content query.

1: SELECT time, location, other
2: FROM events.trento.bars
3: WHERE time > 21 AND date = '21/06/2008'

- *communication*: the middleware manages all the communication tasks. Regarding the communication technology, we choose to rely on Bluetooth technology for achieving localized peer-to-peer data exchanges among mobile nodes. This is due to the large availability of this networking technology on mobile phones and to the available J2ME programming APIs (such as the well-known JSR 82 [1]. Obviously, Bluetooth is not properly designed for opportunistic communications, given the amount of time typically required for establishing a connection between two devices. To shorten up this connection time, we leveraged on some assumptions and on a few properties of the Bluetooth technology. As an example, we implemented a device caching mechanism that inhibits consecutive meetings of nodes. This allows speeding the Bluetooth peer and service discovery phases.

5.6.2   Throwboxes

Throwboxes functionalities are implemented as a Java application running on laptops. As for the U-Hopper implementation, we assumed Bluetooth as the proximity communication technology, utilizing the Avetana Bluetooth library as the Bluetooth stack implementation. This allows us to reuse much of the U-Hopper software developed for the mobile platform.

In the current version of R-P2P, the Throwboxes implementation has been oversimplified: as a first step, in fact, we assumed Throwboxes to be connected to an IP network and to access to a shared centralized MySQL database. Through this database it is possible to (i) store queried content and (ii) store advertisements. Clearly, the spatial/temporal indexing of data is simplified. In the next phase, the DHT functionalities will be implemented replacing the database with a full *DHT* solution and instantiating the related primitives accordingly.[5]

---

[5]The interested reader may refer to `u-hopper.create-net.org` for details and free code download.

5.7   Conclusions

In this chapter we described R-P2P, a system reproducing a request-response model for distributed queries. R-P2P leverages a Delay-Tolerant Wireless Network: queries, tagged with an identifier, are diffused through opportunistic communications and responses are stored and retrieved (by using the query identifier), through interconnected throwboxes equipped with a set of DHTs.

The proposed solution requires several extensions to enrich its usability and reliability. For instance, the language used for queries is rather crude; refinements could be introduced in order to make it more user-friendly and flexible: introduction of semantics support should be considered in order to enlarge the criteria for matching description of requested information/contents and the data available in a user node. The role of throwboxes in the system could be enriched in order to store fragments of ontologies or elements of folksonomies, accessed by the user node for tagging the available information or before formulating a query. A second aspect concerns the improvement of the robustness of the system, by avoiding that user nodes have a "free riding" behavior, i.e., nodes that perform queries but have a limited rate in processing the messages received by their neighbors (e.g., providing the required contents, or forwarding the messages). Moreover, security mechanisms should be introduced in order to limit the damages caused by malicious nodes, e.g., aiming at causing a DoS by forwarding fictitious or corrupted messages. Finally, the U-Hopper prototype will be enriched, by refining the implementation of throwboxes, and organizing them to provide a distributed storage, e.g., based on DHT or on a hierarchical solution supporting locality criteria.

The future technology will be pervasive in our environment. Mobile data mining will be part of the fabric of our life. We will be carrying this data in pods, phones, and ID cards. The patterns we make will all be networked into retrievable data and contents structures that can be sourced for information. The use of this information and data will allow richer new interpretations of the way our world is built, used, and designed, now and in the future. Even if today there are some shortcomings, for example, the limited processing power available in users' devices, in five to ten years the balance will be very different: technology will increase devices, processing power, while reducing dimensions and power consumption. In this scenario, technologies and solutions like R-P2P will provide viable technical solutions in those use-cases where several wireless mobile devices cooperate in order to feed local data to applications and leverage opportunistic communications.

5.8   Acknowledgments

This work has been partially supported by the European Commission within the framework of the BIONETS project IST-FET-SAC-FP6-027748, see www. bionets.eu.

Bibliography

[1] JSR-000082 JavaTM APIs for Bluetooth.

[2] BORREL, V., AMMAR, M. H., AND ZEGURA, E. W. Understanding the wireless and mobile network space: a routing-centered classification. In *Proc. of CHANTS* (New York, New York, USA, 2007), ACM, pp. 11–18.

[3] CARRERAS, I., MIORANDI, D., AND CHLAMTAC, I. A framework for opportunistic forwarding in disconnected networks. In *Proc. of Mobiquitous* (Palo Alto, California, USA, July 17–21, 2006).

[4] CARRERAS, I., AND TACCONI, D. U-hopper: User-centric heteogeneous opportunistic middleware. In *BIONETICS 2007 — SAC Workshop Demo Session* (Budapest, Hungary, December 2007).

[5] CARRERAS, I., TACCONI, D., AND MIORANDI, D. Data-centric information dissemination in opportunistic environments. In *MASS* (Pisa, Italy, October 2007).

[6] CHAINTREAU, A., HUI, P., CROWCROFT, J., DIOT, C., GASS, R., AND SCOTT, J. Impact of human mobility on the design of opportunistic forwarding algorithms. In *Proc. of INFOCOM* (Barcelona, Spain, April 23–29, 2006).

[7] FALL, K. A delay-tolerant network architecture for challenged internets. In *Proc. of ACM SIGCOMM* (Karlsruhe, Germany, March 25–29, 2003).

[8] HUI, P., CHAINTREAU, A., SCOTT, J., GASS, R., CROWCROFT, J., AND DIOT, C. Pocket switched networks and human mobility in conference environments. In *WDTN '05: Proceedings of the 2005 ACM SIGCOMM workshop on Delay-tolerant networking* (New York, New York, USA, 2005), ACM Press, pp. 244–251.

[9] IBRAHIM, M., NAIN, P., AND CARRERAS, I. Analysis of relay protocols for throwbox–equipped dtns. In *Proc. of WiOpt* (June 2009).

[10] KHELIL, A., BECKER, C., TIAN, J., AND ROTHERMEL, K. An epidemic model for information diffusion in MANETs. In *Proc. of ACM MSWiM* (Atlanta, Georgia, Sept. 28, 2002), pp. 54–60.

[11] KLEINBERG, J. The small-world phenomenon: An algorithmic perspective. In *Proceedings of the 32nd ACM Symposium on Theory of Computing* (2000), pp. 163–170.

[12] LAU, W. H. O., KUMAR, M., AND VENKATESH, S. A cooperative cache architecture in support of caching multimedia objects in MANETs. In *WOWMOM* (New York, NY, USA, 2002), ACM, pp. 56–63.

[13] LE, H., HONG, D., AND SIMMONDS, A. A self-organizing model for topology-aware overlay formation. In *ICC* (May 16–20, 2005).

[14] LEBRUN, J., AND CHUAH, C. Bluetooth content distribution stations on public transit. In *MobiShare* (Los Angeles, Sept. 2006).

[15] LEE, U., MAGISTRETTI, E., ZHOU, B., GERLA, M., BELLAVISTA, P., AND CORRADI, A. MobEyes: smart mobs for urban monitoring with vehicular sensor networks. Tech. Rep. 060015, UCLA CSD, 2006.

[16] LEGUAY, J., LINDGREN, A., SCOTT, J., FRIEDMAN, T., AND CROWCROFT, J. Opportunistic content distribution in an urban setting. In *Proc. of ACM Chants* (Florence, Italy, September 15, 2006).

[17] LINDEMANN, C., AND WALDHORST, O. P. A distributed search service for peer-to-peer file sharing in mobile applications. In *P2P* (Washington, DC, USA, 2002), IEEE Computer Society, p. 73.

[18] LUO, J., HUBAUX, J.-P., AND EUGSTER, P. T. Pan: providing reliable storage in mobile ad hoc networks with probabilistic quorum systems. In *MobiHoc* (New York, New York, USA, 2003), ACM, pp. 1–12.

[19] MAGHAREI, N., AND REJAIE, R. Mesh or multiple-tree: A comparative study of live p2p streaming approaches. In *Proceedings of IEEE INFOCOM* (2007), pp. 1424–1432.

[20] MALKHI, D., REITER, M., AND WRIGHT, R. Probabilistic quorum systems. In *PODC* (New York, New York, USA, 1997), ACM, pp. 267–273.

[21] OTT, J. Application protocol design considerations for a mobile internet. In *MobiArch '06* (New York, New York, USA, 2006), ACM, pp. 75–80.

[22] OTT, J., AND PITKNEN, M. Dtn-based content storage and retrieval. In *IEEE WoWMoM Workshop on Autonomic and Opportunistic Communications* (Helsinki, June 18–21, 2007).

[23] PAPADOPOULI, M., AND SCHULZRINNE, H. Effects of power conservation, wireless coverage and cooperation on data dissemination among mobile devices. In *ACM MobiHoc* (Long Beach, New York, 2001).

[24] SAILHAN, F., AND ISSARNY, V. Cooperative caching in ad hoc networks. In *MDM* (London, UK, 2003), Springer-Verlag, pp. 13–28.

[25] SCHLOSSER, M., SINTEK, M., DECKER, S., AND NEJDL, W. Hypercup – hypercubes, ontologies and efficient search on networks. In *LNCS* (2002), Springer, pp. 112–124.

[26] SHAH, R., ROY, S., JAIN, S., AND BRUNETTE, W. Data mules: Modeling a three-tier architecture for sparse sensor networks. In *IEEE SNPA Workshop* (May 2003).

[27] SPYROPOULOS, T., PSOUNIS, K., AND RAGHAVENDRA, C. S. Spray and wait: An efficient routing scheme for intermittently connected mobile networks. In *SIGCOMM WDTN* (2005), ACM.

[28] STOICA, I., MORRIS, R., LIBEN-NOWELL, D., KARGER, D. R., KAASHOEK, M. F., DABEK, F., AND BALAKRISHNAN, H. Chord: a scalable peer-to-peer lookup protocol for internet applications. *IEEE/ACM Trans. Netw. 11*, 1 (2003), 17–32.

[29] YANG, G., CHEN, L.-J., SUN, T., ZHOU, B., AND GERLA, M. Ad-hoc storage overlay system (asos): A delay-tolerant approach in MANETs. In *IEEE MASS* (Vancouver, October 2006).

[30] YIN, L., AND CAO, G. Supporting cooperative caching in ad hoc networks. *IEEE Trans. on Mobile Computing 5*, 1 (2006), 77–89.

[31] ZHANG, R., HU, C., LIN, X., AND FAHMY, S. A hierarchical approach to internet distance prediction. In *ICDCS '06: Proceedings of the 26th IEEE International Conference on Distributed Computing Systems* (Washington, DC, USA, 2006), IEEE Computer Society, p. 73.

[32] ZHANG, R., TANG, C., HU, Y. C., FAHMY, S., AND LINY, X. Impact of the inaccuracy of distance prediction algorithms on internet applications — an analytical and comparative study. In *INFOCOM* (April 23-26 2006).

Chapter 6

Mobile P2P: Peer-to-Peer Systems over Delay Tolerant Networks

Angela Sara Cacciapuoti, Marcello Caleffi, and Luigi Paura

6.1 Introduction 160
6.2 Peer-to-Peer Overlay Networks 161
    6.2.1 Overview 161
    6.2.2 Structured Peer-to-Peer Overlay Networks 162
    6.2.3 Unstructured Peer-to-Peer Overlay Networks 164
6.3 Delay Tolerant Networks 165
    6.3.1 The Store-Carry-Forward Paradigm 166
    6.3.2 MANETs as a Special Case of DTNs 166
6.4 Mobile Peer-to-Peer Overlay Networks For DTNs 168
    6.4.1 Challenges 169
    6.4.2 Unstructured Mobile Peer-to-Peer Overlay Networks 170
        6.4.2.1 Optimized Routing Independent Overlay
                Network 171
        6.4.2.2 Mobile Peer-to-Peer 172
        6.4.2.3 Ad-Hoc Storage Overlay System 173
        6.4.2.4 Peer-to-Peer Swarm Intelligence 173
        6.4.2.5 Prophet-Based Information Retrieval 174
    6.4.3 Structured Mobile Peer-to-Peer Overlay Networks 175
        6.4.3.1 Mobile Ad Hoc Pastry 175
        6.4.3.2 Indirect Tree-Based Routing 176
        6.4.3.3 Virtual Ring Routing 178
        6.4.3.4 Opportunistic DHT-Based Routing 179

      6.4.4   Summary and Open Problems 180

6.5   Conclusions 182

6.1 Introduction

P2P systems are distributed systems able to form self-organizing overlay networks to provide efficient search/distribution of data items [1]. By introducing the concept of *peer*, i.e., an entity that provides and, at the same time, consumes resources/services offered by others entities, P2P systems go beyond the traditional client/server paradigm thanks to its features of self-organization, fault-tolerance, and high scalability. In the last years, the P2P paradigm has gained popularity as a consequence of the diffusion of Internet file sharing applications like Napster [2], Gnutella [3], and Emule [4], which have allowed millions of users to share files in a decentralized manner.

On the other hand, DTNs represent a novel paradigm for wireless multihop networks that aims to provide connectivity also when links on an end-to-end path may not exist contemporaneously and, therefore, intermediate nodes may need to store data waiting for communication opportunities [5].

As pointed out in [6], DTNs and P2P systems share the same key concepts of self-organization and distributing computing, and both aim to work in a completely decentralized environment. Both lack central entities to which to delegate the management and the coordination of the network, and both rely on a time-variant topology. In fact, in P2P networks the time-variability is due to joining/leaving peers, while in DTN ones it is due to both node mobility and wireless propagation condition instability. Finally, both adopt a *store-and-forward* like paradigm: DTN nodes store packets waiting for a chance to deliver them to the destinations, while peers store data items waiting for requests from other peers.

Despite these similarities, the adoption of the P2P paradigm to disseminate and discover information in a DTN raises new and challenging problems [7, 8]. One of the main issues concerns the layer where they operate. P2P systems build and maintain overlay networks at the application layer, assuming the presence of an underlying network layer which assures connectivity among nodes. DTNs focus on providing a multi-hop wireless connectivity among nodes in scenarios where frequent and numerous network partitions would prevent packets from being delivered in a timely fashion. In addition, traditional P2P systems rely on wired infrastructures, characterized by reliable and bandwidth-supplied links. On the other hand, DTNs rely on unreliable and bandwidth-limited wireless links. Finally, DTN nodes usually have hard constraints on the available resources like energy, memory, and computation, while peers are commonly assumed to be resourceful.

For these reasons, trying to couple a P2P overlay network over a DTN is still an open problem. Nevertheless, since the ad hoc network paradigm, which assumes that most of the time an end-to-end connectivity between each pair of nodes exists, can be considered as a special case of DTN [5], we start looking at the P2P solutions proposed for MANETs. However, since DTNs share with P2P systems a store-and-forward like paradigm which requires a unitary approach able to assure the effectiveness of integrated solutions, we describe some interesting examples of such an integrated approach.

The remaining part of the chapter is organized as follows. In Section 6.2 an overview of P2P overlay networks is provided, by distinguishing the unstructured P2Ps from the structured ones. Section 6.3 describes the main features and applications of DTNs. The store-carry-forward paradigm is presented and it is highlighted how MANETs can be considered a sub-class of DTNs. In Section 6.4 we provide the main challenging issues for the design of P2P overlay networks in a DTN, and the main features of some representative Mobile P2P (MP2P) systems are discussed. This section is concluded with the open problems. Finally, conclusions are drawn in Section 6.4.

6.2 Peer-to-Peer Overlay Networks

Peer-to-Peer (P2P) systems represent a recently proposed scalable and fault-tolerant paradigm to disseminate and discover information in a communication network. In this section, we provide an overview of such a paradigm in terms of the main characteristics, some interesting applications, and finally some illustrative examples.

6.2.1 Overview

The Peer-to-Peer (P2P) paradigm is an application level paradigm that aims to share both resources and services and in which the involved entities, namely the *peers*, behave both as resource/service consumers and providers. Such a paradigm assumes that the peers collaborate spontaneously by means of distribute procedures without the necessity of establishing a hierarchy and/or relying on a pre-existent infrastructure. Differently from the traditional client/server paradigm, the lack of hierarchy guarantees the absence of bottlenecks and single point of failures, allowing the P2P paradigm to exhibit properties of fault-tolerance and high scalability.

The P2P paradigm does not deal with the communication issues, since it assumes the presence of an underlying layer, which assures connectivity among nodes. In this sense, it defines an overlay network, i.e., a logical network built on top of the physical one. Figure 6.1 shows an example of an overlay network built upon a physical one.

We note that usually the logical proximity, i.e., the proximity in the overlay network among peers, is not related to the physical one, namely, the proximity in the physical topology. Therefore, two neighboring peers in the overlay network are likely not to be neighbors in the physical one, and so one logical hop usually involves multiple physical hops. Moreover, although traditional

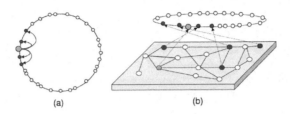

Figure 6.1: Overlay network.

P2P systems are commonly developed over static networks, the related overlay networks are characterized by time-variant topologies due to the peer joining/leaving.

P2P systems can be grouped in two classes, namely *unstructured* and *structured* systems, according to the solution adopted for content dissemination/discovery. More specifically, in unstructured P2Ps, peers are unaware of the resources that neighboring peers in the overlay network maintain. So, they typically resolve search requests by means of flooding techniques, and they rely on resource replication to improve the lookup performance and reliability. Differently, in structured P2P networks peers have knowledge about the resources offered by overlay neighbors, usually by resorting to the Distributed Hash Table (DHT) paradigm, and, therefore, the search requests are forwarded by means of unicast communications. We note that in both the approaches the locations where data items have to be stored are selected regardless of the physical topology of the network.

In the following paragraphs we present the main features of each approach, along with some illustrative examples.

6.2.2   Structured Peer-to-Peer Overlay Networks

The adoption of a structured approach for content dissemination/discovery imposes that data items must be placed at specific peers according to a globally known rule. In this way, the items can be efficiently retrieved with unicast communications, avoiding so the inefficiency of the flooding techniques adopted in unstructured systems.

Usually, structured P2P systems utilize as data structures the Distributed Hash Tables (DHTs), which allow retrieval of data items without the need of a-priori knowledge about the locations where the items are stored. More specifically, each peer has a unique identifier, namely a *peer id*, belonging to the identifier space I, and each data item is univocally identified by a *key* belonging to the key space K.

The core of a DHT is a globally known hash function $h : K \rightarrow I$ able to map data items on live peers. By means of the hash function, a P2P system

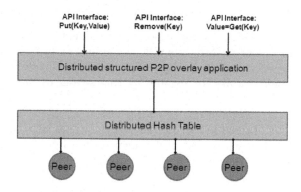

Figure 6.2: Application interface for structured DHT-based P2P systems.

is able to provide a scalable storage and retrieval of data items in the overlay network by means of three common interfaces: *put, remove,* and *get,* as shown in Figure 6.2 derived from [1]. Given the data item and the corresponding key, the put operation *put(key,value)* stores the data item *value* at the peer whose identifier is equal to $h(key)$. The remove operation *remove(key)* simply removes from the hash table the data item corresponding to the key. Finally, the lookup operation *get(key)* retrieves the data item corresponding to the key.

Structured P2P systems require that each peer maintains a table that stores, for each logical neighbor peer, both its identifier and its Internet Protocol (IP) address. The communication among peers exploits the overlay neighborhood: when a peer has to send a message to another one, it forwards the message to the neighbor peer whose identifier is the closest to the destination one according to a certain metric (e.g., numerically closest, shortest Euclidean distance, etc.). In such a way, structured P2P systems impose a *structure* on the overlay network topology, and the defined structure depends on the particular P2P protocol. Typical structures are the ring, the tree, and the butterfly.

In theory, DHT-based systems can guarantee that each data item can be located/retrieved in $O(\log n)$ overlay hops, where n is the number of peers in the system. However, since the underlying network path between two neighbor peers can be composed by several physical links, the latency times for data items dissemination/discovery can be quite long and can affect the overall performances. Moreover, the table maintenance can introduce a considerable overhead.

Widely-known examples of file-sharing applications based on structured P2P overlay networks are Content Addressable Network (CAN) [9], Tapestry [10], Chord [11], Pastry [12], Kademlia [13], and Viceroy [14].

6.2.3 Unstructured Peer-to-Peer Overlay Networks

Unstructured P2P systems do not impose that content has to be placed at a pre-defined peer. In other words, they do not impose a pre-defined structure on the overlay network and, therefore, the peers have to resort to strategies like flooding, random walks, or expanding-ring search to discover the data items.

From an operational point of view, when a peer receives a resource, query, it first locally evaluates the query on its own data items. Then, it replies to the requesting peer with a list of the owned data items corresponding to the query. Such a strategy is easy to implement and, moreover, it natively supports complex keyword-based queries.

Nevertheless, the lack of relation between a data item and its location implies scalability issues. The strategy, in fact, is effective in case of widely replicated data items, while in case of rare contents the queries have to be sent to a large set of peers, thus incurring a considerable overhead. On the other hand, structured P2P systems are able to efficiently locate rare items, but they incur significant overhead in discovering popular content.

Unstructured P2P systems can be classified in three main groups according to the adopted architecture: centralized, de-centralized, and hybrid, as shown in Figure 6.3. In centralized P2P systems some functions are provided by a central entity, which coordinates and provides auxiliary information to peers. Nevertheless, peers communicate directly without any intermediate entity. The advantages of P2P centralized systems are easy management and implementation of security policies. On the other hand, the presence of the central entity limits the scalability and introduces single points of failures. Examples of file-sharing applications based on centralized P2P architectures are Napster [2] and SETI@home project [15]. Decentralized P2P systems, like Freenet [16] and Gnutella [3], are based on a flat peer hierarchy where all

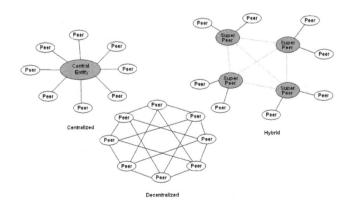

Figure 6.3: Unstructured P2P architectures.

Table 6.1: Characteristics of Different Unstructured P2P Architectures

		Architectures	
Characteristics	Centralized	Decentralized	Hybrid
Manageable	Yes	No	No
Extensible	No	Yes	Yes
Fault-tolerance	No	Yes	Yes
Secure	Yes	No	No
Scalable	No	Maybe	Maybe

peers share the same role. As advantages, the decentralized systems are scalable and they exhibit fault tolerance properties. Finally, hybrid P2P systems try to conjugate both the advantages of centralized and decentralized architectures. In such systems, peers are organized in clusters, and the cluster-heads, namely the *super-peers*, are responsible for forwarding queries received by the peers. The communications among super-peers are decentralized since hybrid P2P systems adopt a two-level hierarchy. Examples of hybrid architecture are KazaA [17] and Morpheus [18]. Table 6.1, derived from [19], summarizes the main characteristics for each architecture.

6.3 Delay Tolerant Networks

Disruption or Delay Tolerant Networks (DTNs) are an emerging class of networks in which the assumption of a persistent end-to-end path between each pair of nodes is relaxed. DTNs are characterized by the following features [20]:

- intermittent connectivity: a DTN exhibits a weak, episodic connectivity as a consequence of unstable end-to-end paths;

- unpredictable end-to-end delays: as a consequence of intermittent connectivity, the end-to-end delays can exceed the requirements of real-time applications or protocols that rely on quick return of acknowledgments or data;

- asymmetric communications: communications exhibit asymmetric characteristics (data rates, loss rates, delays, etc.);

- unreliable communications: DTN routes are characterized by unreliable communications, and end-to-end Automatic Repeat Request (ARQ) strategies cannot be adopted in presence of long delays.

In addition, DTN nodes can have strong resource limitation (power, storage and computation), especially in case of mobile networks.

Several military and civilian applications can benefit from the DTN paradigm. Typical examples are the deep-space networks such as the NASA

JPL's Deep Impact Networking (DINET) [21] (where the delay/disruption tolerance is required due to long delays and high packet loss of the interplanetary communications), networks for satellite communications [22], and networks for rural areas such as Kiosknet[23].

6.3.1   The Store-Carry-Forward Paradigm

A typical networking paradigm for DTNs characterized by sparse topologies is the *store-carry-forward* one, which assumes that a message is stored and carried by the nodes, until an opportunity to deliver the message arises. An example of the process is shown in Figure 6.4 through Figure 6.6. More specifically, in Figure 6.4 node S has to communicate with D, but there is not any connected path between the two nodes. Therefore, S has to store the message, waiting for a communication opportunity. The node R acts as a relay for S by carrying the message as shown in Figure 6.5, and forwarding it to D as depicted in Figure 6.6.

The routing protocols which adopt the store-carry-forward paradigm can be classified in two main groups, according to the assumptions made about the available knowledge of the network topology [24].

The first class of protocols requires a minimal knowledge about the topology. In such a case, the simplest delivery strategy is to replicate the messages in the network. In more detail, the source forwards a copy of the message each time that another node comes into its communication range. The same procedure is followed by the receiving node, by forwarding copies of the same message to nodes which in turn come in contact with it. Clearly, this implies that several copies of the same message are present in the network, wasting the resources. This strategy is the basic idea behind *Epidemic Routing* protocols [25, 26, 27, 28], which try to solve the scalability issues by adopting some limitations on the message replication, i.e., by limiting the number of copies for each message or by using historical encounter-based metrics.

The second class of protocols, namely the *message ferrying ones*, assumes the presence of well-connected islands of nodes that intermittently communicate with each other thanks to node mobility [29, 30, 31, 32]. The mobile nodes responsible for carrying the messages among the islands are called *ferries*. Differently from epidemic routing-like protocols, the message ferrying ones adopt single copy forwarding strategies.

6.3.2   MANETs as a Special Case of DTNs

As mentioned before, DTNs are a class of wireless networks that aims to provide connectivity in the absence of a persistent end-to-end path between nodes, usually by requiring that relays store the messages waiting for connectivity. On the other hand, ad hoc networks are wireless networks in which a persistent end-to-end connectivity between each pair of nodes exists. In these networks it is assumed that if a path fails due to node mobility and/or wireless

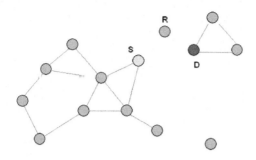

Figure 6.4: S stores the message waiting for a communication opportunity.

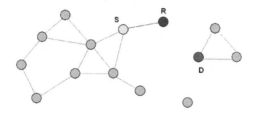

Figure 6.5: S forwards the packet to R.

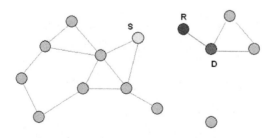

Figure 6.6: R carries the message, and forwards it to D.

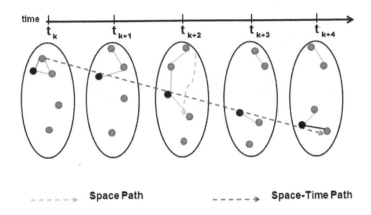

Figure 6.7: Example of a DTN space-time path.

propagation conditions instability, such a failure is temporary since alternative routes are likely available.

According to [5], we define the un-persistent paths of DTNs as *space-time paths* to underline that not all the links belonging to the path exist simultaneously, while ad hoc paths are referred to as *space paths*. Figure 6.7 derived from [5] shows an example of a path in which the links appear at different temporal intervals.

According to this classification, space paths are a special case of space-time paths, and, therefore, ad hoc networks can be considered as a sub-class of DTNs. Clearly, since Mobile Ad hoc NETworks (MANETs) are a special case of ad hoc networks in which nodes are mobile, they are also a special case of DTNs. This classification is depicted in Figure 6.8.

In the following, we will use such a classification to distinguish P2P systems proposed for DTNs from those proposed for MANETs and to analyze the limits of the latter ones when they are applied on mobile DTNs.

6.4 Mobile Peer-to-Peer Overlay Networks for Delay Tolerant Networks

In this section we first present the main challenging issues for the design of P2P overlay networks in a DTN, and then we describe the main features of some representative Mobile P2P (MP2P) systems, that is P2P systems for mobile multi-hop wireless networks. The considered proposals have been selected according to one or more of the following motivations: i) they may be popular choices among the research community; ii) they may be illustrative examples of interesting approaches; iii) they may have unique features that

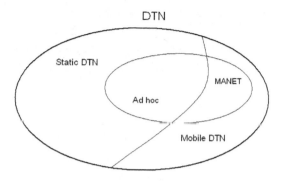

Figure 6.8: Classification of wireless multi-hop networks.

make them appealing. In the following no comparison is carried out among the considered overlay networks, since their citations often provide performance evaluations of the systems.

6.4.1 Challenges

Providing an efficient architecture for information dissemination/discovery in DTNs is an open problem, since there are several challenging issues related to sparse topologies and intermittent connectivity.

DTNs operate with a smaller bandwidth than MANETs since in space-time paths links are available at different temporal intervals. Clearly, this implies that it is necessary to adopt solutions that avoid high overlay maintenance traffic (common in structured P2P overlay networks) or inefficient flooding-based searches (common in unstructured ones) to make them suitable for DTNs.

Moreover, the typical unpredictable delays of DTNs affect the information dissemination/discovery procedures. Some redundancy in queries/content forwarding is necessary to compensate the unreliability of wireless communications. Nevertheless, congestion due to excessive query/content messaging has to be avoided.

As mentioned in Section 6.2.1, traditional P2P systems exploit a logical proximity among peers that is not related to the physical one. More in detail, messages are routed among peers which are neighbors in the overlay network. However, since two logically neighboring peers are likely not to be neighbors in the physical topology, each logical hop usually involves multiple physical hops, thus introducing a considerable overlay route stretch effect. Mobile P2P systems for DTNs should implement a kind of relation between the overlay and the physical topology to avoid such a route stretch effect.

Finally, implementing security policies in decentralized, self-organizing,

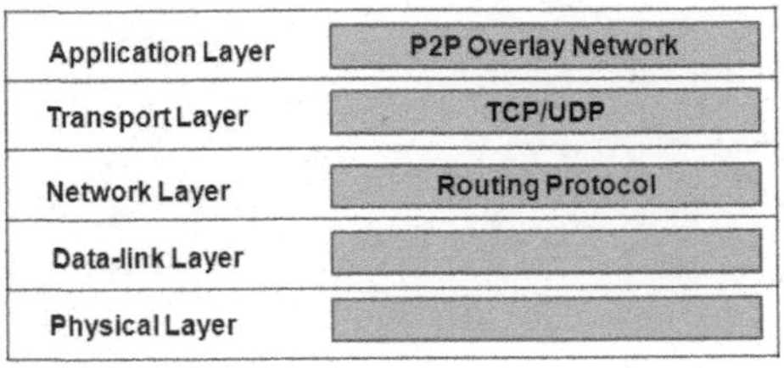

Figure 6.9: Traditional layered approach.

and anonymous systems like the P2P ones is a complex task, which becomes harder in MP2P due to the node mobility and broadcast characteristics of wireless communications.

These challenging issues require new approaches to provide scalable MP2P systems [33]. It has been proved that the traditional layered approach (Figure 6.9), namely simply deploying a P2P overlay network over an unreliable network substrate, causes significant message overhead and redundancy due to the lack of cooperation and communication between the two layers. For these reasons, several proposals exploiting the cross-layer approach (Figure 6.10) have been presented in the last years. In these systems, an inter-layer communication between the network and the application layers is introduced, thus allowing a weak interaction between the routing and the P2P functionalities. However, very recent solutions [34, 6, 35] that integrate the P2P services at the network layer have been proposed (Figure 6.11), thus allowing a more strong interaction between the two layers.

6.4.2 Unstructured Mobile Peer-to-Peer Overlay Networks

Unstructured MP2P overlay networks generally provide flooding-based content discovery using reactive routing protocols as network substrate.

In the following, we describe five unstructured MP2P systems: the Optimized Routing Independent Overlay Network (ORION) [36], the Mobile

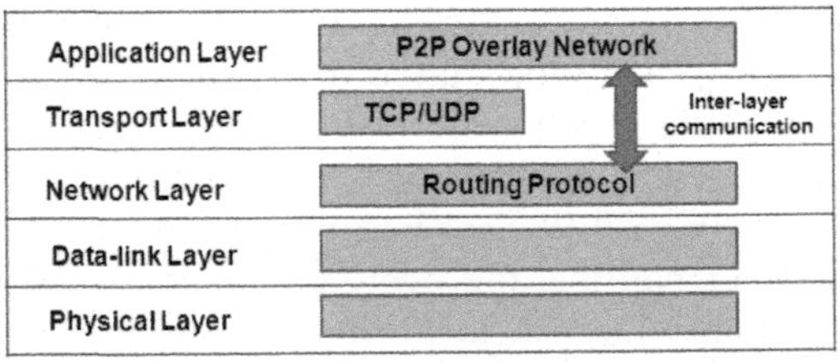

Figure 6.10: Cross-layered approach.

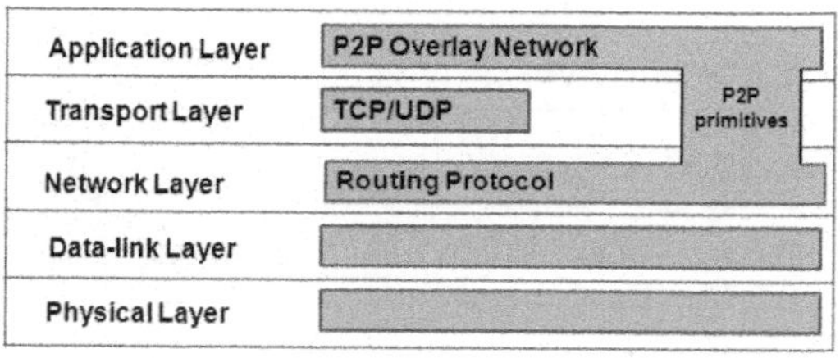

Figure 6.11: Integrated approach.

Peer-to-Peer (MPP) [37], the Ad-hoc Storage Overlay System (ASOS) [38], the Peer-to-Peer file sharing system based on Swarm Intelligence (P2PSI) [39], and a Prophet-based information retrieval system [40]. A brief discussion about the limitations of each system is also provided.

6.4.2.1 Optimized Routing Independent Overlay Network

The Optimized Routing Independent Overlay Network (ORION) [36] offers file-sharing services over MANET scenarios with a layered approach. ORION provides advanced keyword-based content discovery using as network substrate the Ad hoc On demand Distance Vector (AODV) [41] and a Gnutella-like overlay network.

Each node maintains a list of the data items locally stored and an AODV-like routing table for the reverse paths. When a node needs a data item, it floods the network with a query message. Each node that receives the query first stores in the routing table the reverse path towards the source node, i.e., it stores as next hop toward the source the node that has forwarded the query. Then, it broadcasts the query to the (physical) neighbor nodes. Finally, it looks in the local data item list for content matching the query and it replies with a query reply message in case of success.

To reduce the overhead of the discovery process, ORION adopts reduced query replies. The intermediate nodes belonging to the path of a query reply do not forward messages for already discovered data items.

Once the query source acquires the knowledge about the nodes that store the data item, it splits the data item in equal length blocks and sends a data request message for each block toward one of the storing nodes. When the data request is received by the storing node, it replies with a data reply message which contains the requested block and which follows the same reverse path discovered during the query process.

The adoption of ORION in DTNs poses several issues. The main problem is due to the assumption about the persistence of reverse paths, which is clearly unrealistic in DTNs. Moreover, the flooding-like strategy adopted for content discovery is not suitable in bandwidth-limited environments.

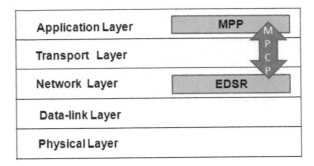

Figure 6.12: MPP cross-layer architecture.

6.4.2.2 Mobile Peer-to-Peer

The Mobile Peer-to-Peer (MPP) [37] offers MANET file-sharing functionalities with a cross-layered approach, by combining MANET routing with flooding-based content discovery. More in detail, MPP uses as network substrate a modified Dynamic Source Routing (DSR) protocol [42], namely the Enhanced Dynamic Source Routing (EDSR). The inter-layer communication between network and application layer is provided by Mobile Peer Control Protocol (MPCP), which allows the P2P application to register itself in the EDSR layer, as shown in Figure 6.12. In such a way, the application can initialize search requests and it can process incoming requests from other nodes.

On startup, the P2P application on the mobile node announces itself to the EDSR layer via MPCP. When a node has to access to a data item, MPCP forwards the P2P application request to EDSR, which in turn transforms it into a search request. Similar to DSR route requests, EDSR floods the search request through the network and when a node receives the request via the EDSR substrate, it forwards such a request to the P2P application via MPCP. In such a way, the application layer can determine if any locally stored data item satisfies the request criteria. If so, the application layer initializes an EDSR data reply, which is sent back to the originating node and contains all necessary information for the data item transfer. Similar to DSR route replies, a data reply includes the complete path between source and destination.

MPP adopts the Hyper Text Transfer Protocol (HTTP) for the data item transfers between peers. Moreover, MPP specifies additional features to overcome the connection break events by allowing peers to continue the transfer from the last received byte.

Besides the adoption of a cross-layer approach, MPP shares the same limitations of ORION when applied to DTNs: flooding inefficiency and persistent path assumption. In particular, the inefficiency of MPP in case of un-persistent paths is made worse by its source routing nature. In fact, the complete ordered list of nodes through which the packets have to pass is singled out at the source side.

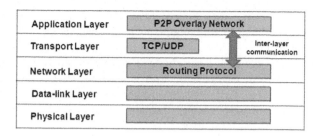

Figure 6.13: P2PSI cross-layer architecture.

6.4.2.3 Ad-Hoc Storage Overlay System

The Ad-hoc Storage Overlay System (ASOS) [38] is a self-organized P2P system specifically designed for MANETs. Nevertheless, the proposed approach is suitable for DTN scenarios, since it tolerates disruption-prone communications.

ASOS assumes the existence of nodes with high memory capabilities, namely ASOS agents, which are exploited to provide reliable communications over unreliable paths. In case of link failures, the ASOS agents cache the data items and deliver them to the original destinations when the connectivity is restored.

From an operational point of view, after the source node submits a data item to its ASOS agent, it becomes the first ASOS peer to hold a copy of such an item. To increase storage reliability, the item is also replicated to other ASOS agents, by selecting among the reachable neighbors of the agent $K-1$ locations to replicate the data to (where K is a configurable parameter). With the assumption that pairwise distances between nodes can be measured, storage locations are selected based on three guidelines: distance from the destination, distance from other ASOS agents, and load of the agent.

ASOS supports both implicit and explicit data deletion and replacement. In the explicit scheme, the original source or destination deletes the item from the system when the data is successfully delivered or it is not useful anymore. In the implicit scheme, instead, the system can accommodate storage scarcity with prioritized storage management, such as the Least Recently Used (LRU) and First-in-First-out (FIFO) algorithms.

Although ASOS is a promising P2P overlay, the authors underline that it is necessary to further investigate the ASOS performances in a DTN environment. Moreover the authors assume the availability of pre-configured agents, which is clearly unrealistic in DTN scenarios.

6.4.2.4 Peer-to-Peer Swarm Intelligence

In [39], a Peer-to-Peer file sharing system based on Swarm Intelligence (P2PSI) has been presented. Swarm intelligence [43] is an artificial intelligence technique based on the study of biological swarms, such as ants or bees. P2PSI applies it to the problem of implementing P2P services in MANETs with a cross-layer approach.

P2PSI adopts as network layer the Ant-colony-based Routing Algorithm (ARA) [44] as shown in Figure 6.13, and it assumes that a large fraction of users are *free-riders*, i.e., they consume resources without providing them. In other words, there are only a small fraction of collaborative nodes, namely *HotSpots*, that store and share files.

P2PSI relies on two processes: advertisement (push) and discovery (pull). In the advertisement process each HotSpot periodically advertises to neighbor nodes the available data items within a limited area. The amount of information about the available items is limited by means of Bloom filtering techniques [45].

Query messages are forwarded at intermediate nodes based on the pheromone tables stored at these nodes. A pheromone table records the pheromone intensity for each neighbor. Intuitively, the pheromone intensity of a neighbor denotes the probability that a query message reached the destination via that neighbor.

When a node receives a resource query, it looks for the requested data item in the cached advertise messages and it eventually replies to the originating node with the identity of the node storing the item. The reply message reinforces the pheromone information along the way. In such a way, subsequent query messages that look for the same data items can follow previously laid pheromone information, without the need of a further route discovery process. Although the swarm intelligence is a very promising research area, more investigations are required to understand how such an approach performs over un-persistent routes. Moreover, the proposed solution requires a careful setting of several parameters, which is not as easy in dynamic environments as in DTNs ones.

6.4.2.5 Prophet-Based Information Retrieval

In a very recent work [40], the authors propose an unstructured content-based information retrieval system for DTNs, by focusing on the aspects related to data caching, query disseminations, and message routing.

For data caching, two schemes have been proposed, namely *random caching* and *intelligent caching*. In the first scheme each node storing a data item creates K tokens, where each token represents the right to make a copy of the data item. Then, the node spreads half of the owned tokens, along with a copy of the data item, to the nodes that it meets. Instead, the intelligent caching scheme requires that the K tokens are spread to nodes selected according to a friendliness metric, which represents a measure of the average number of nodes met during an observation period.

As regards to query dissemination, the authors propose two strategies: *W-copy selective query spraying* (WSS) and *L-hop neighborhood query spraying*

(LNS). The WSS strategy replicates a query to nodes selected according to the friendliness metric, while in the LNS one, each query is replicated to L-hop neighborhood, that is the nodes which are distant L hops from the originating node.

The authors adopt for message routing an enhanced version of the Prophet protocol [26], although different routing strategies can be used. According to the adopted strategy, messages are forwarded to neighbors based on the estimate of the delivery probability to the destination.

The numerical performance analysis conducted by the authors reveals that the best solution is the combined use of intelligent caching with the WSS scheme.

Although the proposal deals with DTNs and therefore it does not suffer from the issues underlined for the above discussed solutions, there are some open problems pointed out by the authors. One problem is related to the design of indices for the cached data items that allow nodes to determine if the newly encountered ones carry any data items that match the queries stored locally. Another problem is the assumption of fixed expiration times for the data items, since realistic applications require strategies to invalidate expired data.

6.4.3 Structured Mobile Peer-to-Peer Overlay Networks

Structured MP2P overlay networks generally adopt the Distributed Hash Table (DHT) paradigm as substrate to provide scalable content management. Although the use of DHTs simplifies the discovery process thanks to the a-priori knowledge of the data items locations, the management of DHT tables is still an open problem in disconnected networks like DTNs ones.

In the following, we describe four proposals: the Mobile AD-hoc Pastry (MADPastry) [46], the Indirect Tree-based Routing (ITR) [6], the Virtual Ring Routing (VRR) [34], and the Opportunistic DHT-based Routing (ODR) [35]. For each system, we underline the main features along with the main limitations.

6.4.3.1 Mobile Ad-Hoc Pastry

Mobile AD-hoc Pastry (MADPastry) [46] is a structured cross-layered P2P overlay network for MANETs in which a Pastry-like [12] application layer is combined with the Ad-hoc On demand Distance Vector (AODV) [41] routing protocol.

In standard DHTs, there is no relation between the overlay distance and the physical one, thus causing large overlay route stretch. To solve this issue, MADPastry utilizes the concept of Random Landmarking [47] to cluster nodes according to the overlay identifiers. Thus, two nodes that are physically close in the physical topology are also likely to be close in the overlay network.

To couple with node mobility, MADPastry does not rely on fixed land-mark nodes. Instead, it uses a set of landmark keys chosen so that they split the overlay id space into equal-sized segments. Nodes whose overlay identifiers are currently closest to one of the landmark keys become temporary landmark nodes. Clusters are formed by imposing that nodes have to associate themselves with the temporary landmark node that is currently closest to them. The association consists of adopting its overlay id as identifier prefix. Since broadcast messages introduce excessive overhead in resource-constrained networks, landmark beacons are only propagated within the landmark's own cluster.

MADPastry routes packets based on the overlay identifiers, i.e., by means of indirect routing. Each route is composed of several overlay hops, and each overlay hop corresponds to a physical path composed of several physical hops. At each overlay hop, the node will consult its Pastry routing table to find the node whose overlay identifier is numerically closest to the key. On the other hand, at each physical hop the node looks for the next physical hop in its AODV routing table.

To avoid unnecessary overhead in case of absence of valid information about the overlay next hop, a node belonging to the target's cluster broadcasts the data item within the confines of its cluster. Otherwise, if the node does not belong to the target cluster, it queues the data item and starts a regular AODV expanding ring broadcast to discover a route to the item's destination.

As pointed out by [48], the main issue with MADPastry is that queries are very sensitive to changes in AODV (physical) routes. In addition, the indirect routing based on overlay identifiers is a form of source routing, which is unsuitable in the absence of persistent paths.

6.4.3.2   Indirect Tree-Based Routing

The authors in [6] propose the Indirect Tree-based Routing (ITR), which extends the Multi-path Dynamic Address RouTing (M-DART)[49, 50], a DHT-based routing protocol for MANETs, by providing fully functional P2P services.

ITR assigns location-dependent identifiers, namely strings of l bits, to peers by means of a distribute procedure and locally broadcasted hello packets. The peer identifier space can be represented as a complete binary tree of $l+1$ levels as shown in Figure 6.14-a, that is, a binary tree in which every vertex has zero or two children and all leaves are at the same level. In the tree structure, each leaf is associated with a peer identifier, and an inner vertex of level k, namely a level-k subtree, represents a set of leaves (that is a set of peer identifiers) sharing a prefix of $l - k$ bits. A level-k sibling of a leaf is the level-k subtree which shares the same parent with the level-k subtree the leaf belongs to.

Indirect Tree-based Routing performs the whole routing by adopting an iterative procedure that explores the topological meaning of the node identi-

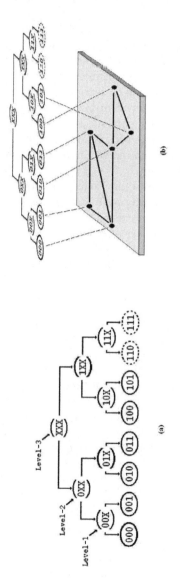

Figure 6.14: Address space and overlay network of ITR.

fiers with a hierarchical form of multi-path proactive distance-vector routing [51]. Each node stores a routing table with l sections, one for each sibling, and the k-th section stores the physical 1-hop neighbor peers which can forward a packet towards peers whose location-dependent identifiers belong to the level-k sibling.

From an operational point of view, ITR performs like traditional P2P systems: namely, when a node stores a resource, it periodically sends a pointer composed of a resource identifier and a storing peer identifier to the rendezvous-point, i.e., the node responsible (according to the hash function) for that resource. When a node has to retrieve a resource, it sends a resource query to such a rendezvous-point. Similarly, for MANET communications, each node periodically sends its current identifier to the rendezvous-point. When a node has to communicate with a node, it will send an identifier query to its rendezvous-point. After the reception of the query reply, the node can start a MANET communication.

The key-feature of ITR is the ability to forward both resource and identifier queries without introducing overlay path stretch, since the overlay distance is strictly related to the physical one, as shown by Figure 6.14-b. Although ITR is one of the first proposals that tries to couple location-dependent identifiers with P2P overlay networks, the address space overlay management introduces a considerable overhead which could not be suitable in DTNs. Moreover, the ITR performances in DTNs have not been evaluated.

6.4.3.3 Virtual Ring Routing

In Virtual Ring Routing (VRR) [34], the authors adopt an integrated approach to provide connectivity in MANETs by exploiting the DHT paradigm. Like ITR, VRR integrates the DHT functionalities directly at the network layer, by providing both direct and indirect routing.

Nodes are identified by means of random location-independent unsigned integers, organized into a virtual ordered ring. Each node maintains a small number of routing paths, say r, to its logical neighbors, namely neighbors in the virtual ring. In more detail, it proactively stores the paths towards the $r/2$ closest neighbors clockwise in the virtual ring and the $r/2$ closest neighbors counter clockwise. Since node identifiers are random and location independent, the virtual neighbors of a node will be randomly distributed across the physical network and each virtual path is composed of several physical hops. Each node also stores a physical neighbor set by means of locally broadcasted hello packets.

The virtual neighborhood is used to route a packet in the network, by forwarding it to the node whose identifier is numerically closest to the destination in the overlay network. In addition, physical neighbors are exploited for packet forwarding to limit the overlay path stretch.

Since VRR routes messages sent to numerical keys to the node whose identifier is numerically closest to the key, it also supports DHT functionalities when the keys identify data items instead of VRR nodes.

Although VRR adopts an integrated approach, the management of the routing tables poses a strong issue about its application in DTNs since the node identifiers are randomly assigned to peers, i.e., there is no relation between logical and physical topology. In addition, like ITR, performances in DTNs have not been evaluated.

6.4.3.4 Opportunistic DHT-Based Routing

The Opportunistic DHT-based Routing (ODR) [35] protocol exploits the broadcast nature of the wireless propagation, by resorting to broadcast communications instead of traditional unicast ones, to provide connectivity in the presence of hostile conditions.

ODR exploits the same location-dependent address space of ITR, and it pushes down the stack the P2P functionalities at the network layer by resorting to indirect key-based routing. Differently from ITR, ODR is explicitly designed for disruption tolerant networks and its performances have been evaluated in such a scenario.

To accomplish the packet routing, each forwarder locally broadcasts the packet to all its neighbors, together with an estimate of its distance from the destination. By means of such a distance, the receiving nodes are able to understand if they are potential forwarders, that is if they belong to the candidate set, by comparing their distances with the one stored in the packet header as shown in Figure 6.15. Clearly, the candidate set is composed by all the neighbors closer than the forwarder to the destination as well as the forwarder.

Each candidate node delays the packet forwarding by an amount of time which depends on its distance estimate from the destination: the closer is a node to the destination, the shorter is the delay. A subsequent reception of the same packet from a neighbor closer to the destination allows the node to discard that packet, while a subsequent reception from a farther neighbor gives rise to an acknowledge transmission.

This proposal, like VRR, does not deal with P2P overlay networks; nevertheless both can be used as P2P overlay networks since they provide all the necessary P2P functionalities directly at the network layer. However, ODR, differently from VRR, has been designed to operate in DTNs. We underline that both ODR and VRR need further evaluations to assess their performances as P2P overlay networks.

6.4.4   Summary and Open Problems

Delay Tolerant Networks and Peer-to-Peer systems are emerging technologies sharing a common underlying decentralized networking paradigm. Nevertheless, the related research activities have been mainly developed by different research communities, nullifying therefore the idea of a unitary approach able to assure effective integrated solutions.

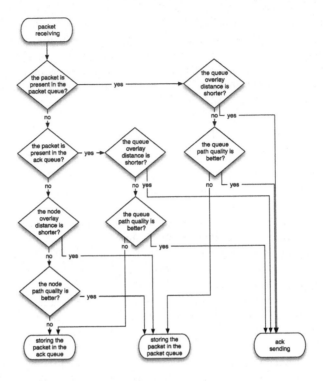

Figure 6.15: ODR packet forwarding process.

In the last years, different proposals based on cross-layered or integrated approaches have been presented to overcome the poor performances due to the lack of cooperation and communication between the two layers, namely the application layer and the networking one. Despite these efforts, the design of Peer-to-Peer overlay networks on wireless multi-hop networks is still an open problem, and DTNs pose additional issues related to the lack of persistent connectivity.

In Table 6.2 we state a comparison among the systems previously described in terms of the following characteristics:

- adopted paradigm: resource-location aware (structured) or not (unstructured);

- adopted architecture: traditional layered, cross-layered, or integrated;

- P2P overlay network;

- routing protocol adopted for the network substrate;

- applicability: explicitly designed for DTNs or not;

Table 6.2: Comparison among the Considered Mobile Peer-to-Peer Overlay Networks

System	Paradigm	Architecture	Overlay network	Routing protocol	Applicability	Performance evaluation
ASOS	Unstructured	Cross-layered	Gnutella-like	AODV-like	Maybe DTN	Simulation
ITR	Structured	Integrated	Tree based	M-DART	MANET	Simulation
MADPastry	Structured	Cross-layered	Pastry	AODV	MANET	Simulation
MPP	Unstructured	Cross-layered	Gnutella-like	EDSR	MANET	Simulation
ODR	Structured	Integrated	Tree based	ODR	DTN	Incomplete
ORION	Unstructured	Layered	Gnutella-like	AODV	MANET	Simulation
P2PSI	Unstructured	Cross-layered	Gnutella-like	ARA	MANET	Simulation
VRR	Structured	Integrated	Pastry-like	VRR	MANET	Incomplete
Yang and Chuang	Unstructured	Cross-layered	Spraying based	Prophet	DTN	Simulation

- performance evaluation: how the proposed P2P system has been evaluated.

We note that most of the unstructured P2P overlay networks adopt a Gnutella-like P2P application layer, modified to couple with node mobility. On the other hand, Pastry is a popular solution for structured P2P overlay networks. In both the classes, reactive routing is often used as a network substrate.

In theory, structured systems are able to efficiently retrieve data items thanks to their content-location awareness. However, they usually suffer from overlay route stretch since the overlay neighborhood concept is not related to the physical one. Such an effect is particularly significant in resource-constrained networks as in DTN ones. Moreover, the stretch effect implies additional latency for data items dissemination/discovery. Finally, the maintenance of the structure in the overlay network can introduce a considerable overhead. On the other hand, unstructured systems are able to react quicker to changes in the network topology since they do not maintain topological information. However, their flooding-like strategies for resource discovery exhibit poor scalability with respect to the number of nodes in the network.

In the future, we expect that a new class of P2P overlay networks able to provide connectivity when both the assumptions of dense network topology and stationary wireless conditions are not verified will be developed. The design of these systems requires the exploration of the similarities of P2Ps and DTNs in terms of the common store-carry-forward paradigm.

6.5 Conclusions

In this chapter we have focused on the issue of allowing the P2P functionalities to operate over a Delay Tolerant Network. More specifically, we have described the P2P system characteristics, capabilities, applications, and design constraints, thus providing an opportunity for beginner readers to acquire familiarity with such a very active research area.

As it has been shown in this chapter, there exists a variety of P2P systems designed specifically for mobile ad hoc networks, but few proposals deal with the problem of providing content information dissemination/discovery in delay tolerant networks.

It is likely that currently a single solution able to satisfy the needs of every conceivable DTN scenario is not available. However, the understanding gained from these first proposals can be used, in the coming years, to improve future designs of Mobile P2P systems, since there still remains much to do in terms of understanding, developing, and deploying a P2P overlay network for DTN scenarios.

Bibliography

[1] E. K. Lua, J. Crowcroft, M. Pias, R. Sharma, and S. Lim, "A survey and comparison of Peer-to-Peer overlay network schemes," *IEEE Communications Survey and Tutorial*, Vol. 7, No. 2, pp. 72–93, March 2004.

[2] "The napster website." [Online]. Available: http://free.napster.com

[3] "Gnutella protocol specification v0.6." [Online] Available: http://rfc-gnutella.sourceforge.net/rfc-gnutella.zip

[4] "The emule website." [Online]. Available: http://www.emule-project.net

[5] V. Borrel, M. H. Ammar, and E. W. Zegura, "Understanding the wireless and mobile network space: a routing-centered classification," in *CHANTS '07: Proceedings of the second ACM workshop on Challenged networks*, 2007, pp. 11–18.

[6] M. Caleffi and L. Paura, "P2p over manet: Indirect tree-based routing," in *PerCom '09: The Seventh Annual IEEE International Conference on Pervasive Computing and Communications*, Mar. 2009, pp. 340–344.

[7] R. Schollmeier, I. Gruber, and M. Finkenzeller, "Routing in mobile ad-hoc and Peer-to-Peer networks: a comparison," in *Revised Papers from the NETWORKING 2002 Workshops on Web Engineering and Peer-to-Peer Computing*, 2002, pp. 172–186.

[8] A. C. Viana, M. D. de Amorim, S. Fdida, and J. F. de Rezende, "Self-organization in spontaneous networks: the approach of dht-based routing protocols," *Ad Hoc Networks*, vol. 3, no. 5, pp. 589–606, Sept. 2005.

[9] S. Ratnasamy, P. Francis, M. Handley, R. Karp, and S. Schenker, "A scalable content-addressable network," in *SIGCOMM '01: Proceedings of the 2001 conference on Applications, technologies, architectures, and protocols for computer communications*. ACM, 2001, pp. 161–172.

[10] B. Zhao, L. Huang, J. Stribling, S. Rhea, A. Joseph, and J. Kubiatowicz, "Tapestry: a resilient global-scale overlay for service deployment," *Selected Areas in Communications, IEEE Journal on Selected Areas in Communication*, vol. 22, no. 1, pp. 41–53, Jan. 2004.

[11] I. Stoica, R. Morris, D. Karger, M. F. Kaashoek, and H. Balakrishnan, "Chord: A scalable Peer-to-Peer lookup service for internet applications," in *SIGCOMM '01: Proceedings of the 2001 conference on Applications, technologies, architectures, and protocols for computer communications*, Aug. 2001, pp. 149–160.

[12] A. Rowstron and P. Druschel, "Pastry: Scalable, decentralized object location and routing for large-scale Peer-to-Peer systems," in *IFIP/ACM International Conference on Distributed Systems Platforms (Middleware)*, Nov. 2001, pp. 329–350.

[13] P. Maymounkov and D. Mazières, "Kademlia: A Peer-to-Peer information system based on the xor metric," in *IPTPS: 1st International workshop on Peer-to-Peer Systems*, 2002, pp. 53–65.

[14] D. Malkhi, M. Naor, and D. Ratajczak, "Viceroy: a scalable and dynamic emulation of the butterfly," in *PODC '02: Proceedings of the Twenty-First Symposium on Principles of Distributed Computing*, 2002, pp. 183–192.

[15] "The seti@home website." [Online]. Available: http://setiathome.berkeley.edu

[16] "The freenet website." [Online]. Available: http://freenetproject.org

[17] "The kazaa website." [Online]. Available: http://www.kazaa.com

[18] "The morpheus website." [Online]. Available: http://www.morpheus.com

[19] F. P. Franciscani, M. A. Vasconcelos, R. P. Couto, and A. A. Loureiro, "(re)configuration algorithms for Peer-to-Peer over ad hoc networks," *Journal of Parallel and Distributed Computing*, vol. 65, no. 2, pp. 234–245, 2005.

[20] F. Warthman, "Delay-tolerant networks (dtns): A tutorial v1.1." [Online]. Available: http://www.dtnrg.org

[21] "Nasa jpl's deep impact networking (dinet)." [Online]. Available: http://deepspace.jpl.nasa.gov/dsn

[22] L. Wood, W. Eddy, W. Ivancic, J. McKim, and C. Jackson, "Saratoga: a delay-tolerant networking convergence layer with efficient link utilization," in *International Workshop on Satellite and Space Communications, 2007. IWSSC '07*, Sept. 2007, pp. 168–172.

[23] S. Guo, M. H. Falaki, E. A. Oliver, S. Ur Rahman, A. Seth, M. A. Zaharia, and S. Keshav, "Very low-cost Internet access using kiosknet," *SIGCOMM Comput. Commun. Rev.*, vol. 37, no. 5, pp. 95–100, 2007.

[24] P. Mundur and M. Seligman, "Delay tolerant network routing: Beyond epidemic routing," in *3rd International Symposium on Wireless Pervasive Computing, 2008. ISWPC 2008*, May 2008, pp. 550–553.

[25] A. Vahdat and D. Becker, "Epidemic routing for partially-connected ad hoc networks," Duke University, Tech. Rep., 2000.

[26] A. Lindgren, A. Doria, and O. Schelén, "Probabilistic routing in intermittently connected networks," *SIGMOBILE Mob. Comput. Commun. Rev.*, vol. 7, no. 3, pp. 19–20, 2003.

[27] T. Spyropoulos, K. Psounis, and C. S. Raghavendra, "Spray and wait: an efficient routing scheme for intermittently connected mobile networks," in *WDTN '05: Proceedings of the 2005 ACM SIGCOMM workshop on Delay-tolerant networking*, 2005, pp. 252–259.

[28] T. Small and Z. J. Haas, "Resource and performance tradeoffs in delay-tolerant wireless networks," in *WDTN '05: Proceedings of the 2005 ACM SIGCOMM workshop on Delay-tolerant networking*, 2005, pp. 260–267.

[29] W. Zhao, M. Ammar, and E. Zegura, "A message ferrying approach for data delivery in sparse mobile ad hoc networks," in *MobiHoc '04: Proceedings of the 5th ACM international symposium on mobile ad hoc networking and computing*, 2004, pp. 187–198.

[30] M. C. Chuah and P. Yang, "A message ferrying scheme with differentiated services," in *Military Communications Conference, 2005. MILCOM 2005. IEEE*, Oct. 2005, pp. 1521–1527, Vol. 3.

[31] S. Jain, R. C. Shah, W. Brunette, G. Borriello, and S. Roy, "Exploiting mobility for energy efficient data collection in wireless sensor networks," *Mob. Netw. Appl.*, vol. 11, no. 3, pp. 327–339, 2006.

[32] K. A. Harras and K. C. Almeroth, "Inter-regional messenger scheduling in delay tolerant mobile networks," in *WOWMOM '06: Proceedings of the 2006 International Symposium on on World of Wireless, Mobile and Multimedia Networks*, 2006, pp. 93–102.

[33] M. Conti, G. Maselli, G. Turi, and S. Giordano, "Cross-layering in mobile ad hoc network design," *Computer*, vol. 37, no. 2, pp. 48–51, 2004.

[34] M. Caesar, M. Castro, E. Nightingale, G. O'Shea, and A. Rowstron, "Virtual ring routing: network routing inspired by dhts," in *SIGCOMM '06: Proceedings of the 2006 conference on Applications, technologies, architectures, and protocols for computer communications*, 2006, pp. 351–362.

[35] M. Caleffi and L. Paura, "Opportunistic routing for disruption tolerant networks," in *AINA '09: the IEEE 23rd International Conference on Advanced Information Networking and Applications*, May 2009, pp. 826–831.

[36] A. Klemm, C. Lindemann, and O. Waldhorst, "A special-purpose Peer-to-Peer file sharing system for mobile ad hoc networks," in *Vehicular Technology Conference, 2003. VTC 2003-Fall. 2003 IEEE 58th*, vol. 4, Oct. 2003, pp. 2758–2763.

[37] I. Gruber, R. Schollmeier, and W. Kellerer, "Performance evaluation of the mobile Peer-to-Peer service," in *IEEE International Symposium on Cluster Computing and the Grid, 2004. CCGrid 2004*, April 2004, pp. 363–371.

[38] G. Yang, L.-J. Chen, T. Sun, B. Zhou, and M. Gerla, "Ad-hoc storage overlay system (asos): A delay-tolerant approach in manets," in *IEEE International Conference on Mobile Adhoc and Sensor Systems (MASS), 2006*, October 2006, pp. 296–305.

[39] C.-C. Hoh and R.-H. Hwang, "P2p file sharing system over manet based on swarm intelligence: A cross-layer design," in *Wireless Communications and Networking Conference, 2007. WCNC 2007. IEEE*, Mar. 2007, pp. 2674–2679.

[40] P. Yang and M. Chuah, "Performance evaluations of data-centric information retrieval schemes for dtns," *Computer Networks*, vol. 53, no. 4, pp. 541–555, 2009.

[41] C. Perkins and E. Royer, "Ad hoc on-demand distance vector routing," in *2nd IEEE Workshop on Mobile Computing Systems and Applications*, 1999, pp. 90–100.

[42] D. Johnson and D. Maltz, "Dynamic source routing in ad hoc wireless networks," in *Mobile Computing*, Kluwer Academic Publishers, 1996, vol. 353, pp. 153–181.

[43] E. Bonabeau, M. Dorigo, and G. Theraulaz, *Swarm Intelligence: From Natural to Artificial Systems*, Oxford University Press, 1999.

[44] M. Gunes, U. Sorges, and I. Bouazizi, "Ara-the ant-colony based routing algorithm for manets," in *International Conference on Parallel Processing Workshops, 2002. Proceedings*, 2002, pp. 79–85.

[45] B. H. Bloom, "Space/time trade-offs in hash coding with allowable errors," *Commun. ACM*, vol. 13, no. 7, pp. 422–426, 1970.

[46] T. Zahn and J. Schiller, "Madpastry: A dht substrate for practicably sized manets," in *Proc. of 5th Workshop on Applications and Services in Wireless Networks (ASWN2005)*, June 2005.

[47] K. Xu, X. Hong, and M. Gerla, "Landmark routing in ad hoc networks with mobile backbones," *Journal of Parallel and Distributed Computing*, vol. 63, no. 2, pp. 110–122, Feb. 2003.

[48] K. Takeshita, M. Sasabe, and H. Nakano, "Mobile p2p networks for highly dynamic environments," in *PERCOM '08: Proceedings of the 2008 Sixth Annual IEEE International Conference on Pervasive Computing and Communications*, 2008, pp. 453–457.

[49] M. Caleffi, G. Ferraiuolo, and L. Paura, "Augmented tree-based routing protocol for scalable ad hoc networks," in *MASS '07: the IEEE Internatonal Conference on Mobile Adhoc and Sensor Systems*, Oct. 2007, pp. 1–6.

[50] M. Caleffi and L. Paura, "M-dart: Multi-path dynamic address RouTing," *Wireless Communications and Mobile Computing*, vol. 11, no. 3, pp. 392–409, March 2011.

[51] M. Caleffi, G. Ferraiuolo, and L. Paura, "A reliability-based framework for multi-path routing analysis in mobile ad-hoc networks," *International Journal of Communication Networks and Distributed Systems*, vol. 1, no. 4-5-6, pp. 507–523, 2008.

Chapter 7

Delay Tolerant Monitoring of Mobility-Assisted WSN

Abdelmajid Khelil, Faisal Karim Shaikh, Azad Ali, Neeraj Suri,[1]
Christian Reinl

"You only see what you know." J.W. von Goethe

7.1 Introduction 190
7.2 State of the Art 192
    7.2.1 Node-Centric Delay-Critical Monitoring Techniques 193
    7.2.2 Region-Centric Delay-Critical Monitoring Techniques 193
    7.2.3 Mobility-Assisted Delay-Tolerant Monitoring
          Techniques 194
    7.2.4 Our Contributions Compared to Existing Approaches 196
7.3 System Model 197
7.4 gMAP: Mobility-Assisted Monitoring Using Global Maps 198
    7.4.1 Overview of Approach 198
    7.4.2 Scenario Classification 198
    7.4.3 Path Planning of ANs 199
          7.4.3.1 Structured mWSN Scenarios 199
          7.4.3.2 Semi-Structured mWSN Scenarios 200
          7.4.3.3 The Integrated Path Planning Algorithm 203
    7.4.4 Data Collection 204
    7.4.5 eMap Construction 204

[1]Research supported in part by DFG GRK 1362 (TUD GKMM), EC INSPIRE, EC CoMiFin, MUET, and HEC.

7.5   Evaluation 208
    7.5.1   Path Planning Performance 208
    7.5.2   Data Collection and Map Construction Performance 209
        7.5.2.1   Simulation Settings 209
        7.5.2.2   Evaluation Metrics 210
        7.5.2.3   Data Collection Performance 211
        7.5.2.4   Regioning Performance 213
    7.5.3   Comparison to Related Work 214
7.6   Conclusions 215
7.7   Open Issues 215

Wireless Sensor Networks (WSN) usually are composed of fragile sensor nodes equipped with short range radios. Despite its elements fragility, the system is expected to deliver dependable services. This is mainly ensured through the inherent node redundancy. However, being battery-powered, the question is not whether the desired dependability cannot be maintained anymore but when. Therefore, frugal monitoring techniques are vital for a WSN. Global maps of the sensor field, such as residual energy maps, are of high acceptance for both system users and designers. However, the map construction can become very inefficient if it requires an extensive intervention of the resource-limited sensor nodes. In this chapter, we present gMAP, an extremely efficient mobility-assisted approach to constructing global maps. In gMAP (a) sensor nodes do not need to process readings of other nodes and (b) need to communicate a minimal number of messages compared to the existing map-based approaches. This is achieved by opportunistically exploiting node mobility to collect data of interest, letting sensor nodes transmit only their own readings on-demand to a mobile node in their transmission area. Our approach is designed for generalized scenarios from unstructured to semi-structured and unstructured. If node mobility is controllable, gMAP includes an integrated scalable path planning algorithm for mobile nodes.

7.1   Introduction

Wireless Sensor Networks (WSN) are seeing increasing usage in several applications, such as military, rescue, and surveillance scenarios. Typical for such scenarios is that mobile nodes cooperate side-by-side with stationary sensor nodes to monitor the area of interest and to support the core network operations such as data collection [60, 52]. One dilemma of WSNs is that while they are often composed of low-cost and basic elements (hardware and software) they are still expected to provide reliable services, such as detecting fire. Another dilemma of WSNs is that they are supposed to be long-life systems despite the fact that they rely on a finite energy source and operate under stress in harsh environments. Consequently, monitoring the operational health (especially energy levels) of the WSN elements aids the overall WSN in providing effective services. Accordingly, utilities for network monitoring and diagnosis are required.

Most of the existing monitoring techniques operate on sensor node level and ignore the inherent redundant node deployment in WSNs, which wastes valuable battery and network resources. An emerging diagnostic schema entails creating WSN wide maps of interested attributes [18] called global maps (gMaps), which operate on a region-level and less on a node-level. Region-level approaches group neighboring nodes that show correlated attribute values and cope with the group as a single node. gMaps present the spatial distribution of such relevant attributes. For example the energy map (eMap) depicts the spatial distribution of residual energy of the WSN elements. gMaps provide elementary utilities for the assessment of the WSN dependability. They support network designers and administrators to monitor and optimize the network in operation. An example of supporting the network management is to utilize the eMap to detect and predict important vulnerabilities such as network partitioning [19].

A variety of approaches for map data collection have been developed [20, 21, 22, 23, 24, 25, 26]. However, all these approaches rely on multi-hop communication and/or in-network aggregation, which overstrains the sensor nodes through the use of the limited energy and processing resources of the stationary sensor nodes. In [27], the authors demonstrated that node mobility can increase the capacity of ad hoc networks, if the mobile nodes transport the data closer to the destination instead of immediately using multi-hop communication. This comes at the cost of higher end-to-end delays for communication. Fortunately, several WSN applications and network management tasks can tolerate delays in the range of minutes, hours or even days [28, 60, 52, 55]. For instance, the energy spatial distribution usually evolves only on a relatively large time scale, therefore, the data collection for constructing and updating the eMap can last longer to increase efficiency.

In [29], we have presented gMAP, an extremely efficient mobility-assisted approach to constructing global maps. In gMAP (a) sensor nodes do not need to process readings of other nodes and (b) need to communicate a minimal number of messages compared to all existing approaches. This is achieved by opportunistically exploiting node mobility to collect data of interest, letting sensor nodes transmit only their own readings on-demand to a mobile node in their transmission area.

While [29] focused on the construction of eMaps for two extreme deployment scenarios, in this chapter we synthesize our previous results [18, 29] and extend the gMAP approach to support scenarios of generalized mobility and deployment. In particular, we present a novel path planning algorithm for mobile nodes to efficiently collect map data and its updates from the WSN. In order to determine an optimal path w.r.t. completeness of collected data and energy overhead, we solve an adapted traveling salesman problem with "flexible cities."

The remainder of this chapter is organized as follows. After the review of the state of the art in Section 7.2, Section 7.3 clarifies our system model. Section 7.4 presents our novel gMAP approach using the example of eMaps. In Section 7.5, we evaluate our approach through simulations. We conclude the chapter in Section 7.6 and discuss open issues in Section 7.7.

7.2 State of the Art

Given the particular properties of WSNs such as energy-constraints and node fragility it is difficult to simply customize the existing monitoring techniques for WSNs. Subsequently, many monitoring techniques tailored for WSNs have been developed. Concerning latency of data collection, we identify two main classes of network monitoring techniques in WSN: Delay-critical and delay-tolerant. Regarding the monitoring level, we distinguish between approaches that focus on node properties and diagnose the network on node-level and those that group nodes into regions (Figure 7.1).

A few of the time-critical techniques exploit the spatial correlation of node properties. These are known as region-centric or (map-based) monitoring techniques. They do not consider node mobility and rely on pure multi-hop communication. Delay-tolerant techniques exploit mobility to collect the data. However, they do not exploit the spatial correlation of monitoring attributes to optimize collection and presentation of the monitoring data. Our gMAP approach profits from the spatial correlation of monitoring data as well as the high data's lifetime (Figure 7.1).

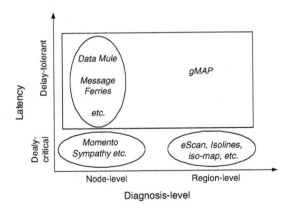

Figure 7.1: State of the art in problem space.

7.2.1 Node-Centric Delay-Critical Monitoring Techniques

Sympathy [42] is a tool for detecting and debugging node-level failures in WSN. The authors show that it is possible to detect and diagnose failures by collecting and analyzing a minimal set of metrics regarding connectivity, traffic flow, and node properties at a centralized sink. Sympathy includes an algorithm that root-causes failures and localizes their sources in order to point the user to a small number of probable causes: Node, path, or the sink. *Memento* [43] provides failure detection and symptom alerts. Memento provides an energy-efficient protocol to deliver state summaries and a distributed failure detector module. The authors show that distributed monitoring of a subset of well-connected neighbors using a variance-bound-based failure detector is suitable for use in practice. The *Sensor Network Tomography (SNT)* [44, 45] provides techniques to compute global predicates (e.g., the total number of nodes) using in-network processing and aggregation.

Sympathy, Memento and SNT waste bandwidth and energy resources. In addition, the monitoring performance suffers from network perturbations. Therefore, (non-intrusive) approaches such as [47] have been presented to passively monitor existing traffic and deduce network health indications. Unfortunately, these approaches suffer from volatile and high inaccuracies as they need to reduce the cost of this supplemental deployment. For instance, in [47], the authors suggest temporarily deploying an additional monitoring network on top of the WSN. Being temporary, the monitoring does not provide for an acceptable accuracy over the entire WSN lifetime.

7.2.2 Region-Centric Delay-Critical Monitoring Techniques

The region-centric (or map-based) techniques collect data from the entire network and construct a map out of this data on the fly. In cartography [33], isolines (also isopleth or contour) and choropleths are the common types of maps. For the varied map types, different data collection techniques have been developed. The naive approach to collect raw data for map construction would be if each node reports its value to the sink using multi-hop communication. This naive approach is obviously inefficient. Consequently, more efficient approaches have been developed based on techniques such as in-network aggregation [20, 21, 22]. Other approaches use suppression mechanisms to reduce the number of nodes reporting their raw readings to the sink [24, 25, 26].

Aggregation-based Approaches: *eScan* [20] and *isobar* [21] are approaches based on polygon aggregation. First, a request for energy values is flooded to all network nodes. This constructs an aggregation tree that can be used to aggregate the energy values while being reported by each node. The aggregation consists of grouping sensor readings that meet a certain criteria (being geographically adjacent and in the same value range). The outcome of the aggregation is a list of (spatial) regions. A region is a polygon that is defined by the line spanning its border nodes. At the sink the aggregation results in an energy map delivered to the user. Each sensor node propagates its position along with the energy values for aggregation purposes. Furthermore, the sensor nodes (even those that have critical residual energy level and especially those closer to the sink) are main actors in map construction, leading to higher processing and communication activities and subsequently to a serious degradation of the network lifetime and disturbance of the core functionality. *INLR* [22] is an aggregation-based approach that focuses on small scale WSNs. A sensor node sends its reading or the calculated aggregate not only to its parent in the aggregation tree but to all its neighbors that are 1 hop closer to the sink. Therefore, nodes possess a partial map and the sink has the global map (choropleth). While using more than one parent increases the accuracy of the map, the efficiency is sacrificed.

Suppression-based Approaches: *Isoline* [24] is an approach based on localized isocluster aggregation. The map building is reduced to the detection of isolines. Neighboring nodes share their readings. A node compares its reading with the readings of all neighbor nodes and detects an isoline, when the readings lie in different sides of a globally defined isoline. The detection of an isoline needs to be reported to the sink by the closest neighbor to the sink. The isocluster aggregation outperforms polygon aggregation in terms of accuracy with minor energy savings. Meng et al. [25] motivate the use of contour (isoline) maps for efficient continuous monitoring in sensor networks. The main contribution of this paper is the design of a temporal and spatial local suppression mechanism that prohibits some nodes from reporting their readings. The number of saved reports highly depends on the spatial correlation between sensor readings. Sensor nodes report their readings using multi-hop routing without any in-network processing. The map is constructed on the sink using interpolation and smoothing techniques. *Iso-Map* [26] also does not rely on in-network processing. It uses a suppression mechanism to reduce the number of nodes that report their readings to the sink using multi-hop communication. This approach is very similar to that of Isoline [24]. However, nodes need to report the gradient direction of the isolines, which requires excessive processing on sensor nodes.

7.2.3   Mobility-Assisted Delay-Tolerant Monitoring Techniques

Mobility-assisted data collection achieves high bandwidth and energy efficiency at the cost of a high end-to-end delay. In the following, we survey the techniques to collect data from the WSN in a mobility-aided fashion. Next, we review the existing approaches to controlling mobility.

Mobility-Aided Data Collection: There exist different mobility-assisted data collection techniques in the literature. These techniques have been developed to collect user data. However, they are easy to adapt to collect monitoring data. In the following, we review these techniques. For further readings we refer to the excellent survey [28].

Data Mule [60, 59, 58, 57, 56] is a mobile node that collects data from sensor nodes as it passes by. In [60], a basic theoretical analysis has been developed. Hereby, simplistic deployments and communication protocols have been considered. Sensor nodes are deployed randomly on a grid and multiple data mules move according to the random walk model. Sensor nodes buffer their data until the mule can receive it through direct communication. The mule buffers the data until it can deliver it to the sink. In [59, 58, 57, 56], the movement paths are fixed and the coverage is not guaranteed. Therefore, some sensor nodes may need to have a multi-hop path to reach the mule. Nodes not on the movement path establish a local routing tree to send data to sensor nodes on the path, which collect the data till the data mule passes by.

Message Ferrying [52] is similar to data mules, but designed for mobile ad hoc networks (MANET) and mainly to overcome partitioning. Mobile nodes act like ferries to reach disconnected sensor nodes. [52] uses controlled movement to deliver data. Ferries can either follow a predefined movement path or change their movement path on-demand. To "order" the ferry, nodes can spontaneously increase their communication range to reach a message ferry nearby and inform it about communication need so that the ferry can change its path. In order to allow for differentiated end-to-end latencies in [53], the authors suggested prioritizing messages. They present a forwarding strategy for the ferry route and discuss the buffer requirements to deal with this proposal. In [54, 55], the authors improve message ferrying by introducing a power management framework. If nodes know when they are going to encounter the ferry they can sleep to conserve power.

Also, [63] is designed for MANET. This work analyzes two simple data delivery schemes, namely, the direct transmission and flooding, based on (1) the likelihood that a sensor node can deliver data messages to the sink, and (2) the message fault tolerance, i.e., the probability that at least one copy of the message is delivered to the sink by other sensor nodes.

Path Planning: Path planning for coverage problems has been studied for many years. Many centralized and distributed approaches, for various indoor and outdoor scenarios (e.g., [5, 4]), have been proposed. Many of these approaches are basically built on studying traveling salesman problems (TSP [6]) and vehicle routing and scheduling problems [7]. For the case of unknown or only roughly estimated node positions that have to be covered, a distributed planning on the mobile node based on its local information is more adequate [2].

A suitable approach for the problem of planing an optimal path based on the nodes' topology is using hybrid optimal control theory, which results in solving mixed-integer nonlinear programming (MINLP) in practice. By considering communication ranges and optimality w.r.t. a specific physical motion dynamics of the mobile nodes [12, 16]. Therefore, an adjusted objective function describes desired features to be minimized like overlappings or path-length. This solution is well suited to small scale networks but becomes very inefficient for common WSN settings. This is mainly caused by the underlying discrete structure, which results in the highly combinatorial character of the entire problem.

Even the basic problem (known as the traveling salesman problem with neighborhoods – TSPN [17]) of planning the shortest path is NP-hard, without considering a certain locomotion dynamic. Thus, many approximating algorithms for the TSPN under mild assumptions were proposed in recent years [15, 11]. For small scale WSN scenarios (a few dozen nodes) a solution based on TSP-path-planning has recently been presented in [61]. The TSPN-problem is only briefly discussed there without presenting a solution for larger scenarios with a dense setting of nodes. For larger WSN scenarios a problem similar to TSPN is solved in [3], but also disregarding overlaps within the transmission areas of subsequent breakpoints.

7.2.4 Our Contributions Compared to Existing Approaches

Our monitoring approach, gMAP, is a mobility-assisted delay-tolerant technique. Efficient data collection is achieved by opportunistically exploiting node mobility to collect data of interest, letting sensor nodes transmit only their own readings on-demand to a mobile node in their transmission area. gMAP benefits from the spatial correlation of collected data and provides for monitoring at region-level, i.e., in the form of maps.

Compared to node-centric monitoring techniques, gMAP collects data at comparable high accuracy (node level). Moreover, gMAP outperforms these techniques w.r.t. load balancing on sensor nodes and the map-based presentation of monitoring data.

Compared to map-based techniques, our gMAP approach uses a *minimal* number of messages without sacrificing the completeness of sensor information. This provides for high efficiency with respect to both energy and bandwidth consumption. In gMAP, we *decouple* the collection of the sensor values from the construction of the map, which results in minimal processing on sensor nodes, reducing the energy consumption on them. Furthermore, gMAP charges all sensor nodes similarly and contributes to the desired energy balancing in WSNs. Our approach is *resilient to network partitioning*, which increases the dependability of the WSN, since monitoring tasks can continue reporting the health of the network even if critical failures/situations occur.

Data collection in gMAP is similar to that of data mule and message ferrying approaches, but we do not rely on multi-hop communication. In addition, most of the existing delay-tolerant approaches have been developed for specific scenarios such as sparsely deployed and structured WSNs (data mule) or MANET (message ferrying). We focus on WSN and provide techniques that are for generalized scenarios (from structured to unstructured). We compare our gMAP approach to the existing approaches and provide the results in the evaluation section.

7.3 System Model

In this chapter, we consider the established *mobile* Wireless Sensor Network (mWSN) model. This model is used in a variety of WSN deployments, in particular in emergency and military scenarios. The main functionality of the mWSN is implemented by a large number of stationary resource-limited *sensor nodes* (SN) that are deployed following either an arbitrary or structured spatial distribution in the area of interest. Also, one dedicated stationary sink is selected as the interface to the user. Additionally, a few mobile *assist nodes* (AN) are deployed with generalized support roles such as (1) application support (e.g., additional interface to users), (2) functionality support (e.g., delay-tolerant data transport), and (3) network support (e.g., diagnosis). The mobile nodes cover a functional capability spanning robots, unmanned aerial vehicles (UAV), etc. Hereby, we assume that the AN is able to move to and stop at any position in the sensor field. In this chapter, we consider an mWSN composed of N_N SNs, with one sink and one mobile AN. We are using one single mobile AN for simplicity of communicating the idea whereas a real implementation can consider multiple nodes or some primary/secondary arrangements.

We consider two major classes of mobility: Structured mobility, i.e., predictable and controllable, and unstructured mobility, i.e., unpredictable and uncontrollable. The AN possesses high processing, storage, and energy capabilities compared to SNs. Furthermore, it has no energy limitations because it can recharge its batteries by means of on-board renewable energy resources [34] or through moving to recharging energy-stations. We assume that SNs use the batteries as a main energy source. These batteries continuously discharge following a long-running process in the range of months or even years. We consider that SNs as well as the AN knows its position. We assume that all deployed nodes are cooperating and that no misbehaving nodes may exist.

For simplicity, we consider all nodes (AN and SNs) are equipped with a conformal level of communication technology and are able to communicate if they are in each other's transmission range R. We use a CSMA/CA-based MAC layer, where communication links are symmetric and bidirectional, and collisions may occur. Furthermore, we assume that network can get partitioned, i.e., some SNs may not be able to communicate with the sink. We allow for the use of duty cycles for SNs. However, we assume that the magnitude of the movement distance covered by the AN during the time period of a duty cycle is negligible and that the duty cycles scheduler assures that all SNs in the AN's transmission area eventually receive the messages sent by the AN.

7.4 gMAP: Mobility-Assisted Monitoring Using Global Maps

We now present our novel gMAP approach comprising new algorithms to collect samples in a mobility-assisted way and a new technique to construct maps. We use the eMaps as an example. However, our methodology is generic and can be easily adopted for other functionality maps. We refer to our gMAP approach for eMaps as eMAP.

7.4.1 Overview of Approach

The main reasoning behind the eMAP approach to construct eMaps is that battery depletion occurs over an extended period of time and it is sufficient to check the battery level on a daily or weekly basis. This shows that collection of energy-based health indications is a *delay-tolerant* process, which allows us to deploy established concepts from the delay-tolerant networking research. Accordingly, the main design principle for the eMAP approach is to exploit the mobility of nodes to transport messages and collect information in a delay-tolerant way, thus reducing the communication overhead.

We let the mobile AN scan the sensor field and collect the energy information from each node it encounters. The AN sends a short beacon, on which nodes reply with their energy value and optionally their position. We proceed progressively, by first considering a structured scenario, then a semi-structured one, and finally an unstructured one. For scenarios with controllable node mobility we design an integrated path planning algorithm. For each scenario we design an appropriate algorithm to collect energy information.

We also present an efficient technique for the mobile AN to locally construct an appropriate eMap from the collected energy samples. The technique is based on measuring inequalities between neighboring samples and grouping similar values into a region. Therefore, we refer to our technique as regioning (Section 7.4.5).

7.4.2 Scenario Classification

In this work, we focus on three important types of scenarios that provide basic features to build realistic scenarios.

1. In a structured scenario, we assume that the spatial deployment of SNs is known a priori and that the mobility of the AN is controllable (Figure 7.2(a)).

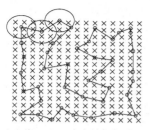 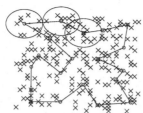 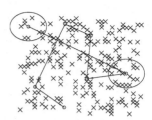

(a) Structured (grid deployment of SNs with controllable AN)

(b) Semi-structured (random SNs deployment with controllable AN)

(c) Unstructured (unknown topology and unpredictable mobility)

Figure 7.2: Basic scenarios (nodes are represented by × and breakpoints by ◦), with results (in (a) and (b)) from the path planing algorithm as detailed in Section 7.4.3.2.

2. In a semi-structured scenario with an a priori known (or reliably estimated) spatial deployment of SNs, we assume the mobility of the AN to be controllable (Figure 7.2(b)).

3. In a scenario with an unknown topology (e.g., random spatial deployment), the mobility of the AN is assumed to be unpredictable and uncontrollable (Figure 7.2(c)).

Our main driver for the scenario selection is the proof of concept in extreme scenarios. Furthermore, in a realistic scenario the spatial deployment of SNs can be structured or known only partially. The mobility of the AN can be either controllable or uncontrollable and may follow varied patterns.

7.4.3 Path Planning of ANs

Path planning is required for the structured and semi-structured scenarios.

7.4.3.1 Structured mWSN Scenarios

In such scenarios, the SNs are deployed according to a specific uniform scheme (e.g., on a grid). The AN knows the accurate positions of SNs and accordingly plans its movement.

Planning an optimal tour to collect the data with a minimal number of messages corresponds to determining a minimal number of breakpoints with a minimal overlap of their corresponding transmission areas. The knowledge of the uniformly structured spatial distribution of the SNs can be used to simplify the complex problem of the optimal AN path planing, e.g., by applying an adopted pattern. Especially if there are repeatedly occurring parts, this structure may allow combining solutions from a simpler path planing problem determined for these smaller parts.

Further optimizations of the tour are possible, e.g., concerning tour time, length, and the energy overhead for an AN. As our main goal is the proof-of-concept, we do not further consider these optimizations in this work. We also assume a simple structure for SN deployment, i.e., grid topology.

For the grid topology, the work [5] suggests a zigzag movement of nodes. Accordingly, the AN crosses the full length of the sensor field in a straight line, turns around, and then traces a new straight line path adjacent to the previous one and $2 \cdot R$ far from it. By repeating this the AN covers the entire WSN field.

7.4.3.2   Semi-Structured mWSN Scenarios

In such scenarios, the AN knows the accurate positions of SNs and accordingly plans its movement. However, nodes deployment is not structured, which complicates an intuitive path planning of the AN. In the following, we investigate the approaches to plan the movement of one controllable AN to efficiently collect network health information from SNs. Our approach consists on stepwise decomposition of the problem into different, less complex subproblems: finding suitable breakpoints (STEP 1), reducing overlaps in the communication range (STEP 2), and planing a shortest-path-tour (STEP 3). Subsequently, we integrate the sub-solutions into a single path planning algorithm. Our main contribution is in STEP 1. However, for the sake of completeness, we detail our optional STEP 2 and the TSP-inspired STEP 3.

We are looking for a set of points where the AN stops and communicates with the SNs within the AN's communication range. The goal is to get a set of points, considerably fewer than the number of SNs and fewer than the number of points needed for a whole coverage of the area with communication ranges. We point out here that it is not necessarily desired to get the smallest possible number of breakpoints, but to get a small set of breakpoints that minimizes the useless overlapping in their coverage. The reasoning behind this is that sensor nodes in coverage overlap will waste energy to listen to redundant beacons.

Figure 7.3 depicts our proposal for path planning. First, we get candidates for a set of AN-positions by solving a nonlinear problem (NLP) and removing unnecessary points. Then, we repeat solving an approximative mixed-integer linear program (MILP) to finally get the optimal sequence TSP-solution.

STEP 1: Finding a reduced number of breakpoint candidates: To avoid solving a large mixed-integer NLP we are proposing a new basic algorithm (Alg. 1) for a given set of sensor positions $P_S := \{(\xi^j, \eta^j) | j = 1, ..., N_N\}$. It works fine, even for the case of a very dense spatial deployment of SNs and thus strongly overlapping transmission ranges of the SNs.

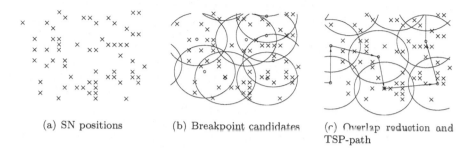

(a) SN positions   (b) Breakpoint candidates   (c) Overlap reduction and TSP-path

Figure 7.3: Visualization of the basic steps of our path planning.

Minimizing the non-linear penalty function

$$\min_{\substack{(x^j,y^j)\\ j=1,...,N_B}} \varphi((x^j,y^j)) = \min_{\substack{(x^j,y^j)\\ j=1,...,N_B}} \sum_{i \in I_j^{BS}} \log(\|(x^i,y^i) - (x^j,y^j)\| + 1) \qquad (7.1)$$

subject to the constraints $\forall k \in I_j^{BS} : \|(x^j,y^j) - (\xi^k,\eta^k)\| \le R$ effects that closely adjacent breakpoints converge towards a common position and ideally coincide (see Alg. 1). In the context of solving a TSP for determining the shortest round trip for visiting all these points, we finally refer to them as "flexible cities."

The efficiency of solving Eq. (7.1) depends decisively on the number of adjacent SNs and the number of breakpoints N_B. Thus it is desirable to start with a set P_B as small as possible. The algorithm can be adjusted by the adjacency parameter k_1.

Intermediately maximizing the penalty function effects changes in the set of adjacent SNs. This has been shown to improve local minima of N_B again. Eventually, this will result in a reduced set of N_B^* breakpoints (Figure 7.3 (b)).

STEP 2: (Optional optimization) Reducing overlapping between breakpoints: An overlap of breakpoints occurs if some nodes are covered by the transmission area of multiple breakpoints. The reduction of overlappings for a given number of N_B^* breakpoints can also be achieved by solving a MINLP:

$$\min \quad \sum_{(i,j)\in I_{BN}} b_{ij} \qquad (7.2)$$

$$\text{subject to} \quad \forall (i,j) \in I_{BN} : \quad b_{ij} = 1 \iff \|(x^i,y^i) - (\xi^j,\eta^j)\| < R \qquad (7.3)$$

$$\forall j \in 1,...,N_N : \quad \sum_{i=1}^{N_B^*} b_{ij} \ge 1 \qquad (7.4)$$

Algorithm 1 Reducing breakpoint candidates

1: Initialize a set of breakpoints $\mathbf{P}_B := \{(x^j, y^j)|j = 1, ..., N_B\}$ that guarantees connectivity to all SN

2: **while** (number N_B of breakpoints decreases) **do**

3:    For each breakpoint $(x^j, y^j) \in \mathbf{P}_B$ determine the indexes of SN lying inside the Voronoi-cell of (x^j, y^j) : $I_j^{BS} = \{i|j = \arg\min_{k=1,...,N_B}\{\|(x^k, y^k) - (\xi^i, \eta^i)\|\}\}$

4:    For each (x^j, y^j) determine the set of indexes of breakpoints in its adjacency: $I_j^{BB} = \{i|\|(x^j, y^j) - (x^i, y^i)\| \leq k_1 \cdot R\}$

5:    Minimize penalty function $\varphi((x^1, y^1), ..., (x^{N_B}, y^{N_B}))$ satisfying connectivity constraints

6:    Inspect for redundant transmission areas and delete breakpoints from $\mathbf{P}_B$.

7:    **if** ($\mathbf{P}_B$ could not be reduced) **then**

8:       Maximize (approximatively) penalty function $\varphi((x^1, y^1), ..., (x^{N_B}, y^{N_B}))$ satisfying connectivity constraints.

9:       Inspect for redundant transmission areas and delete breakpoints from $\mathbf{P}_B$.

10:   **end if**

11: **end while**

12: Return $\mathbf{P}_B^* := \mathbf{P}_B = \{(x^j, y^j)|j = 1, ..., N_B^*\}$

---

with $I_{BN} = \{(i, j) \in \{1, ..., N_B^*\} \times \{1, ..., N_N\} \,|\, \|(x^i, y^i) - (\xi^j, \eta^j)\| < 2 \cdot R\}$ and $b_{ij} \in \{0, 1\}$. Therefore, logical constraints like (7.3) have to be transformed to (in-)equalities by using, e.g., Big-M- or Convex-Hull-formulations [13].

To increase efficiency and robustness of path planning for larger settings, we suggest a scalable linearized standard approximation for (7.3): ($\forall(i, j) \in I_{BN}$)

$$\forall k = 1, ..\gamma : (x^i - \xi^j)\sin(k\frac{2\pi}{\gamma}) + (y^i - \eta^j)\cos(k\frac{2\pi}{\gamma}) \leq R + (1 - b_{ij})M_j$$

$$(7.5)$$

$$\forall k = 1, ..\gamma : (x^i - \xi^j)\sin(k\frac{2\pi}{\gamma}) + (y^i - \eta^j)\cos(k\frac{2\pi}{\gamma}) > R + \hat{b}_{k,(i,j)}m_j \quad (7.6)$$

$$\sum_{k=1}^{\gamma} \hat{b}_{k,(i,j)} \leq b_{i,j} + \gamma - 1 , \quad\quad\quad\quad\quad (7.7)$$

where $M_j = max_{x^i, y^i}\{(x^i - \xi^j)\sin(k\frac{2\pi}{\gamma}) + (y^i - \eta^j)\cos(k\frac{2\pi}{\gamma}) - R\}$, $\hat{b}_{k,(i,j)} \in \{0, 1\}$ and $m = max_{x^i, y^i}\{(x^i - \xi^j)\sin(k\frac{2\pi}{\gamma}) + (y^i - \eta^j)\cos(k\frac{2\pi}{\gamma}) - R\}$.

Compared to a MINLP problem, the MILP has an increased discrete structure. On the other hand, solving an MILP has some strong advantages: It depends less on the quality of solution estimations, results in a global minimum, and is much more robust and more efficient so that it is also easier to handle time limits with the solver (i.e., the best feasible solution so far is returned to the user, when a given time limit is exceeded).

STEP 3: Finding the optimal path: Assuming that the AN has to stop to communicate, the costs to go from breakpoint i_1 to breakpoint i_2 are constant and independent from the sequence of breakpoints visited before i_1 and after i_2. Thus, one has to solve an Euclidean TSP. This can be efficiently achieved for hundreds of positions using an existing solver such as [10]. In [1], we present an extension to a cooperative and synchronized movement of multiple ANs. Each AN moves and collects data in a specified region of the sensor field. Deploying multiple ANs additionally allows for higher efficiency in solving TSP and lower latencies in collecting data.

7.4.3.3 The Integrated Path Planning Algorithm

As an example we now propose a scalable path planning algorithm based on solving the subproblems above. Depending on R, N_N, and on the spatial distribution, the steps can be scaled by time constants $t_{max,1}$, $t_{max,2}$ and $t_{max,3}$ and by parameters in the implementation like the size of the adjacency in the set I_{NN} and I_{BN} or γ in (7.5)–(7.7).

Algorithm 2 Optimization-based path-planning for mWSN

1: /***** On sink in order to plan the mobile data collection ******/
2: /***STEP 1***/
3: $[\mathbf{P}_B]$=determine_breakpoints_NLP$(R, \mathbf{P}_S, t_{max,1})$;
4: $[\mathbf{P}_B]$=remove_unnecessary_points$(R, \mathbf{P}_B, \mathbf{P}_S)$;
5: **while** $t \leq t_{max,2}$ AND $\mathbf{P}_B$ have been improved in last run **do**
6:    /***STEP 2***/
7:    $\mathbf{P}_B$=reduce_overlapping_MILP$(R, \mathbf{P}_B, \mathbf{P}_S, t_{max,3})$;
8:    $[\mathbf{P}_B]$=remove_unnecessary_points$(R, \mathbf{P}_B, \mathbf{P}_S)$;
9: **end while**
10: /***STEP 3***/
11: solve_TSP$(\mathbf{P}_B)$;

Interrupting the MILP optimization after $t_{max,3}$ and then looking for unnecessary breakpoints on the best solution so far may reduce the combinatorial character of the MILP before the minimization starts again. Because of the MILPs' structure here with many minima of equal value, this was shown to be more efficient than waiting until the solver gives back a proven global optimum.

We also apply Alg. 2 to the structured scenario and compare the performance of our path planning to the optimal zigzag tour.

7.4.4   Data Collection

For the structured and semi-structured scenarios, we present the following data collection algorithm. The AN performs a first *snapshot* by sending a REQ-beacon to all SNs in its transmission area using a MAC broadcast. A SN replies by sending a message containing its node-ID, location (loc), and energy level E_{lev}. In order to reduce collisions, nodes schedule their reply for a random time t_{rand} between 0 and a maximum value T_{max}. The AN performs the subsequent snapshot after visiting the next breakpoint according to the path planning algorithm. The optimal result of the collection operation is a set of N_N elements with the following structure: {node-ID, loc, E_{lev}}.

Now, we present our algorithm to collect energy information (Alg. 3) for unstructured scenarios, where the movement of the AN is neither controllable nor predictable. If the AN performs a snapshot, moves $2 \cdot R$ away without changing the direction, and performs a second snapshot, then both snapshots are covering disjointed areas. Subsequently, we let the AN perform a second snapshot, only after moving $2 \cdot R$ from the location of the previous snapshot. The data collection completes when the total WSN area is covered by all snapshots. We note that if the AN changes its movement direction, then the snapshots overlap and some nodes may receive redundant REQ beacons. The major concern for SNs is to minimize the number of messages to be sent or received. The AN is powerful enough to send REQ beacons frequently. However, the REQ beacons are received by energy precious SNs. Therefore, we have to minimize the number of unnecessary REQ messages sent by AN. To avoid unnecessary snapshots, the AN maintains a history of snapshots {$snapshot_{id}$, $snapshot_{loc}$}. After moving $2 \cdot R$ from the location of the previous snapshot and before performing a second snapshot, the AN uses the history to calculate if the second snapshot has an additional coverage higher than a fixed threshold coverage $COV_{th}\%$. Only in this case the AN performs a snapshot. The value of $COV_{th}\%$ allows investigation of the trade-off between the number of redundant REQ beacons and the sampling latency. Once the AN scans the whole sensor field, the history of snapshots will be flushed and a new round will be initiated by the AN. To avoid unnecessary transmissions, SNs send information only once in a round as presented in Alg. 3.

7.4.5   eMap Construction

The prime goal of the map construction is to identify inequalities of energy density. Expected is an eMap that divides the sensor field into regions, which are indicators of similar energy-densities. The input of the construction algorithm is the collected residual-energy information and the output is the map's regions. The eMap is a geometrical/spatial data structure (e.g., tree) which is easy to evaluate. The construction operation has to satisfy some crucial requirements. First, it should be easy to evaluate the AN. Second, two neighboring regions should have two "sufficiently" different energy densities. The

---

Algorithm 3 Data Collection Algorithm for Unstructured mWSN

1: /***** On Assist Node (AN) ******/
2: var $HIST_{snapshot}$
3: Initiate a new round $round_{id}$ for sampling
4: AN: Do first snapshot: SEND REQ beacon with $round_{id}$
5: AN: STORE $\{snapshot_{id}, AN_{loc}\}$ in $HIST_{snapshot}$
6: $snapshot_{id} + +$
7: If AN has moved a distance of 2*R since previous snapshot do:
8: AN: CHECK $HIST_{snapshot}$
9: AN: Compute $COV_{additional}$ from current AN_{loc} and $HIST_{snapshot}$
10: **if** $COV_{additional} \geq COV_{th}$ **then**
11:     AN: SEND REQ beacon with $round_{id}$
12:     AN: STORE $\{snapshot_{id}, AN_{loc}\}$ in $HIST_{snapshot}$
13:     $snapshot_{id} + +$
14: **else**
15:     AN: Suppress REQ beacon
16: **end if**
17: AN: RECEIVE E_{msg}
18: AN: GOTO 7
19: /***** On Sensor Node (SN) ******/
20: SN: RECEIVE REQ beacon
21: **if** SN: new *round* **then**
22:     SN: Schedule transmission between $0 < t_{rand} < T_{max}$
23:     SN: SEND E_{msg} $\{ID, SN_{loc}, E_{lev}\}$
24: **else**
25:     SN: Suppress SEND E_{msg}
26: **end if**

---

map construction process is composed of the spatial partitioning of the sensor field (*space partition*) and the fusion of the regions of similar residual energy values (*regioning*).

For space partitioning, Voxel grid, triangulation (e.g., Voronoi or Delaunay), octree, k-d tree and BSP tree [37] can be used. All these schemes except the Voxel grid are dependent on the input data. For this reason we select the simple Voxel grid for space partitioning. The primitive parameter to divide the sensor field is the size of smallest fragment of area, i.e., grid-cell size or the partitioning *resolution* (r). The energy density in the cell is the basis to form a region. Selecting r is a crucial decision for creating the eMap. Depending on this resolution a cell may contain more than one SN. We refer to the residual energy value of one cell by the sum of the energy values of all the nodes in that cell.

In order to merge the cells into regions (regioning), we need to ascertain if neighboring cells have similar values for residual energy. For this, we need a technique to decide if two neighbor cells can be merged or not. A first possible technique is to use a metric to measure the inequality between two neighboring cells. In the literature, we identify several inequality indices [38] that measure the inequality of a set: Variance, entropy coefficients, Hoover coefficient, Coulter coefficient, Gini coefficient, etc. A second technique is to use global classes. In the eMAP approach we rely on the class-based technique for its simplicity and easy evaluation on ANs. Furthermore, we are investigating the suitability of other indices in ongoing work. The cells are classified into a fixed number of classes depending on their energy density. Neighboring cells are merged into the same region if they belong to the same class.

In Alg. 4, we propose the pseudocode of our regioning algorithm. This algorithm is based on searching and is inspired by the region growing algorithm for image segmentation [39]. We assign a region ID to any cell to start regioning. Then, we check if neighbor cells can be merged with this cell. When we merge cells we assign the same region ID to them. Once the neighbor cells are checked for merge, we will repeat the process of cell merging for the neighbors that have been successfully merged to the current region. After completing regioning for the starting cell, all the other cells (which are not assigned to a region) will form regions in the same way. Hence, we complete regioning for the whole WSN field.

We observe a trade-off between the *accuracy* and comprehensibility of the constructed map. The accuracy of the eMap depends on the accuracy of energy information collection and on the accuracy of regioning. Regioning accuracy is important to comprehend node distribution. If the model provides a map such that only the neighboring cells with the same energy-level form regions, it becomes the most accurate region map. It would be a worse map if regions consisted of cells with highly different energy densities.

Abstraction is far more meaningful for human comprehension than sheer raw numbers. For example, *hot water* is more meaningful than telling the exact temperature of water as $83.51^{o}C$. Therefore, the selection of the number of energy classes is crucial since it allows for tuning the trade-off between map accuracy (exact numeric values) and comprehensibility (abstraction for perception). It should take into account the range of possible values and the level of inequality tolerated for regioning. Selecting a higher number of classes provides for higher map accuracy on the one hand but hardens regioning and subsequently the comprehensibility of the map on the other hand. A lower number of classes sacrifices accuracy in order to provide for a better comprehensibility. However, if the number of classes is too low, we merge cells with high difference in energy densities. Thus, regioning weakly reflects the energy spatial distribution. This results in an erroneous map. Summarizing, the number of classes should be appropriately selected to provide for the required trade-off between accuracy and comprehensibility.

Algorithm 4 Regioning

1: cell := structure{cellID, neighborList[], regionID=-1, energyClass}
2: grid := array of all cells
3: var currentRegionID
4: var currentRegion:= array of all cells with regionID=currentRegionID
5: /********* regioning () *********/
6: regioning()
7: **for** each $cell_i \in$ grid **do**
8:   **if** $cell_i.cellID > -1$ **then**
9:     next iteration
10:   **end if**
11:   $cell_i$.regionID=currentRegionID;
12:   regionMaking($cell_i$)
13:   **for** each $cell_j$ merged with the region of $cell_i$ **do**
14:     regionMaking($cell_j$)
15:   **end for**
16:   currentRegionID++
17: **end for**
18: /***** regioning with the 8 neighboring cells ******/
19: regionMaking(myCell)
20: **for** each neighborCell **do**
21:   **if** (myCell.energyClass = neighborCell.energyClass) **then**
22:     neighborCell.regionID = myCell.regionID
23:   **end if**
24: **end for**

7.5 Evaluation

In this section, we provide simulation-based analysis of the gMAP approach. In particular we evaluate the performance of the path planning framework, the data collection accuracy and efficiency, as well as the map construction accuracy and comprehensibility. We further compare the message complexity, processing complexity, data completeness, and latency of gMAP to that of the existing approaches.

7.5.1 Path Planning Performance

The performance of the proposed algorithm strongly depends on the spatial deployment of the SNs and particularly on the nodes' local density. The more neighbors within the communication range of SNs, the more discrete structures in the neighborhood of each breakpoint occur, which arises as sets of constraints in the NLP or MILP to solve.

Obviously, it is easier to solve a coverage problem over the whole area for scenarios with a high density of SN. For scenarios with very low node density

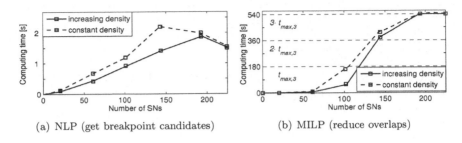

(a) NLP (get breakpoint candidates) (b) MILP (reduce overlaps)

Figure 7.4: Time to solve the NLP (STEP 1) and the approximating MILP (STEP 2).

a TSP solution is recommended, followed by a minimization that shortens and smoothes the path again by moving the breakpoints ("flexible cities") within a certain neighborhood of the nodes.

For the proof of concept, the proposed algorithm in Section 7.4.3.2 has been designed and tested for the representative scenarios discussed in Figure 7.2. We solved STEP 1 using the IPOPT [9] NLP-solver on the NEOS server [8]. The MILP STEP 2 was solved by using CPLEX [14] running on a PC (Intel® Pentium® M processor 1.86GHz; 1024 MB RAM). The TSP in STEP 3 was solved with CONCORDE [10] on the same platform.

For a single AN, we investigated scenarios with an increasing number of SNs but: (1) In a fixed area with an increasing density and (2) with a constant density. As time limits $t_{max,1} = \infty$, $t_{max,2} = 540$s, $t_{max,3} = 180$s and as MILP-approximation parameter $\gamma = 8$ were used. Resulting computing times for a setting with randomly distributed nodes is depicted in Figure 7.4, and a resulting path is shown in Figure 7.2(b).

For the setting with 225 randomly distributed SNs, each node was covered by an average of 1.3 breakpoints, when the arbitrary time limit $t_{max,2} = 540$s was reached. For the harder problem with 225 nodes on a fixed grid, each node was covered by an average of 1.5 breakpoints after the same time limit $t_{max,2}$.

7.5.2 Data Collection and Map Construction Performance

We first describe our simulation settings. Then, we define the evaluation metrics, based on which we present our results.

7.5.2.1 Simulation Settings

We use Tossim [41] and its Tython extension for network simulations, and Matlab for map construction. Tossim is an event-driven simulation tool widely used in the WSN community. We have used the disc radio model provided by

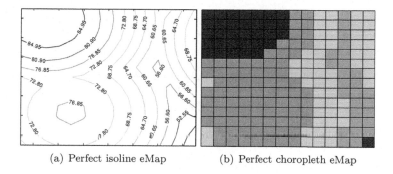

(a) Perfect isoline eMap   (b) Perfect choropleth eMap

Figure 7.5: Perfect maps.

Tossim with 5 units communication range. All nodes lying in this communication range communicate without errors and have symmetric links, although collisions may occur. We have considered 225 SNs either generated randomly in the area of 42 units x 42 units or in a grid topology of 15 x 15 (cell size $c = 3$ units) including sink. The AN moves either in controlled fashion (with pause time of $t_0 = 3sec$) or according to the random waypoint model using a constant speed of 1 $unit/sec$. We selected the commonly used random waypoint mobility model as its high randomness maximizes the unpredictability and assume that the mobility of the AN eventually covers the entire sensor field. SNs use $T_{max} = 500msec$ as a maximum time to schedule their replies to REQ beacons.

Nodes initially have energy values following an arbitrary distribution. In this work, we use the distribution depicted in Figure 7.5. The choice of space partitioning resolution (r) is critical. Intuitively, a good choice is $D \leq r \leq R$, where D is the average distance between two neighboring SNs. For the structured as well as the unstructured scenarios (225 nodes and 42 x 42 units area), simulations with different r values showed that $r = 3$ $units$ allows for the most comprehensive eMap. This is about the average distance between two neighboring SNs in the structured scenario. The simulation results are the average of 5 runs, i.e., the collected data map is the average of 5 collected maps via AN, and then we compare it with the original perfect map.

7.5.2.2   Evaluation Metrics

The performance of constructing global maps is commonly measured with respect to its completeness, efficiency, and regioning accuracy.

- *Collection Completeness*: The ratio of nodes whose values are collected by the AN to the total number of SNs.

- *Collection Efficiency*: To measure collection efficiency we consider the number of energy messages per SN, i.e., the ratio of total number of energy messages sent by SNs to the number of SNs that received a REQ-beacon. We also consider the number of snapshots as overhead since it implies the reception of beacons by the SNs.

- *Region Accuracy*: To evaluate the regioning accuracy we compare the eMap constructed by the AN with the perfect eMap, i.e., the map constructed from complete energy information.

To compute the regioning accuracy we compare both maps either cell-wise or region-wise:

(1) Cell Misclassification Percentage (CMP): We count the number of grid cells in the map that have been misclassified due to incompleteness of collected data. To evaluate CMP we define the original perfect map as the ideal map and the test map that is collected as the data map acquired through AN. We have one reference map for each scenario, i.e., one for the structured scenario and one for the unstructured scenario, and we have two test maps for each scenario. The CMP can be defined by the following equation:

$$Ce = \sum_i \sum_j count(Cr_{ij} - Ct_{ij}) \div C_o, \qquad (7.8)$$

Ce is the total count of class cells that differ between the reference Cr and test class map Ct. C_o refers to the number of occupied cells. The count() function returns '1' if the two cells do not belong to the same class, or else it returns '0'. Ce is the direct measure of correct classification of the grid cells into the classes and indirect measure of the accuracy of area and perimeter of the detected event area.

(2) Regional Percentile Accuracy (RPA): RPA is the percentage of correctly classified cells for each region. RPA assesses the accuracy of regions level and is computed as follows. The regioning algorithm (Alg. 4) is used to detect various regions in both the original perfect map and the collected data map. Afterwards, each region from the collected map is compared to the perfect region to find out the percentile accuracy of each region.

7.5.2.3   Data Collection Performance

The results of our simulations are summarized in Table 7.1. We observe that the completeness of energy information is close to but lower than 100% in all scenarios. In structured and semi-structured scenarios, the incompleteness is due to MAC collisions. We also observe that our path planning has improved the completeness of the collected eMap, though it requires more snapshots in structured scenario. Furthermore, we observe a lower number of

Table 7.1: Simulation Results for gMAP

	Structured		Semi-structured	Unstructured	
	Zigzag	TSP-Grid	TSP-Random	COV = 70%	COV = 90%
Completeness [%]	94.7	98.2	95.1	88.4	81.3
Efficiency	1.0	1.0	1.0	1.0	1.0
#snapshots	25	41	20	61	43
Latency [min]	5.0	7.94	4.14	40.71	39.25

snapshots in semi-structured scenarios, which is a direct result of optimizations in the proposed path planning algorithm for uniform random distribution of nodes. In unstructured scenarios, besides collisions, mobility leads to fast topology changes and therefore to additional message loss. We observe that lower COV_{th} values provide for higher completeness at the cost of a higher number of snapshots. This is due to the fact that if COV_{th} is increased, higher overlaps between snapshots is tolerated. After a sufficient number of snapshots, the additional coverage will not be able to be higher than COV_{th} and no further snapshots are possible although some nodes have not received a REQ beacon. This results in higher efficiency (limited number of snapshots) but the completeness of collected information may suffer. The efficiency is 1.0 in all scenarios, given the fact that nodes that receive a REQ beacon from the AN respond with a single message per collection round, irrespective of whether the AN received it or not. The latency of gMAP is as expected in the range of minutes to hours. The latency for data collection in unstructured scenarios is higher than that in structured scenarios. This is due to the fact that the movement of the AN is random implying that more time is needed to cover the sensor field, whereas in a semi-structured scenario we observe the latency to be optimal corresponding to the lower number of snapshots and resulting shorter path from the path planning algorithm. Generally, the simulation results confirm the gMAP efficiency and utility in the practical scenarios.

7.5.2.4   Regioning Performance

Given that energy levels are between 0 and 100%, and as it is likely to tolerate 10% difference within the single region, we use for regioning 10 classes of energy levels. Figure 7.5(a) shows the isolines of the considered energy distribution. We use this map as a reference and compare different eMaps generated by our approach. In Figure 7.5(b) we show the perfect choropleth eMap of the structured scenario. Obviously, choropleth is more expressive and comprehensive than the isolines.

Given the 10 possible energy classes, the considered energy distribution results in an eMap of 9 regions. Figure 7.6 shows the accuracy of each region (RPA) formed from the collected data along with the overall accuracy (CMP) of the eMap. Map and region accuracies are calculated according to the CMP and RPA metric, respectively. As the structured scenario has the highest level of completeness, its eMap is almost perfect and regions have a very high accuracy. Apart from 2 regions all other regions are perfectly formed. This high accuracy level proves the high performance of our path planning algorithm. In the semi-structured scenario the achieved accuracy is lower in comparison to the structured scenario since nodes are randomly deployed in the former. However, our path planning algorithm helped to collect enough data for each region to be considerably accurate with more than 90% of accuracy in each of 7 regions out of 9 (Figure 7.6). From the map accuracy in the unstructured

Table 7.2: Comparison with Existing Approaches

Approach	Message complexity	Processing complexity	Max data completeness	Latency
isobar [21]	$O(N)$	$O(N)$	$< 100\%$	$\sim sec$
eScan [20]	$O(N^{1.5})$	$O(N^4)$	100%	$\sim sec$
INLR [22]	$O(N^{1.5})$	$\Omega(N^{1.5})$	100%	$\sim sec$
Isoline [24]	$O(N^{0.5})$	$O(N)$	$<100\%$	$\sim sec$
Meng [25]	$O(N)$	$\Omega(Nd)$	$<100\%$	$\sim sec$
Iso-Map [26]	$O(N^{0.5})$	$O(N)$	$<100\%$	$\sim sec$
gMAP	N	$O(1), = CP$	100%	$L_1, \sim min - hr$
Data Mule [56]	$\geq N$	$\geq CP$	100%	$L_2 < L_1, \sim min - hr$
Message Ferrying [55]	$> N$	$> CP$	100%	$L_2 \leq L_3 \leq L_1,$ $\sim min - hr$

scenario, it is clear that data completeness is the lowest among all scenarios. However, gMAP achieves 88.4% map accuracy (CMP) for $COV_{th} = 70\%$ and 81.3% map accuracy for $COV_{th} = 90\%$. Though the selection of $COV_{th} = 70\%$ increases the number of snapshots, the data completeness increases as highlighted in Figure 7.6. We observe there that all regions are more than 80% accurately formed in comparison to $COV_{th} = 90\%$ where we have a worst case situation of a region formed with just 50% accuracy.

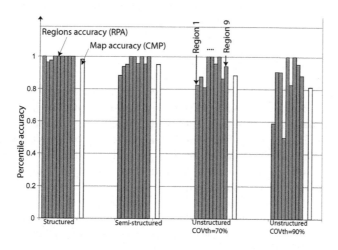

Figure 7.6: Regional Percentile Accuracy (RPA), Cell Misclassification Percentage (CMP).

7.5.3   Comparison to Related Work

In this section, we compare the performance of gMAP to that of the existing approaches. We are mainly interested in the message and processing overhead on SNs, since this determines their energy consumption concerning map construction. In addition, we compare the achievable maximal data completeness (ratio of maximal collected data samples to the total number of possible samples or number of sensor nodes $N = N_N$) as this determines the highest accuracy performed by the corresponding monitoring technique. We denote the processing complexity of gMAP ($O(1)$) by CP. We refer by L_1, L_2, and L_3 to the latency of gMAP, data mule, and message ferrying, respectively.

In [29][26][56][55], analytical results for gMAP as well as related approaches have been developed. We summarize these results in Table 7.2.

The comparison of gMAP to delay-critical monitoring approaches proves that the gMAP approach provides for the minimal processing complexity and for a relatively low message complexity while keeping the data completeness maximal. This comes at the cost of higher end-to-end latency, which is not a concern for the delay-tolerant information (e.g., energy) collection. gMAP shows comparable message/processing complexities, data completeness, and latencies compared to the delay-tolerant monitoring approaches. As in data mule and message ferrying SNs may need to use multihop communication to reach the mobile AN, both approaches usually show higher message and processing complexities. This comes with the benefit of a slightly smaller latency. gMAP as well as data mule and message ferrying can achieve maximal completeness of data collection.

7.6   Conclusions

We have presented gMAP, an extremely energy-efficient methodology that collects data of interest from the WSN and presents its geographical distribution as a map. Our approach is opportunistic as it exploits existing node mobility to collect data. Being mobility-assisted the collection process lasts for the time that mobile entities need to scan the whole sensor field. Therefore, data should be of high time relevance, i.e., do not change suddenly or radically in magnitude. As an example we focussed on the map of residual energy since the battery depletion is a long-running process. Considering three representative scenarios, i.e., structured, semi-structured, and unstructured, we showed the efficiency and the accuracy of our gMAP approach.

7.7   Open Issues

Though there exists a wide effort to design mobility-assisted collection of data of longer lifetime, there is still a need for further research work that we discuss in the following.

We provided an efficient, approximative, and scalable path planning algorithm that performs in the range of seconds on a conventional PC platform. Unfortunately, our optimization to reduce overlapping of breakpoints decreases the performance of path planning to the range of minutes on the same platform. As ANs may be much more restricted in resources, it is interesting to investigate further optimizations of our algorithm in order to sacrifice a tolerable accuracy for increased efficiency. This will be useful to achieve for different conventional platforms from laptops to PDAs to motes. An implementation of the algorithm on resource-constraint devices is a challenge for path planning algorithms. Therefore, a distributed computation of path planning on motes is an interesting open issue.

In this work, we focussed on controlling the mobility in space, which is important to provide for high data collection completeness. Further investigations of mobility control in time such as suggested in [59] will allow AN to autonomously adjust their motion to run time dynamics of the system and to the evolving tolerable data lifetime. The dynamic adaptation of mobility control in both time and space on resource-constraint ANs is a crucial research to allow for autonomous and pro-active reconfiguration.

The coordination between multiple ANs to optimize data collection is also a crucial research field that should be extended beyond the few works existing in the literature [64, 65]. In particular, open issues include how to fix the number of optimal AN in dependency of data lifetime, the WSN coverage area, movement properties of AN, and load balancing policies (buffer etc.). The data mule projet suggested a few of these optimizations such as identifying a set of congested nodes (with high data rates or lossy link, etc.) so that the next tour is improved by adjusting the node speed, for example [56]. In [62], the authors present a technique to adapt the node speed to the varied data rates of different sensor nodes.

Another interesting adaptation issue for path planning is the adaptation to realistic network conditions at run-time. For example, path planning is assuming a disc model for communication (i.e., fixed communication range), however, realistic radio propagation models while running the path planning algorithm are more appropriate. We believe that the ANs should react locally while traversing the path if communication perturbations occur.

In order to provide for continuous monitoring it is important to investigate update strategies to keep monitored status accurate. Usually, the collected samples become obsolete and need to be updated after varied collection time points in different spatial regions. A natural optimization of the path planning is to adapt it to the location of the next needed updates. We believe that AN can optimize their path if they can predict future profiles. One possibility is that ANs can weight the nodes depending on the urgency of the status update from the different nodes. However, we believe that for efficiency reasons ANs should weight the regions of the map and accordingly do a path planning on top of regions and then on node level.

Bibliography

[1] A. Khelil, C. Reinl, B. Ayari, F. K. Shaikh, P. Szczytowski, A. Ali, N. Suri. Sensor Cooperation for a Sustainable Quality of Information. *Pervasive Computing and Networking*, New York: John Wiley & Sons, 2011.

[2] W. Burgard, M. Moors, C. Stachniss, and F.E. Schneider. Coordinated Multi-robot Exploration. *IEEE Transactions on Robotics*, vol. 21, pp. 376–386, 2005.

[3] R. Sugihara and R. K. Gupta. Improving the Data Delivery Latency in Sensor Networks with Controlled Mobility. *Distributed Computing in Sensor Systems*, Springer Berlin/Heidelberg, vol. 5067 of Lecture Notes in Computer Science, pp. 386–399, 2008.

[4] M. Bosse, N. Nourani-Vatani, and J. Roberts. Coverage Algorithms for an Under-actuated Car-Like Vehicle in an Uncertain Environment. *IEEE International Conference on Robotics and Automation (ICRA)*, 2007.

[5] C. L. Bloebaum and Ulrich Faigle. Coverage Path Planning: The Boustrophedon Cellular Decomposition. In *International Conference on Field and Service Robotics*, 1997.

[6] J. Beardwood, J.H. Halton, and J.M. Hammersley. The Shortest Path Through Many Points. In *Proc. Cambridge Philosophical Society*, 1959.

[7] M.M. Solomon. Algorithms for the Vehicle Routing and Scheduling Problems with Time Window Constraints. *Operations Research*, 1987, 35(2), pp. 254–265.

[8] http://neos.mcs.anl.gov

[9] A. Wächter and L. T. Biegler. On the Implementation of a Primal-Dual Interior Point Filter Line Search Algorithm for Large-Scale Nonlinear Programming. In *Mathematical Programming* 106(1), pages 25–57, 2006.

[10] D. Applegate, R.E. Bixby, V. Chvátal, and W.J. Cook. Concorde tsp-solver. Technical report, http://www.tsp.gatech.edu/concorde.html, 2006.

[11] K. Elbassioni, A.V. Fishkin, and R. Sitters. On Approximating the TSP with Intersecting Neighborhoods. In *Algorithms and Computation*, volume 4288 of *Lecture Notes in Computer Science*, pages 213–222. Springer Berlin/Heidelberg, 2006.

[12] M. Glocker, C. Reinl, and O. von Stryk. Optimal Task Allocation and Dynamic Trajectory Planning for Multi-vehicle Systems Using Nonlinear Hybrid Optimal Control. In *Proc. 1st IFAC-Symposium on Multivehicle Systems*, 2006.

[13] I.E. Grossmann and S. Lee. Generalized Convex Disjunctive Programming: Nonlinear Convex Hull Relaxation. *Computational Optimization and Applications*, 26(1):83–100, 2003.

[14] ILOG. *ILOG CPLEX 11.0, User's Manual*, 2007.

[15] Joseph S. B. Mitchell. A PTAS for TSP with Neighborhoods among Fat Regions in the Plane. In *Proc. of the eighteenth annual ACM-SIAM symposium on Discrete algorithms (SODA)*, pages 11–18, 2007.

[16] C. Reinl and O. von Stryk. Optimal Control of Multi-vehicle Systems under Communication Constraints Using Mixed-integer Linear Programming. In *Proc. of The First International Conference on Robot Communication and Coordination (RoboComm)*, 2007.

[17] B. Yuan, M. Orlowska, and S. Sadiq. On the Optimal Robot Routing Problem in Wireless Sensor Networks. *IEEE Transactions on Knowledge and Data Engineering*, vol. 19, Iss. 9, pp 1252–1261, 2007

[18] A. Khelil, F. K. Shaikh, B. Ayari, and N. Suri. MWM: A Map-based World Model for Event-driven Wireless Sensor Networks. In *The 2nd ACM International Conference on Autonomic Computing and Communication Systems (AUTONOMICS)*, 2008.

[19] A. Ali, A. Khelil, F. K. Shaikh, and N. Suri. MPM: Map-based Predictive Monitoring for Wireless Sensor Networks. In *The 3rd International Conference on Autonomic Computing and Communication Systems (AUTONOMICS)*, 2009.

[20] Y. Zhao, R. Govindan, and D. Estrin. Residual Energy Scan for Monitoring Sensor Networks. In *IEEE Wireless Communications and Networking Conference (WCNC)*, 2002.

[21] J.M. Hellerstein, W. Hong, S. Madden, and K. Stanek. Beyond Average: Toward Sophisticated Sensing with Queries. In *Second International Workshop on Information Processing in Sensor Networks (IPSN)*, page 553, 2003.

[22] W. Xue, Q. Luo, L. Chen, and Y. Liu. Contour Map Matching for Event Detection in Sensor Networks. In *ACM SIGMOD*, 2006.

[23] R.A.F. Mini, M.d.V. Machado, A.A.F. Loureiro, and B. Nath. Prediction-based Energy Map for Wireless Sensor Networks. *Elsevier Ad-hoc Networks Journal (Special Issue on Ad Hoc Networking for Pervasive Systems)*, 2005.

[24] I. Solis and K. Obraczka. Isolines: Energy-efficient Mapping in Sensor Networks. In *10th IEEE Symposium on Computers and Communications (ISCC)*, 2005.

[25] X. Meng, T. Nandagopal, L. Li, and S. Lu. Contour maps: Monitoring and Diagnosis in Sensor Networks. *Computer Networks, 50(15)*, 2006.

[26] Y. Liu and M. Li. Iso-Map: Energy-Efficient Contour Mapping in Wireless Sensor Networks. In *The 27th International Conference on Distributed Computing Systems (ICDCS)*, 2007.

[27] M. Grossglauser and D. Tse. Mobility Increases the Capacity of Ad Hoc Wireless Networks. *IEEE/ACM Transactions on Networking, 10(4)*, 2002.

[28] Y. Wang, H. Dang, and H. Wu. A Survey on Analytic Studies of Delay-Tolerant Mobile Sensor Networks. *Journal of Wireless Communications and Mobile Computing (WCMC) SI on Disruption Tolerant Networking for Mobile or Sensor Networks, 7(10)*, 2007.

[29] A. Khelil, F. K. Shaikh, A. Ali, and N. Suri. gMAP: Efficient Construction of Global Maps for Mobility-Assisted Wireless Sensor Networks. In *Conference on Wireless On-demand Network Systems and Services (WONS)*, 2009.

[30] Y.J. Zhao, R. Govindan, and D. Estrin. Sensor Network Tomography: Monitoring Wireless Sensor Networks. In *Student Research Poster, ACM SIGCOMM*, 2001.

[31] R.A.F. Mini, A.A.F. Loureiro, and B. Nath. The Distinctive Design Characteristic of a Wireless Sensor Network: The Energy Map. *Computer Communications, 27(10)*, 2004.

[32] E. Souto, R. Gomes, D. Sadok, and J. Kelner. Sampling Energy Consumption in Wireless Sensor Networks. In *IEEE International Conference on Sensor Networks, Ubiquitous, and Trustworthy Computing (SUTC)*, 2006.

[33] A.H. Robinson, J.L. Morrison, P.C. Muehrcke, A.J. Kimerling, and S.C. Guptill. *Elements of Cartography*. John Wiley & Sons, New York, 1995. 6th Edition.

[34] P. Dutta, J. Hui, J. Jeong, S. Kim, C. Sharp, J. Taneja, G. Tolle, K. Whitehouse, and D. Culler. Trio: Enabling Sustainable and Scalable Outdoor Wireless Sensor Network Deployments. In *The Fifth International Conference on Information Processing in Sensor Networks (IPSN)*, 2006.

[35] S.M. LaValle. *Planning Algorithms*. Cambridge University Press, Cambridge, U.K., 2006. Available at http://planning.cs.uiuc.edu/.

[36] L. Lima and J. Barros. Random Walks on Sensor Networks. In *the 5th International Syposium on Modeling and Optimization in Mobile, Ad hoc, and Wireless Networks (WiOpt)*, 2007.

[37] J. Nievergelt and P. Widmayer. Spatial Data Structures: Concepts and Design Choices. In *Handbook of Computational Geometry*, Elsevier Science Publishers, 2000.

[38] B.A. Portnov and D. Felsenstein. *Regional Disparities in Small Countries*. Springer, Berlin/Heidelberg, 2005. ISBN 978-3-540-24303-8.

[39] L.G. Shapiro and G.C. Stockman. *Computer Vision*. Prentice-Hall, Upper Saddle River, NJ, 2001. 978-0130307965.

[40] B. Krishnamachari, D. Estrin, and S. Wicker. Modelling Data-Centric Routing in Wireless Sensor Networks. In *USC Computer Engineering TR CENG 02-14*, 2002.

[41] P. Levis, N. Lee, M. Welsh, and D. Culler. Tossim: Accurate and Scalable Simulation of Entire Tinyos Applications. In *Proc. of The ACM Conference on Embedded Networked Sensor Systems (SenSys)*, 2003.

[42] N. Ramanathan, K. Chang, R. Kapur, L. Girod, E. Kohler, and D. Estrin. Sympathy for the Sensor Network Debugger. In *Proc. of The ACM Conference on Embedded Networked Sensor Systems (SenSys)*, 2005.

[43] S. Rost and H. Balakrishnan. Memento: A Health Monitoring System for Wireless Sensor Networks. In *Proc. of The IEEE Communications Society Conference on Sensor, Mesh and Ad Hoc Communications and Networks (SECON)*, 2006.

[44] Jerry Zhao, Ramesh Govindan, and Deborah Estrin. Tomography: Monitoring Wireless Sensor Networks. In *Student Research Poster, ACM SIGCOMM*, 2001.

[45] Jerry Zhao, Ramesh Govindan, and Deborah Estrin. Computing Aggregates for Monitoring Wireless Sensor Networks. In *Proc. of The First IEEE International Workshop on Sensor Network Protocols and Applications (SNPA)*, 2003.

[46] M. Woehrle, C. Plessl, R. Lim, J. Beutel, and L. Thiele. EvAnT: Analysis and Checking of Event Traces for Wireless Sensor Networks. In *IEEE International Conference on Sensor Networks, Ubiquitous and Trustworthy Computing (SUTC)*, 2008.

[47] M. Ringwald, K. Römer, and A. Vitaletti. Passive Inspection of Sensor Networks. In *Proc. of The 3rd IEEE International Conference on Distributed Computing in Sensor Systems (DCOSS)*, 2007.

[48] G. Xing, T. Wang, Z. Xie, and W. Jia. Rendezvous Planning in Mobility-Assisted Wireless Sensor Networks. In *Proc. 28th IEEE International Real-Time Systems Symposium (RTSS)*, 2007.

[49] G. Xing, T. Wang, Z. Xie, and W. Jia. Rendezvous Planning in Wireless Sensor Networks with Mobile Elements. In *IEEE Transactions Mobile Computing 7(12)*, 2008.

[50] G. Xing, T. Wang, W. Jia, and M. Li. Rendezvous Design Algorithms for Wireless Sensor Networks with a Mobile Base Station. In *Proc. The ACM International Symposium on Mobile Ad Hoc Networking and Computing (MobiHoc)*, 2008.

[51] G. Xing, J. Wang, K. Shen, Q. Huang, X. Jia, and H.C. So. Mobility-Assisted Spatiotemporal Detection in Wireless Sensor Networks. In *Proc. The International Conference on Distributed Computing Systems (ICDCS)*, 2008.

[52] W. Zhao, M. Ammar, and E. Zegura. A Message Ferrying Approach for Data Delivery in Sparse Mobile Ad Hoc Networks. In *Proc. The ACM International Symposium on Mobile Ad Hoc Networking and Computing (MobiHoc)*, 2004.

[53] M.C. Chuah and P. Yang. A Message Ferrying Scheme with Differentiated Services. In *Proc. of Military Communications Conference (MILCOM)*, 2005.

[54] H. Jun, W. Zhao, M. Ammar, E.W. Zegura, and C. Lee. Trading Latency for Energy in Wireless Ad Hoc Networks Using Message Ferrying. In *Proc. Third IEEE International Conference on Pervasive Computing and Communications Workshops (PERCOMW)*, 2005.

[55] H. Jun, M. Ammar, and E. Zegura. Power Management in Delay-Tolerant Networks: A Framework and Knowledge-Based Mechanisms. In *Proc. of The IEEE Communications Society Conference on Sensor, Mesh and Ad Hoc Communications and Networks (SECON)*, 2005.

[56] A.A. Somasundara, A. Kansal, D. Jea, D. Estrin, and M.B. Srivastava. Controllably Mobile Infrastructure for Low Energy Embedded Networks. In *IEEE Transactions on Mobile Computing*, 2006.

[57] A.A. Somasundara, A. Ramamoorthy, and M.B. Srivastava. Mobile Element Scheduling for Efficient Data Collection in Wireless Sensor Networks with Dynamic Deadlines. In *Proc. IEEE Real Time Systems Symposium (RTSS)*, 2004.

[58] A. Kansal, M. Rahimi, W.J. Kaiser, M. Srivastava, G.J. Pottie, and D. Estrin. Controlled Mobility for Sustainable Wireless Networks. In *Proc. of The IEEE Communications Society Conference on Sensor, Mesh and Ad Hoc Communications and Networks (SECON)*, 2004.

[59] A. Kansal, A. Somasundara, D. Jea, M. Srivastava, and D. Estrin. Intelligent Fluid Infrastructure for Embedded Networks. In *Proc. of The International Conference on Mobile Systems, Applications, and Services (MobiSys)*, 2004.

[60] R.C. Shah, S. Roy, S. Jain, and W. Brunette. Data MULEs: Modeling a Three-tier Architecture for Sparse Sensor Networks. In *Proc. of The First International Workshop on Sensor Network Protocols and Applications (SNPA)*, 2003.

[61] O. Tekdas, V. Isler, J.H. Lim, and A. Terzis. Using Mobile Robots to Harvest Data from Sensor Fields. In *Proc. IEEE Wireless Communications*, 2009.

[62] Y. Gu, D. Bozdag, E. Ekici, F. Ozguner, and C.-G. Lee. Partitioning-Based Mobile Element Scheduling in Wireless Sensor Networks. In *Proc. of The IEEE Communications Society Conference on Sensor, Mesh and Ad Hoc Communications and Networks (SECON)*, 2005.

[63] Y. Wang and H. Wu. DFT-MSN: The Delay/Fault-Tolerant Mobile Sensor Network for Pervasive Information Gathering. In *Proc. of The IEEE Conference on Computer Communications (INFOCOM)*, 2006.

[64] W. Cheng, M. Li, K. Liu, Y. Liu, X.-Y. Li, and X. Liao. Sweep Coverage with Mobile Sensors. In *Proc. of The 22nd IEEE International Parallel and Distributed Processing Symposium (IPDPS)*, 2008.

[65] D. Jea, A. Somasundara and M. Srivastava. Multiple Controlled Mobile Elements (Data Mules) for Data Collection in Sensor Networks. In *Proc. of The IEEE International Conference on Distributed Computing in Sensor Systems (DCOSS)*, 2005.

... ...
... ...
... ...

... ...
... ...
... ...

Chapter 8

Message Dissemination in Vehicular Networks

Shabbir Ahmed and Salil S. Kanhere

8.1	Introduction		224
8.2	Characteristics of VANET		226
	8.2.1	Network Characteristics	226
8.3	Innovative Applications		229
	8.3.1	Low Cost Digital Connectivity in Rural Areas	231
	8.3.2	Data Collection from Sensor Nodes	232
	8.3.3	Cooperative Downloading	233
	8.3.4	Miscellaneous Applications	234
8.4	Message Dissemination		234
	8.4.1	MANET Forwarding Schemes and Their Limitations	235
	8.4.2	Forwarding Schemes Suitable for VANET	236
		8.4.2.1 Position-Based Forwarding	236
		8.4.2.2 Trajectory-Based Forwarding	238
		8.4.2.3 Opportunistic Forwarding	239
8.5	Medium Access Control for VANET		247
	8.5.1	Cooperative ARQ	247
	8.5.2	802.11p MAC	249
8.6	Open Issues		251
	8.6.1	Security Challenges	251
	8.6.2	Acknowledgments	252
	8.6.3	Addressing	253
	8.6.4	Emerging New Application	253
8.7	Conclusion		253

A Vehicular Ad Hoc Network (VANET) is a special kind of Disruption Tolerant Network (DTN) which is characterized by its intermittent connectivity and high velocity of nodes. Besides road safety applications which are meant to avoid injuries, comfort and innovative applications are increasingly becoming popular which leverages the power of vehicular networks. Accessing the Internet and offering P2P services are some of those. The main challenges for data dissemination in these kinds of services are: a) route messages among highly churned and disconnected nodes, b) scalability to tackle varying traffic densities, c) not to saturate the channel, and d) adaptive to the environmental factors (e.g., tall buildings, detours, etc.). In this chapter, we discuss the problems and possible solutions of message dissemination at different protocol layers of disruption tolerant VANET. The primary focus is to propagate messages reliably and efficiently.

8.1   Introduction

Advances in computing and wireless communication technologies have increased the interest in *smart* vehicles — vehicles equipped with computing, sensing, and communication capabilities. The network of these smart vehicles (i.e., VANET) have the potential to greatly lower system operating costs by reducing the dependence on public infrastructures. They can be used to extend the coverage of costly roadside infrastructures. In-vehicle sensors have the potential to access much detailed, accurate information about on-road vehicle activities that would otherwise be impossible. Vehicles can exchange information (e.g., detour, traffic accident, congestion information, etc.) and provide early warnings to nearby vehicles in order to reduce traffic jams near the affected areas. Vehicular applications range from road-safety to Internet access to online vehicle entertainments (e.g., gaming and file sharing, etc.). Some innovative applications are discussed in Section 8.3.

The authors [32] mention some of the important characteristics of VANET which affect communications: a) *High velocity* of the vehicles, b) *Environmental factors* such as: obstacles, high-rise buildings, tunnels, traffic jams, etc., c) *Regular mobility patterns* due to the fact that vehicles move only on roadways and mobility depends on the source to destination path and on traffic conditions, d) *Intermittent communications* due to highly churned and isolated networks of cars, and e) *High Congestion Channel* because of high density of vehicles in some areas. These factors result in a network with frequent fragmentation, rapid change in topology due to high node speeds, small effective diameter, and limited redundancy, which render message dissemination a challenging task in these networks. The intermittent connectivity of VANET suggests that it can be readily modeled as DTN. Routing is one of the most challenging aspects of DTN due to the inherent intermittent connectivity. To deal with this episodic connectivity in DTN, the proposed routing strategies rely on the inherent mobility of the participating nodes to *store-carry-and-*

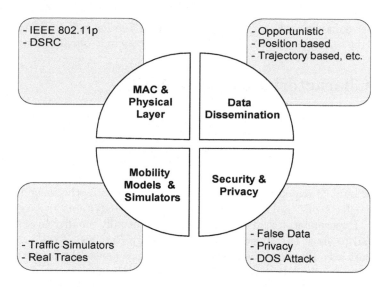

Figure 8.1: VANET research: the big picture [42].

forward [12] the messages for delivery to the destination. Since VANET shares properties of DTN, the store-carry-and-forward paradigm of DTN routing also applies to VANET.

Unlike other types of disruption tolerant networks, the mobility pattern of vehicles on highways is often predictable due to the restrictions imposed by roads, traffic, intersections etc. Another unique characteristic of VANET is the existence of roadside infrastructure which can be leveraged in order to improve the efficiency of routing mechanisms [81, 62]. Other than the network characteristics, the applications (e.g., safety related applications, traffic monitoring, etc. [72]) which are expected to run on top of VANET also make it an unique environment.

The research on VANET can be broadly categorized [42] as (Figure 8.1): a) message dissemination, b) security and privacy, c) addressing, and d) MAC/Physical layer enhancements. In this chapter, we primarily focus on a robust message dissemination system on VANET. We shall only scratch the surface of other aforementioned research areas of VANET. Since most safety critical applications have stringent delay requirements, those schemes will not be discussed in this chapter because those are beyond the scope of the book. Instead we limit our discussions on the challenges and possible solutions of message dissemination schemes for non-safety and comfort related applications.

The organization of this chapter is as follows: We begin our discussion by highlighting the characteristics of VANET in Section 8.2. In Section 8.3, we discuss some commercial applications of VANET. Section 8.4 discusses potential problems and solutions of message dissemination in VANET. The topic of

MAC layer enhancements is briefly mentioned in Section 8.5. In Section 8.6, we highlight some open research challenges and conclude our discussion.

8.2 Characteristics of VANET

It is important to understand the key characteristics of VANET in order to develop efficient and robust message dissemination schemes. In this section, we briefly talk about salient characteristics of VANET which are gathered from simulation and statistical studies of the mobility models.

The characterization of VANET can be carried out either by simulation studies or by field experiments. Simulation studies give us the insight to observe the effects of topology changes and network partitions in large scale networks. Experimental studies on VANET are usually smaller in scale and usually fail to show the effects of global topology changes. However, actual experiments accommodate the effect of interference caused by many factors, which is usually difficult to simulate. A summary of different mobility models and characteristics of VANET can be found in [32]. Mobility models in VANET can be classified as macroscopic and microscopic [20]. The macroscopic model considers streets, buildings, traffic lights etc. as the principal driving constraints, whereas microscopic models consider the individual movement of each vehicular node. The driving constraints of each node in the microscopic model is neighboring cars, drivers' behavior, acceleration/deceleration, braking, etc.

In this section, instead of delving into the details of different aspects of the mobility models, we focus our discussion on the characteristics of VANET which are gathered from simulation and statistical studies of the mobility models.

8.2.1 Network Characteristics

In [7], the authors develop a microscopic simulator to investigate the effects of free-flow traffic on connectivity in a VANET. An imaginary road is divided into fixed length segments where exactly one or zero cars can travel at a velocity (all vehicles follow similar velocity and lane changing restrictions). The output of the traffic simulation model is used to construct a connectivity graph during each simulated time step. The nodes of the graph represent the vehicles and the links between two nodes represent the paths which the vehicles follow. The study shows that vehicle density, relative velocity, and number of lanes are the key factors which influence node connectivity. However, the study did not consider the effect of interference and signal fading.

In [64], the authors develop a mobility model based on actual map data and evaluate the performance of DSR [38] using their model. A vehicle starts on a random point for a random destination in the graph. The driving behavior is simulated by assuming that the driver will always take the shortest path to go to the destination. The study shows that in many cases, the constructed mobility model produces similar results as the statistical Random Waypoint model. Also, the studies show that a transmission range of less than 500 meters produces severe network fragmentation.

The authors in [73] use a commercial microscopic traffic simulator VISSIM to simulate traffic movement. Their road setup is a closed (i.e., without entry and exit) three lane highway of 26 Km in total. The other behaviors including acceleration/deceleration, lane-changing, and car-following are simulated using common models found in literature on transportation. The key findings of the setup is that 90% of all unicast paths have a lifetime of less than 50 seconds.

Blum et al. [9] simulate road geometry which corresponds to 9.2 miles of highway with 10 exits and 10 ramps by using a commercial microscopic simulator CORSIM. Their results indicate that network layers need to be used efficiently due to limited available bandwidth and rapidly changing topology. Most of the links are short-lived (about 1 minute on average) even for the vehicles traveling in the same direction with a long transmission range (i.e. 500ft). Their study also shows that sending even a relatively small message over 3 to 4 hops is likely to suffer a route error.

The work presented in [16] studies how traffic density can affect the delay experienced in VANET by comparing the simulation results of different routing algorithms using different highway directions (unidirectional vs. bidirectional) and varying number of lanes (1-5 lanes). In traditional MANET routing protocols (e.g., AODV, DSR etc.), messages are dropped when no next hop exists whereas in opportunistic routing (i.e., store-carry-forward paradigm in DTN) messages are buffered at the intermediate nodes when next hop is unavailable. The results show that opportunistic routing decreases the message delivery latency. Another important result is that three or more lanes decrease the delay which implies that freedom of movement is essential to prevent vehicle clustering due to a few slow moving vehicles.

A large scale simulation study using real vehicular mobility traces can be found from [36]. The mobility traces are collected from city buses in Seattle, Washington which is composed of around 1200 vehicles covering a 5100 square kilometer area. One of the key findings is that the maximum path length is 48 hops if access points are not considered. This large number of hops raises the question whether it is realistic to expect two far-apart nodes to communicate (using typical MANET routing protocols) without access points since the previous research [9] shows that the first message will not be acknowledged before the link expires even with only 9 hops.

Below is a summary of the physical characteristics of VANET as mentioned in [32].

- **Roadside Infrastructure:** Availability of fixed roadside communication infrastructure (e.g., fixed access points) is unique to VANET. Researchers [81] have shown that the fixed nodes with storage can increase the capacity of the network. Also there are several routing schemes which utilize the fixed roadside access points in order to improve the message delivery ratio. For example, the GVGrid [69] routing protocol uses these fixed infrastructures to create routes on demand to the vehicles in the destination area. These fixed nodes are also responsible for rebuilding links that are broken due to mobility. A more detailed discussion on routing protocols can be found in Section 8.4.2.

- **Predictable Mobility:** VANET has its own unique mobility pattern. The motion of a vehicle is quite predictable as it is constrained by the roads, traffic lights, road intersections, etc. Also, mobility is predictable because humans are creatures of habit. We drive along the same routes at approximately the same time of day (for example, to and from work). Further, public transport buses follow fixed routes and schedules. The velocity, location, and trajectory information of a vehicle can be used to estimate its projected movement in a point in time. For example, the authors in [34, 59] introduce a mobility pattern aware routing scheme for VANET, which leveraged the position information and speed of vehicle. However, the high mobility generates short-life contacts that make routing challenging. Also, the studies in [7, 16] suggest that multi-lane roads ensure greater connectivity and experience less delay than those of single lane roads.

- **Vehicle Density:** Partitions and congestion are common in VANET because vehicles are often distributed unevenly. Rapid topology change and frequent fragmentation are also common phenomena in VANET. Usually as the density of the vehicle increases, connectivity also increases. But after reaching a critical value, an increase in vehicle density results in a decrease in connectivity (analogous to a traffic jam) [7]. Increasing density may also increase delay if there is not a large enough number of lanes to avoid slow moving vehicles [16].

- **Transmission Range:** In general, connectivity drops sharply beyond the transmission range and at least 500 meters of transmission range (simulation scenario) are required to stay well-connected [64]. At a range of 500 ft, a node can only reach 37% of the other nodes [9].

- **Connection Lifetime:** Probability distribution of connection lifetime (i.e., contact duration) and interconnection lifetime (i.e., time between two meetings of nodes) resemble that of a power law [7, 77]. Driving direction and transmission range directly affect connection lifetime. Simulation shows that 90% of all paths have less than 50 seconds of lifetime [73, 77].

- **Path Length:** Connectivity decreases with an increase in path length (i.e., number of hops). Even a small message, sent over 3 to 4 hops away, is likely to suffer route errors using traditional MANET protocols [9].

- **Latency:** Opportunistic routing and bi-directional vehicular traffic increases message delivery ratio while decreasing delivery latency [16].

- **Unbounded Network:** Unlike MANET, large-scale network and large-scale movement are the major implications of the unbounded nature of VANET [50]. Since it is almost impossible to trace the identity of the destination vehicles in an unbounded network prior to data transmission, the traditional identity-based addressing models hardly achieve 100% application delivery ratio with low overhead.

- **Node Availability:** Unlike other types of wireless networks, nodes in VANET can frequently join or leave the network. The reliability of a vehicular node as a message forwarder diminishes even further due to security and privacy issues. This implies that the network should not depend on a single vehicle to forward messages [11].

- **Environmental Factors:** Vehicles can move into many different environments (e.g. city environments, disaster situation, extreme weather conditions etc.) which can interfere with wireless communications. In urban environments, many tall buildings may obstruct and interfere with the wireless signals. Multi-path fading is a common phenomenon in such built-up environments. Vehicles usually remain closer together than in highway scenario, hence can create interference if transmission range is increased by increasing the power of radio communication equipments.

8.3 Innovative Applications

Many interesting applications have been proposed for VANET. An understanding of their requirements would certainly benefit the design of successful network protocols. Typical VANET applications can be categorized as: a) Transportation related including drive-safety applications, and b) Convenience related (i.e., the applications which increase the comfort of the driver and passengers) [20]. When the applications establish connection between vehicles, it is known as V2V (Vehicle to Vehicle) or C2C (Car to Car) communication. And connection between a vehicle and a road infrastructure is known as V2R (Vehicle to Road) or C2R (Car to Road) communication. A classification [15, 20] of VANET applications is shown in Figure 8.2.

The authors [20] mention some of the common problems which VANET applications face:

- *Timeouts:* Getting a response is usually longer than foreseen by timeouts. Usually timeouts are used to re-initiate a transaction or to warn the user that the connection could not be established. Long delay of VANET often renders most network protocols unusable because of this.

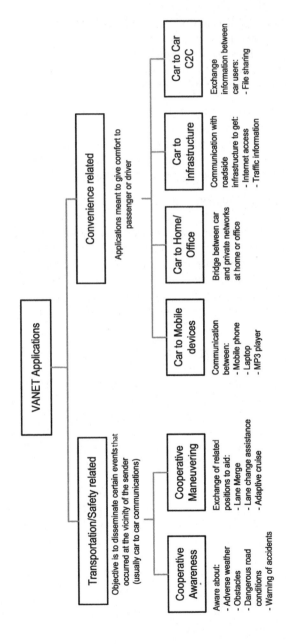

Figure 8.2: VANET applications classification.

- *Synchronous Programming Style:* In common applications, it is assumed that the total execution time of an application is longer than its connection establishments and that both ends be synchronized or running at the same time which is not the case in VANET.

- *Chatty Application Protocols:* With long end-to-end round trip times (which is common for VANET), one cannot afford to have several information exchanges to establish a communication.

In this section, we mention some innovative and useful applications that leverage the power of VANET and circumvent the hindrance mentioned above. We have omitted the discussion of safety related applications because most safety related applications have stringent delay requirements and hence are beyond the scope of this chapter.

8.3.1   Low Cost Digital Connectivity in Rural Areas

The *DakNet* [63] project uses wireless technology to provide asynchronous digital connectivity (non-realtime data) in rural areas. It takes advantage of the existing communication and transport infrastructure to distribute digital connectivity to remote villages which lack digital communication infrastructure. Figure 8.3 shows the DakNet architecture. Instead of trying to relay data over a long distance using a high power transmitter which can be expensive and prone to interference, the DakNet transmits data over short point-to-point links between kiosks and portable storage devices, termed as mobile access points (MAPs). MAP, which is installed on and powered by a bus, a motorcycle, or even a bicycle with a small generator, physically transports data among public kiosks and private communication devices (e.g. intranet) and between kiosks and a hub. Low-cost WiFi radio transceivers automatically transfer the data stored in MAP at high bandwidth for each point-to-point

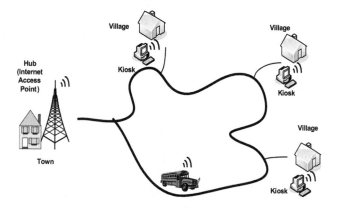

Figure 8.3: DakNet architecture [63].

connection. Every vehicle carrying a MAP unit performs the following two major steps:

- When the MAP-equipped vehicle is within the coverage of a village WiFi-enabled kiosk, it senses the wireless connection and then uploads and downloads data.

- Whenever the MAP-equipped vehicle is within the range of an Internet access point (the hub), it synchronizes the data from all the rural kiosks.

In addition to its tremendous cost reduction, a critical feature of DakNet is its ability to provide a seamless method of upgrading to always-on-broadband connectivity. As the economic conditions of the village improves, its inhabitants can use the same hardware, software, and user interface to enjoy realtime information access.

8.3.2   Data Collection from Sensor Nodes

Using sensor networks to monitor road traffic has become an emerging research area. In [22, 58], researchers use VANET to collect and disseminate traffic information to cars. In the PATH project [1], the researchers have investigated a number of issues related to smart transportation systems, including the use of sensor networks for on-road monitoring [23]. The authors in [46] propose an efficient lightweight protocol for proactive urban monitoring based on the primary idea of exploiting vehicle mobility to opportunistically diffuse summaries about sensed data.

CarTel [33] is a mobile sensor computing system designed to collect, process, deliver, and visualize data from sensors located on mobile units. A CarTel node is a mobile embedded computer in a vehicle coupled to a set of sensors. One example of the use of a CarTel node is to monitor traffic delays on roadways in a distributed manner. CarTel nodes, equipped with GPS devices, can opportunistically obtain information about observed traffic delays as cars move and use that information in traffic monitoring and route planning applications. This type of mobile sensor network has the potential to cover a much larger geographical area with a smaller number of sensors, leveraging the movement of vehicular nodes. Besides traffic monitoring, the authors [33] also mention some other applications of CarTel:

- *Environmental Monitoring:* by using mobile chemical and pollution sensors.

- *Civil Infrastructure Monitoring:* by attaching vibration and other sensors to monitor the state of the roads [24].

- *Automotive Diagnostics:* by obtaining information from a vehicle's onboard sensors, which can help in preventive and comparative diagnostics. This information is also useful to monitor bad driving tendencies.

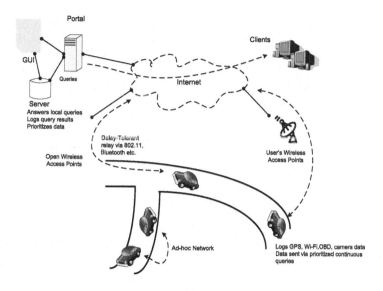

Figure 8.4: Cartel architecture [33].

- *Geo-Imaging:* by attaching cameras on vehicles and using mobile phone cameras, one can capture location-tagged images for various applications, including landmark-based route finding.

- *Data Muling:* by using cars and people as delivery networks for remote sensor networks and sending data from those networks to Internet servers.

8.3.3 Cooperative Downloading

Cooperative downloading is an inexpensive and effective alternative to mirror servers and content distribution networks. It is based on the principle of swarming, wherein the desired file is downloaded in parallel from a number of cooperating peers.

For VANET, where the nodes move at a high speed and topology changes rapidly, vehicles can only download partial data from the gateways before disconnection. Therefore it is an ideally suited candidate for cooperative content sharing systems. However, most existing peer-to-peer swarming protocols are designed for fixed topology-based networks [19]. Due to the changing topology and high mobility, implementing these schemes in VANET is very challenging. In particular, devising peer and content selection strategies for sharing is quite complex. In SPAWN [60], the authors use a new piece selection strategy and a communication efficient swarming protocol called SPAWN to tackle these problems. The SPAWN protocol uses a gossip mechanism to advertise the piece list each node possesses and takes proximity into account when selecting contents among peers. Moreover, leveraging the broadcast nature of wireless media enables SPAWN to reduce redundant transmissions. However, the peer discovery method, which is based on gossip mechanism, incurs message overhead; and the proximity driven piece selection strategy induces delay which might have negative impact on the performance of a VANET with high speed moving vehicles.

In order to simplify the problem of finding cooperator nodes and data block selection strategy, the authors [4, 45] propose network coding-based [3] cooperative downloading schemes for VANET. The basic principle the network coding-based scheme is discussed in Section 8.4.

8.3.4   Miscellaneous Applications

Virtual Marketplace: The authors [44] propose a virtual marketplace concept where a mixture of mobile and stationary users can carry out buy/sell transactions using a vehicular network.

Discovery of Parking Space: Parking in urban areas is often a painful and time consuming experience, especially during peak hours and popular events. The authors [8] have explored the use of mesh networks of wireless parking meters operating in conjunction with mobile devices in automobiles. With the help of such a system, the authors have also shown many variants on the parking paradigm, such as, spot reservation, pricing models, and automatic ticketing.

8.4   Message Dissemination

Robust message dissemination is one of the most indispensable requirements of various VANET applications, which affects the message delivery ratio, transmission delay, and message communication overhead. Since VANET shares, many characteristics of MANET, it would be tempting to apply the routing protocols designed for generic MANET in VANET. In fact, these protocols form the basis of VANET routing protocol research. Based on algorithmic perspective and the use of knowledge in deciding routing paths, these routing protocols can be categorized as: a) *Topological routing*, b) *Energy-aware routing*, c) *Position-based routing*, d) *Trajectory-based routing*, e) *Opportunistic routing*. However, it should be noted that many routing protocols actually belong to more than one of the above mentioned classes. Before discussing the message dissemination techniques, we first list the key performance metrics [50] that are widely used for evaluating the effectiveness of these protocols.

- *In-time delivery ratio:* This indicates the percentages of unique messages delivered within a specified period. Delivery ratio is the single most important metric in DTN routing protocols.

- *Delivery Delay:* It provides the indication of how soon the message can be delivered to the intended receiver. Though a large message delay is common in non-safety applications of VANET, it is always desirable to receive a message as early as possible.

- *Message Overhead:* Two kinds of message overhead should be considered: a) Data overhead and b) control overhead. Data overhead involves the redundant message transmissions and control overhead originates from the task of maintaining topology information or advertising some information to neighbors (using beacons).

Next, we discuss why some MANET routing schemes are not suitable for VANET and also discuss some promising classes of routing protocols for VANET.

8.4.1 MANET Forwarding Schemes and Their Limitations

In the following, we explain why some classes of MANET routing protocols fail to perform well in VANET.

Topological Forwarding: These routing protocols use information about the links (metrics such as hop counts, bandwidth, link quality, etc.) to perform routing. *Proactive* protocols (e.g., DSDV, OLSR etc.) build and update the routing table continuously by maximizing the link metrics even when no routing request is made. On the other hand, *Reactive* protocols (e.g., AODV, DSR etc.) build the routing table dynamically on demand. Depending on the specific parameters (e.g., mobility patterns, traffic load, etc.), any of these approaches may behave better or worse [27].

In a VANET, the rate at which the nodes roam in the network is very high. The network topology changes rapidly due to high speed of the nodes (typical vehicle speed is 60km/h). Because of the rapidly changing topology, the proactive routing schemes create enormous overhead to keep track of the global routing table. Also, the short-lived paths discovered by the protocols often become stale and thus messages may follow incorrect routing paths. Not only does the network topology vary rapidly in VANET but also nodes are often out of coverage of others (or APs) in highways. These protocols are not designed to cope with network disruption. For example, AODV returns a route error message (indicating that the destination is unreachable) in the face of disruption. For these reasons, topological routing mechanisms designed for MANETs are not generally suitable for VANET.

Energy Aware Forwarding: The main objective of energy aware routing is to find a routing path where energy (i.e., lower number of transmissions, low power transmission, etc.) consumption is least. These types of routing protocols [31, 51] are particularly suitable for the nodes that need to conserve battery power to last longer (e.g., sensor nodes in battle field or in remote areas). However, in VANET, energy consumption is not a problem. As a result, these types of routing protocols are not suitable for VANET.

8.4.2   Forwarding Schemes Suitable for VANET

Intermittent connectivity, rapidly changing topology, stringent application requirements (e.g., safety applications have strict requirements not only on guaranteed delivery but also on delivery delay) make routing in VANET challenging. In this section, we discuss several classes of routing schemes which are suitable for VANET (Figure 8.6).

8.4.2.1   Position-Based Forwarding

In position-based schemes, routing decision is based on the physical location of the routing node, source, and destination [32]. The source first defines a routing area between the source and the destination. The nodes within the routing area forward the message, while nodes outside the area drop it [50, 41]. In greedy [10, 39] variants, instead of defining routing area, the intermediate node forwards the message to the neighbors which are nearest to the destination or furthest from the intermediate node. A survey of position-based routing schemes can be found in [71].

A-STAR [52] is a position-based routing protocol which adopts an anchor-based approach with position awareness. The protocol uses city bus information to utilize certain highly connected paths in order to improve delivery ratio. The authors also mention possible improvements which involve utilizing the information about bus schedule.

In general, position-based routing schemes do not keep global network information, hence are more suitable for large networks like VANET than *proactive* routing protocols. However, they suffer from *stale position* problem in highly churned VANET. Usually nodes broadcast their position information in beacons. Due to beacon losses and interval between consecutive beacons, the position information of neighbor nodes often becomes stale. As a result, the routing protocol may drop messages. Figure 8.5a shows an example [14] of the problem. Vehicle A knows the position of B, which is in the transmission range of A. After a while, A generates a data message and selects B as the next hop. But B has already moved out of the transmission range of A and hence the message becomes lost.

Next, in Figure 8.5b, it is shown how outdated information can cause routing loops. Assume each vehicle knows the current position of its neighbors and that the nodes employ greedy forwarding. Vehicle A generates a data message and forwards it to B (since B is closer to the destination than A). Since B has no neighbors closer to the destination, it buffers the message. Now, A is moving towards the next destination while B moves away from it.

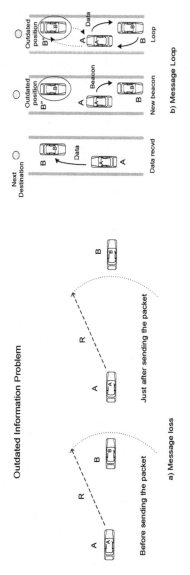

Figure 8.5: Problem with stale position [14].

Assume that B receives A's beacon while it crosses A. In this situation, B sends the message to A because A becomes closer to the destination. Since A has not received B's beacon yet, it still thinks that B is in its old position B'; therefore, A tries to forward the message to B. It creates a routing loop until A gets a beacon from B (or TTL expires). This problem can be solved if the vehicles include their current velocity vectors in beacons, so that the routing vehicle can calculate the estimated position of the neighbor.

As mentioned above, the beaconing mechanism is the heart of many position-based routing protocols since it allows the exchange of various information among nodes. However, the authors [26] have identified that due to highly churned nodes in VANET, the beacon information rapidly becomes stale. Therefore, a beaconless approach is adopted in the FleetNet project [26] which is based on Content-Based routing (CBF). In this scheme, initially the message is broadcast to all of its one hop neighbors. In order to determine the next hop, the CBF uses a biased timer-based contention process. The selected node then issues suppression commands to prevent rest of the nodes from routing the message to avoid duplication.

8.4.2.2 Trajectory-Based Forwarding

A trajectory-based routing scheme is a hybrid scheme combining the ideas from source-based routing and greedy routing [50, 53, 49]. The source calculates the approximate trajectory and each intermediate node makes a greedy routing decision along the trajectory based on local position information. The idea is to deliver the message near the destination, according to the trajectory of the intermediate nodes. The authors [53] propose Geographic Source Routing (GSR) which combines position and topological information in making routing decisions. It is assumed that the vehicular nodes are equipped with an on board GPS and contain digital maps of current areas. The sender computes a sequence of junctions the message has to traverse to reach the destination using the underlying map of streets. By using Dijkstra's shortest path algorithm, the sender selects the route from the street map.

Greedy Perimeter Stateless Routing (GPSR) [39] is a greedy routing scheme which selects the next hop among the nodes which are geographically closer to the destination. However, this scheme fails if the current incumbent is geographically nearer to the destination but unable to forward the message to the destination (e.g., in case of unreachable radio communication). In these circumstances, GPSR switches to a perimeter routing algorithm where messages are forwarded using the right hand rule by traversing the graph model of the network. However, this fallback method was shown to be inappropriate [53] in city environments where interference is a common problem due to the large number of radios. Greedy Perimeter Coordinator Routing (GPCR) [54] is an attempt to ameliorate the situation in which the geographical map is utilized and routing is based on streets and junctions. Nodes situated at the junction, termed as *coordinators*, make routing decisions as to which street the message should be forwarded to in order to reach the next junction. The message is then forwarded to the furthest neighbor of the routing node.

Another prediction-based routing scheme which utilizes the spatial and trajectory information of vehicles is the Predictive Directional Greedy Routing (PDGR) [28]. Next hop is chosen by calculating a weighted score based on this information. This scheme computes the weighted scores of 2-hops away neighbors and forwards messages in greedy manner.

An infrastructure aware prediction-based routing scheme (PBR) is presented in [59] which predicts how long routes will last and preemptively creates new routes to replace old ones before they break, thereby reducing the number of route failures. This protocol utilizes the fact that in spite of dynamic mobility, the mobility patterns of vehicles on highways is quite predictable. This deterministic motion of vehicles is exploited to predict the route lifetime between a vehicle and a stationary gateway. By creating a new route prior to the expiration of predicted route lifetime, this protocol preempts route failures and utilizes most of the connectivity available with the gateway. The simulation results show significant reductions in route failures and improved performance compared to the schemes which do not employ preemptive routing.

In general, the trajectory-based routing schemes have less data overhead than position-based schemes due to the utilization of digital maps. However, the position-based schemes are more robust in face of network disruption because even if there is no vehicle on the path, message can be routed via other paths and message delivery is not confined to a single trajectory [50].

8.4.2.3 Opportunistic Forwarding

Opportunistic routing is a new class of routing paradigm that can cope with sparse and partitioned networks (often termed as *opportunistic* networks) where nodes communicate opportunistically. In opportunistic networks (e.g. VANET), no assumption is made on the existence of a complete end-to-end path from source to destination. Source and destination might never be connected during the lifetime of the network. Moreover, nodes do not require the global knowledge about network topology to forward a message. Routes are built dynamically wherein each node decides the next hop based on local information and may carry messages until a suitable next hop is found. Unlike traditional *store-forward* routing methodology, the routing of messages is usually based on the *store-carry-and-forward* paradigm. This communication paradigm resembles that of e-mail applications. However, this comes at a price of additional delay in message delivery due to the effect of message buffering. An overview of opportunistic routing techniques for MANET can be found in [61]. The class of opportunistic routing schemes is especially suitable for partitioned networks like VANET. In fact, most of the non-safety related VANET applications incorporate the idea of opportunistic routing schemes in order to tolerate disruption in networks. In the next few subsections, we limit our discussions to opportunistic routing in the context of VANET.

Opportunistic Forwarding in VANET:

The opportunistic routing algorithms can be classified according to their dependence on the roadside infrastructure. A taxonomy [42] of opportunistic routing in VANET is shown in Figure 8.6.

It should be mentioned that it may be possible that certain routing protocols may simultaneously belong to different classes.

Infrastructure-Based Forwarding Schemes

Algorithms which rely on the roadside infrastructure (V2I/I2V) can be categorized as: *push* type and *pull* type. In push type methodology, the infrastructure pushes out the message to every vehicle in its reach. These schemes are good for dissemination of popular data (e.g., traffic alert, weather report, etc.) with minimal contention of the physical channel. However, the major problem of this model of communication is that all nodes may not be interested in the messages pushed by the infrastructure. On the other hand, the pull type schemes follow the traditional *request-response* model where the vehicle sends a query and the infrastructure responds back with it. These schemes are suitable for routing user specific data, but channel contention, interference, and collisions are likely to occur when vehicle density becomes larger.

There are few message routing schemes for VANET which leverage the existence of roadside infrastructure. In [62], the authors have proposed the Roadside-Aided Routing (RAR) which utilizes road constraints and roadside fixed communication infrastructure to improve message routing efficiency. Their scheme reduces overhead by employing one-hop agent broadcasting (instead of multi-hop) and eliminates the need for hierarchical addressing. The simulation results show that their approach can provide high delivery rate with low and constant overhead.

In the following paragraphs, we discuss a network coding-based content dissemination system, called VANETCode, which is a hybrid model of both infrastructure-based and infrastructure-less schemes.

VANETCode [4] is a *push* type cooperative content dissemination scheme for VANET. The concept of cooperative downloading scheme for VANET is briefly discussed in Section 8.3. VANETCode extends the idea further by incorporating random linear coding mechanisms in dissemination of messages in order to simplify piece selection strategy.

It is assumed that the files are present in their entirety at the static gateways. These could be proactively distributed by the content providers amongst the gateways, similar to a content distribution network (CDN) or could be downloaded on demand. The gateways, which act as the servers, split the original file into k blocks. The servers (situated at the gateways) produce a linear combination of the k blocks using randomly selected coefficients. By using random linear network coding on the data blocks, this scheme eliminates the need of peer selection, content selection, and neighbor discovery, which take up a significant amount of time and resources in many other cooperative

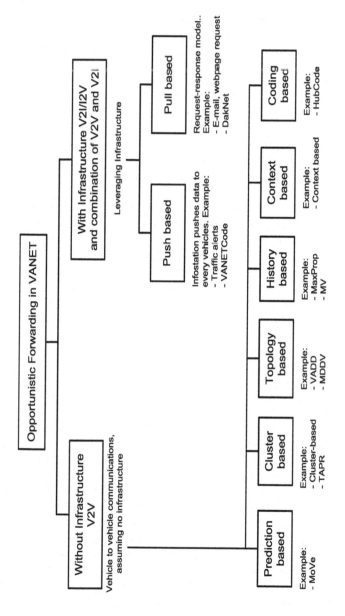

Figure 8.6: A Taxonomy of opportunistic message dissemination in VANET.

downloading mechanisms proposed for VANET [60].

In order to explain the operation of the method [4], a simplified scenario has been illustrated in Figure 8.7, where four vehicles (A, B, C, and D) pass by a stationary gateway along a freeway. The gateway has a file to share that has been split into two blocks, X_1 and X_2. Now, assume that all the vehicles request this file from the gateway and the time a vehicle stays in the range of AP is barely enough to receive one such block. The gateway then randomly selects coefficients and encodes all the constituent blocks using random linear encoding (Eq. 8.1) to form a single encoded block. For example, in Figure 8.7, the gateway picks two random coefficients a_{11} and a_{12} to encode its blocks as $F_1 = a_{11}X_1 \oplus a_{12}X_2$ and sends it to vehicle D together with the coefficient vector $\langle a_{11}, a_{12} \rangle$.

$$F = \sum_{i=1}^{k} a_i X_i \; [where, \; k = total \; number \; of \; blocks] \qquad (8.1)$$

The above operation is repeated for all other requesting nodes, but with different randomly selected coefficients. Since the coefficients are chosen randomly from a very large space (typically 2^8 or 2^{16}) all the encoded blocks contain information important to all the interested nodes.

When all vehicles are out of range of the gateway, instead of waiting for the next gateway, the vehicles cooperatively share their data blocks amongst each other [4]. However, unlike SPAWN where the nodes need to request for specific blocks that are missing, in this scheme all blocks are linearly independent and hence a node does not need to explicitly request specific blocks. Each block that a node receives from its neighbors is beneficial. This scheme also takes advantage of the broadcast characteristics of the wireless medium with each node employing a MAC layer broadcast to transmit its encoded blocks. Similar to the gateway, the intermediate nodes also transmit the new coefficients associated with the encoding along with the block. This allows other nodes in its vicinity to passively listen to the broadcast and receive the transmitted block.

Decoding of the data requires the nodes to capture a sufficient number of blocks with linearly independent coefficients in order to solve a set of linear equations. In this scenario, a vehicle can decode and retrieve all the original blocks (X_1 and X_2) if it receives at any two linearly independent encoded blocks. It should be mentioned that the computation overhead of decoding n blocks using matrix inversion is $O(n^3)$. Further, combining the blocks to reconstruct the original file is an $O(n^2)$ task. However, this is not a problem in VANET because the computing capabilities of vehicular nodes are usually quite high and it has virtually no energy constraint.

Infrastructure-Less Forwarding Schemes

In the next few paragraphs, we discuss some important classes of infrastructure-less opportunistic message routing schemes for VANET.

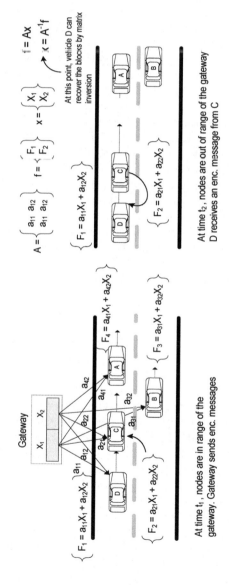

Figure 8.7: Encoding and distribution by the gateway [4].

Predictive Forwarding: The motion of vehicles on highways is constrained by road and traffic conditions. As a result, utilizing the knowledge about location, velocity, trajectory etc., the mobility patterns of VANET are quite predictable. In [43], the authors studied several opportunistic routing strategies for VANET including *MoVe*, an algorithm which leverages velocities of nodes to make intelligent routing. The approaches are: a) NoTalk (i.e., two-hop relay), in which nodes accept data from a source and carry the data until it meets the destination; b) Broadcast (i.e., Epidemic) in which nodes exchange data with every other node they meet; c) Location-based, in which nodes forward data to neighbors only if that neighbor is closer to the destination than its current position; d) MoVe, in which nodes leverage relative velocities and predict the closest forwarder to the destination; and e) a variation of MoVe which takes the node trajectory into account. Their simulation-based study shows how the auxiliary information provides a trade-off between resource overhead and performance of routing.

The Greedy Traffic Aware Routing (GyTAR) [35] protocol is a location aware movement predication-based routing scheme which routes message at road intersection. GyTAR chooses road intersections on the fly, considering the number of vehicles in between the intersections. This protocol assumes that each vehicle is equipped with a navigation system and is able to gather real-time traffic data. The next hop is chosen by calculating vehicles' travel direction, speed, and last known position. This scheme incorporates the store-carry-and-forward mechanism as a fallback strategy when the next hop is unreachable.

Cluster-Based Forwarding: By organizing mobile nodes into clusters managed by cluster heads, the routing scheme can achieve high delivery ratio and low message overhead. The authors in [65] propose a cluster-based location aware routing algorithm for VANET which divides the network into multiple clusters. Each cluster has a cluster-head and a group of members within the transmission range of the cluster-head. The cluster-head is responsible for routing inside the cluster. At least one bridge takes care of the communication between two cluster-heads. Though the scalability issue of large VANET can be addressed by cluster-based routing schemes, the management of cluster members and the election of cluster heads in a rapidly changing topology as in VANET often offset its advantage. For this reason, a distributed and self-maintained cluster structure is preferred in VANET.

Traffic Adaptive Packet Relaying (TAPR) [2] is another trajectory-based routing scheme which considers VANET as a disconnected network of many clusters. A message is gradually forwarded to an edge node of the current cluster which is headed towards the destination. At this point the message is carried until two clusters are connected. This scheme is another example of store-carry-and-forward mechanism.

Topology-Based Forwarding: This class of routing protocols utilizes geographical maps in addition to location information in order to aid routing. A Vehicle Assisted Data Delivery (VADD) [80] protocol has been proposed for VANET to forward messages to the best road (in urban scenarios) with lowest data delivery delay. VADD assigns a cost to each road (segment between intersections) based on expected routing delay (d_{ij}) between intersections I_i and I_j. It is assumed that vehicles arrive at the intersections following Poisson distribution. If the vehicle density on that road is high so that multi-hop routing is possible, then d_{ij} is set to the Euclidean distance between the intersections. If the vehicle density on that road is low, vehicles carry the messages. The authors propose a stochastic model to estimate the data delivery delay (D_{ij}). Upon arriving at an intersection, the vehicle may choose that road with smallest D_{ij}. Several enhancements to the routing strategies have been proposed based on vehicle movement and direction. The protocol performs better in comparison to DSR (topological), GPSR (geographic), and epidemic routing protocols.

A complete routing architecture, MDDV [76], for VANET is proposed which combines the ideas of opportunistic routing, trajectory-based routing, and geographical routing. The roads can be viewed as a directed graph where the nodes represent the intersections and the edges represent road segments. However, the shortest road distance from source to destination does not necessarily mean least delay because vehicle density often leads to faster propagation. So, the authors calculate a *dissemination length* which is used as weight for the corresponding links in the graph (see Eq. 8.2):

$$d(A, B) = r(A, B)(m - (m - 1)(i^p + cj^p)) \qquad (8.2)$$

where, $d(A, B)$ is the dissemination length of the road segment from node A to node B, $r(A, B)$ is the road length from A to B, and i/j is the number of lanes from A/B to B/A. The constants p and c take values between 0 and 1 and m is set to 5. Messages are forwarded opportunistically along the routing trajectory through the intermediate nodes. Each intermediate node decides whether it will participate in routing or not based on the local information (i.e., traffic density) in the area. Some of the design issues are: decision to store/drop messages, identification of routing group (similar to concept of clusters in cluster-based routing), etc. The authors provide simulation results to show the efficacy of their approach.

History-Based Forwarding: This class of protocols utilizes history movement patterns to aid routing. Routing can benefit from the fact that most vehicular networks exhibit some sort of periodicity in their mobility patterns. For example, public transportation networks follow periodic schedules. Even most individuals have fairly repetitive movement patterns, for example, driving to and from work at approximately the same time every day. The authors in [6] propose a Bayesian classifier-based routing scheme which incorporates the periodic behavior of mobility patterns in order to choose the best forwarder. In the context of public transport networks, which are characterized by their regular mobility patterns, the simulation results suggest that this approach performs better than MaxProp in terms of delivery ratio.

MaxProp [12] is a DTN routing protocol that utilizes nodes' encounter history and is based on prioritizing both the schedule of packets transmitted to other peers and the schedule of packets to be dropped. These priorities are based on the path likelihoods to peers according to historical data. In addition, MaxProp uses acknowledgments that are propagated throughout the network and not just to the source, in an effort to reduce stale data from buffers. Also, MaxProp stores a list of previous intermediaries to prevent data from propagating twice to the same node. The authors have performed their experiments on real mobility of the bus-based DTN testbed, called *UMassDieselNet*, which consists of 30 buses. The experiments show a significant improvement of the delivery rate and latency in a wide variety of scenarios compared to some opportunistic routing approaches.

The Meets and Visits (MV) [13] protocol is somewhat similar to Maxprop, but presents an alternative method to estimate the likelihood of routing. MV utilizes past meeting frequencies of nodes to rank each message according to the delivery predictability through a specific path. MV estimates the probability, $P_n^k(i)$, that the current node k can successfully deliver a message bundle to a region i within n transfers, as:

$$P_n^k(i) = 1 - \prod_{j=1}^{N} \left[1 - m_{j,k} P_{n-1}^j(i) \right]$$

where $P_O^k(i) = t_i^{(k)}/t$ and $t_i^{(k)}$ is the number of times node k visited region i during the previous t rounds. The meeting probability between nodes j and k in the same region is given by:

$$m_{j,k} = \frac{t_{j,k}}{t}$$

where, $t_{j,k}$ is meeting frequency between nodes j and k. MV operates in *pull* type fashion in that it only forwards the message bundle upon request from next-hop neighbors. Under the assumption of an infinite buffer at each node, MV outperforms a FIFO scheme by 50 percent [78].

Priority/Content-Based Forwarding: As the name suggests, the content-based routing schemes make their routing decision by examining the content of the message [17]. If the message is valuable for traffic safety, the receiver accepts the message and rebroadcasts it. Speed, traffic density, and pool of interested neighbors are also taken into account in the decision process to rebroadcast the message [50]. The intended receiver and the effective range is calculated on the fly. For example, only vehicles which are traveling fast should get a speed warning message on the slippery highway.

Coding-Based Forwarding: With network coding, a node may recombine several input messages into one or several output messages. The benefits of this approach is: a) potential throughput improvements and resilience to data loss [25]. In [5], the authors propose a network coding-based routing strategy called *HubCode*, which seeks to use the *hubs* (a fraction of the nodes, which are collectively connected to the rest of the nodes) as message relays. Analysis of the temporal connectivity graph of vehicular networks (and many other people centric networks) reveals the existence of hubs. The hubs employ random linear network coding to encode multiple messages addressed to the same destination, thus eliminating the disadvantages imposed by the *Coupon Collector's Problem*. Further, the use of the hubs as relays, ensures that most messages are delivered to the destinations. Routing is a simple three step process: 1) Source nodes forward messages to a hub, 2) a hub encodes (using random linear encoding as in Eq. 8.1) multiple messages headed to the same destination and disseminates the encoded messages among other hubs and 3) a hub delivers the encoded message together with coefficient vector to the destination. A simplified operating principle of HubCode is illustrated in figure Figure 8.8.

Once the destination receives a sufficient number of linearly independent coefficient vectors, it can recover original messages by solving the set of linear equations. The authors have shown the efficacy of their protocol by simulating empirically collected movement traces of a large citywide public transport network.

8.5   Medium Access Control for VANET

A good VANET MAC protocol should be concerned with fast varying topology rather than focusing on power constraints or time synchronization problems. Moreover, in order to utilize the short contact durations of highly churned vehicles effectively, a VANET MAC must have ways to reduce medium access delay. In this section, we give an overview of *cooperative ARQ*, an important link layer enhancement for VANET, and also discuss briefly the 802.11p standardization effort for VANET MAC.

8.5.1   Cooperative ARQ

Automatic Repeat ReQuests (ARQs) are *de facto* parts of wireless link layer protocols to avoid expensive re-transmissions of erroneous messages by the transport layer. Cooperative ARQs are the schemes which increase link reliability in data link protocols by using node cooperation. The main idea behind these approaches in wireless networks is to exploit the broadcast nature and spatial diversity of wireless transmission.

The basic operation of C-ARQ is as follows [57, 79, 70, 21]: let node x need to transmit a frame to node y. Also assume that node y has designated M

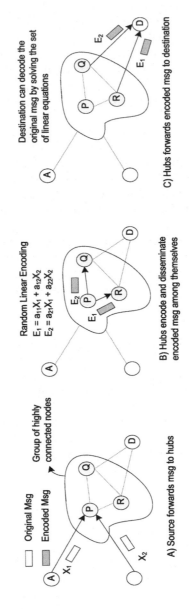

Figure 8.8: Routing scheme using hub-based network coding.

neighbor nodes as *cooperator nodes*. When x transmits the frame addressed to y, the cooperators of y will also receive a copy of the frame due to the broadcast nature of wireless medium. In typical ARQ protocol, if node y receives a frame with erroneous bits (can be determined by calculating checksum), it will ask node x for a frame re-transmission. In C-ARQ, node y will first request its cooperator nodes to re-transmit the frame. Since the cooperator nodes are nearby nodes of y, it is more likely that this re-transmission will succeed. In case no cooperator nodes have the correct copy of the frame, node y will fall back to generic ARQ protocol (i.e., node y will request the frame from node x). The authors [56, 20] have effectively applied a delayed C-ARQ protocol (DC-ARQ) in a delay tolerant VANET scenario where vehicles accessing an AP have a few seconds to download information in a high loss environment. The key idea of the scheme [20] is to delay re-transmissions coming from cooperators until the platoon of vehicles is out of coverage of the AP. By restricting re-transmissions by the vehicle nodes in the AP coverage area, the AP can transmit more data to cars and thus can increase throughput. Here, cars in a platoon recover messages from other cars that they have failed to receive from an access point. By evaluating their scheme through an experimental testbed, the authors show that message losses in transmissions from an access point to cars can be reduced by half.

8.5.2   802.11p MAC

The IEEE 802.11p [75] standardization process for vehicular environment originates from the allocation of the Dedicated Short Range Communications (DSRC) spectrum band when FCC allocated 75MHz DSRC spectrum at 5.9GHz to be used exclusively for V2V and V2I communications. The primary goal of this spectrum allocation is to enable public safety applications and improve traffic flow. IEEE 802.11p for the vehicular environment, which is a part of the DSRC system, operates in the licensed 5.9GHz band in United States.

The IEEE 802.11p draft specifies a new PHY/MAC amendment of the 802.11 standard designed for VANET: the Wireless Access in Vehicular Environment (WAVE). This amendment mostly comes from the requirements of Active Safety concept of VANET where reliability and low latency are extremely important. At the PHY layer, 802.11p will work in the 5.850-5.925 GHz spectrum in North America, and by using the OFDM system it provides both V2V and V2I communications over distances up to 1000m considering the effects of the surrounding environment (absolute and relative velocities up to 200km/h) and fast multi-path fading.

In an overly simplified manner, as mentioned in [37], the IEEE 802.11 MAC is about how to arrange for a set of radios in order to establish and maintain a cooperation group (known as Basic Service Set BSS). Within the BSS, radios can freely communicate among themselves and all the transmissions from outside are filtered out. The key purpose of the 802.11p amendment at the MAC level is to enable a very efficient communication group setup without much overhead as typically needed in 802.11 MAC. Even for non-safety applications, the overhead of traditional *association* and *authentication* may be too expensive for highly churned vehicles at highways. The IEEE 802.11p introduces a new mode of operation, called *WAVE* mode. A node in WAVE mode can transmit or receive data frames without performing any *authentication* and *association* procedures before participating in a WAVE BSS (termed a WBSS), which greatly increases actual data communications opportunity. Below is a summary for WAVE operations [37]:

- A WBSS is a type of BSS which consists of a set of cooperating stations using common BSSID. A WBSS is initialized when a radio (could be an AP or a node) in WAVE mode sends a WAVE beacon which includes all necessary information for the beacon receiver to join.

- A radio joins a WBSS when it is configured to send and receive data frames with BSSID defined for that WBSS. On the other hand, it ceases to belong to WBSS when its MAC stops sending and receiving frames that use the BSSID of that WBSS.

- A station cannot be a member of two WBSS's at the same time. A station does not join an infrastructure BSS and it should not use active or passive scanning. And it does not use MAC authentication and association procedures.

- WBSS dissolves when it has no members. The initiating radio is no different from any other member after the establishment of WBSS. So a WBSS can continue even if the initiating station disappears.

Problems of 802.11p Draft: As pointed out in [18], there are a couple of problems in the operations of 802.11p which need further discussion:

- *Stateless Channel Access:* Since there is no authentication or association in a WBSS, it is not clear how WBSS providers keep track of connectivity with WBSS users. When a WBSS user moves out of a WBSS provider's coverage, the WBSS provider cannot know whether the WBSS user exits or not due the the absence of de-association phase. In this case, if the WBSS provider has a frame destined for that user, it continues to transmit the frame until re-transmission reaches a pre-threshold value and thus wastes valuable bandwidth and channel allocations. The problem becomes worse in multi-rate WBSS where the absence of ACK frames prompts the WBSS provider to use lower bit rate since it assumes that the distance between WBSS provider and user has increased, or radio condition becomes poorer, which results in a severe waste of wireless channel.

- *Caching for Handoff:* Due to rapid movement of vehicular nodes, a WBSS user frequently moves from one WBSS to another WBSS without finishing data exchange. Therefore, the IEEE 802.11p needs to support fast handoff among multiple WBSS providers. Although in [74], a proactive caching is suggested to reduce signaling delay in re-association process, it cannot be used directly because there is no association process in 802.11p.

- *Opportunistic Frame Scheduling:* Due to frequent disconnections experienced by vehicular nodes, data transmissions have to take place depending on channel state in an opportunistic manner. In addition, bitrate should be determined in a timely manner in order to accommodate rapidly varying channel state.

8.6 Open Issues

In this section we have listed some of the open research challenges which need to be addressed in order to achieve robust message dissemination in VANET.

8.6.1 Security Challenges

Given that VANET are open self-organized networks, ensuring security is challenging since a malicious node may be able to launch a range of attacks. Because of this, several threats potentially exist in VANET, which include traffic disruption by launching a denial of services attack using fake messages, compromising drivers' private information, etc. Detailed discussions about potential security issues in VANET can be found in [68, 48, 11]. The issues to be addressed include trust (vehicles are able to trust the messages they receive), real-time message authentication, etc. The following paragraphs discuss some of the security issues.

Authentication: The constantly varying topology and network disruption make the use of PKI and certificates challenging in VANET. Since vehicular nodes roam at a high speed and contact time between two nodes is on the order of a few seconds [64], the cumulative latency due to cryptographic calculations and key agreements limits the speed of data exchange.

Privacy: Privacy is also a major issue that needs to be addressed. For example, in many location-based routing protocols, the location and a persistent vehicle ID are constantly broadcast to neighboring vehicles [55, 32]. Besides, most safety-related applications broadcast messages to all neighbors and do not contain secrets to be processed by a specific destination. If each vehicular node would carry a unique permanent MAC address, then it could be possible to trace such a car and its driver (the introduction of dynamically assigned MAC address in IEEE 802.11p might solve the problem).

Incorrect Location Information: Most VANET routing schemes utilize some sort of position information in order to aid message routing. However, the impact of false position information distribution by malicious or defective nodes has serious impact on the performance of those position aware routing schemes. Researchers [47, 40] have shown that even 10% of the nodes which distribute false position information can severely degrade the delivery ratio in a highway scenario.

Geocast Vulnerability: Most safety critical information is disseminated using geocasting, which refers to the distribution of messages within a geographic destination region. Consequently, the geocasting protocols must be very robust against accidental or intentional failures. However, it has been shown [66] that selective jamming can completely disrupt geocast flows along a highway. This is a potentially dangerous attack because anyone with suitable hardware can initiate it and it is also hard to identify the attacker in the jammed network.

Non-cooperative behavior: It is pointed out in [67] that most of the proposed ad hoc routing protocols assume an implicit trust relationship with neighboring nodes in which it is expected that all nodes behave properly. Non-cooperative behavior of some vehicular nodes, such as message modification and dropping, could significantly degrade the performance of message delivery schemes of VANET and further affect the safety of other vehicles [50, 32]. Therefore enforcement of cooperative communications among VANET must be ensured in a secure way. The authors in [67] propose location service for position-based ad hoc routing to prevent message tampering, dropping, and location table tampering attack by malicious nodes. However, security measures for store-carry-forward mechanisms of opportunistic routing schemes need to be explored further.

The requirements for somewhat contradictory properties, such as authentication versus anonymity, and robustness of cryptographic algorithms versus real-time requirements of those, exacerbate the security problems in VANET [72]. Most of the above research studies usually try to solve only a specific problem. The simultaneous solutions of many issues or the integration of the solutions in order to develop a general architecture for VANET is required to assess whether the specified system is capable of satisfying the rather contradicting requirements of VANET.

8.6.2   Acknowledgments

Some form of acknowledgments are required for robust message dissemination and for reducing resource utilization (e.g., buffer space at each node, re-transmissions etc.) in VANET. In the dynamic VANET environment, the probability of receiving an acknowledgment is often low [50]. In [29, 30], the authors use the concept of passive cure in which the destination sends a cure-ack to the last forwarder to indicate that the message has been delivered successfully. This healing propagates through the network to prevent future re-transmission of the same message. But in VANET, where it is important to maximize the use of each contact opportunity for transferring data messages, these types of anti-message broadcasts become a burden.

8.6.3 Addressing

Traditional identity-based addressing models of vehicular nodes impose a huge overhead in maintaining the identities of the nodes in an unbounded network. Moreover, many VANET applications do not require the specific address of the receivers, rather concern about the characteristics of them [50]. Depending on applications, a more precise addressing model is desirable in order to reduce control overhead and increase throughput.

8.6.4 Emerging New Application

We have seen in Section 8.3 that the current VANET applications have rather loose requirements. But we can envision that many future applications (e.g., instant messaging, conferencing etc.) will emerge. These innovative applications of VANET impose newer challenges to robust message dissemination. The requirements of these applications, such as long-duration session, low latency, and high throughput, demand a more complex application aware message dissemination mechanism. [50].

8.7 Conclusion

The integration of hundreds of millions of automobiles attached to wireless communication media and over a billion mobile phone-equipped people in the world may turn out to be the world's largest and most dynamic network in the coming years. However, robust message dissemination is one of the most challenging aspects of VANET due to its highly churned vehicular nodes and intermittent connectivity. It is presumed that once the robust message dissemination can be achieved, a plethora of innovative applications will emerge. This chapter discusses the unique features and characteristics of VANET and their implications towards robust message dissemination. A range of innovative applications which unleash the power of VANET have been mentioned in this chapter. Research challenges and existing schemes for robust message dissemination from the perspective of different protocol layers is the main focus of this chapter. Some open research issues, such as security, privacy, addressing models, etc., have also been outlined.

Bibliography

[1] Path project, [online] http://www.path.berkeley.edu.

[2] M. Abuelela and S. Olariu. Traffic-adaptive packet relaying in vanet. In *Proceedings of the 4th ACM International Workshop on Vehicular Ad Hoc Networks*, pages 77–78, 2007.

[3] R. Ahlswede, N. Cai, S. Y. R. Li, and R. W. Yeung. Network information flow. In *IEEE Transactions on Information Theory*, volume 46, pages 1204–1216, 2000.

[4] S. Ahmed and S. S. Kanhere. Vanetcode: Network coding to enhance cooperative downloading in vehicular ad-hoc networks. In *Proceedings of the International Wireless Communications and Mobile Computing Conference, IWCMC'06*, pages 527–532, July 2006.

[5] S. Ahmed and S. S. Kanhere. Hubcode: Message forwarding using hub-based network coding in delay tolerant networks. In *Proceedings of the 12-th ACM International Conference on Modeling, Analysis and Simulation of Wireless and Mobile Systems (MSWiM'09)*, pages 288–296, October 2009.

[6] S. Ahmed and S. S. Kanhere. A bayesian routing framework for delay tolerant networks. In *Proceedings of the IEEE Wireless Communications and Networking Conference, WCNC'10*, April 2010.

[7] M. M. Artimy, W. Robertson, and W. J. Phillips. Connectivity in inter-vehicle ad hoc networks. In *Proceedings of the Canadian Conference on Electrical and Computer Engineering*, pages 293–298, May 2004.

[8] P. Basu and T. D. C. Little. Networked parking spaces: architecture and applications. In *Proceedings of the Vehicular Technology Conference, VTC'02*, volume 2, pages 1153–1157, 2002.

[9] J. Blum, A. Eskandarian, and L. Hoffman. Performance characteristics of inter-vehicle ad hoc networks. In *Proceedings of the IEEE 6th International Conference on Intelligent Transportation Systems*, pages 114–119, 2003.

[10] L. Briesemeister and G. Hommel. Integrating simple yet robust protocol layers for wireless ad hoc inter-vehicle communications. In *Proceedings of Communication Networks and Distributed Systems Modeling and Simulation Conference, CNDS'02*, pages 186–192, January 2002.

[11] I. Broustis and M. Faloutsos. Routing in vehicular networks: Feasibility, modeling, and security. In *International Journal of Vehicular Technology*, 2008.

[12] J. Burgess, B. Gallagher, D. Jensen, and B. N. Levine. Maxprop: Routing for vehicle-based disruption-tolerant networks. In *Proceedings of the 25th IEEE International Conference on Computer Communications, IN-FOCOM'06*, pages 1–11, Barcelona, Spain, April 2006.

[13] B. Burns, O. Brock, and B. N. Levine. Mv routing and capacity building in disruption tolerant networks. In *Proceedings of IEEE INFOCOM*, volume 1, pages 298–208, 2005.

[14] V. Cabrera, F. J. Ross, and P. M. Ruiz. Simulation-based study of common issues in vanet routing protocols. In *Proceedings of the IEEE 69th Vehicular Technology Conference, VTC'09*, Spring 2009.

[15] W. Chen and S. Cai. Ad hoc peer to peer network architecture for vehicle safety communications. In *IEEE Communications Magazine*, pages 100–107, April 2005.

[16] Z. D. Chen, H. Kung, and D. Vlah. Ad hoc relay wireless networks over moving vehicles on highways. In *Proceedings of the 2nd ACM international symposium on Mobile ad hoc networking and computing*, October 2001.

[17] I. Chisalita and N. Shahmehri. A context-based vehicular communication protocol. In *Proceedings of the 15th IEEE International Symposium on Personal, Indoor and Mobile Radio Communications*, volume 4, pages 2820–2824, September 2004.

[18] N. Choi, S. Choi, Y. Seok, T. Kwon, and Y. Choi. A solicitation-based IEEE 802.11p mac protocol for roadside to vehicular networks. In *Proceedings of Mobile Networking for Vehicular Environments*, pages 91–96, May 2007.

[19] B. Cohen. Incentives build robustness in bittorrent. In *P2P Economics Workshop*, Berkeley, CA, 2003.

[20] O. T. Cruces. Master thesis: Applying delay tolerant protocols to vanets, June 2008.

[21] M. Dianati, X. Ling, K. Naik, and X. Shen. A node cooperative arq scheme for wireless ad-hoc networks. In *Proceedings of the IEEE Wireless Communications and Networking Conference*, 2005.

[22] M. D. Dikaiakos, S. Iqbal, T. Nadeem, and L. Iftode. Vitp: an information transfer protocol for vehicular computing. In *Proceedings of the Workshop on Vehicular Ad Hoc Networks*, pages 30–39, 2005.

[23] S. C. Ergen, S. Y. Cheung, P. Varaiya, R. Kavaler, and A. Haoui. Wireless sensor networks for traffic monitoring. In *Proceedings of the ISPN*, 2005.

[24] J. Eriksson, L. Girod, B. Hull, R. Newton, S. Madden, and H. Balakrishnan. The pothole patrol: using a mobile sensor network for road surface monitoring. In *Proceedings of the 6th international conference on mobile systems, applications, and services*, pages 29–39, June 2008.

[25] C. Fragouli, J.-Y. L. Boudec, and J. Widmer. Network coding: An instant primer. In *Proceedings of the ACM SIGCOMM Computer Communication Review*, volume 36(1), pages 63–68, January 2006.

[26] H. Fubler, H. Hartenstein, J. Widmer, M. Mauve, and W. Effelsberg. Contention-based forwarding for street scenarios. In *Proceedings of the 1st International Workshop on Intelligent Transportation, WIT'04*, pages 155–160, 2004.

[27] A. Goldsmith. *Wireless Communications*. Cambridge University Press, 2005.

[28] J. Gong, C.-Z. Xu, and J. Holle. Predictive directional greedy routing in vehicular ad hoc networks. In *Proceedings of the 5th International Workshop on Mobile Distributed Computing, MDC'07, in conjunction with 27th International Conference on Distributed Computing Systems Workshops, ICDCS'07*, pages 2–2, June 2007.

[29] Z. J. Haas and T. Small. A new networking model for biological applications of ad hoc sensor networks. In *ACM Transactions on Networking*, volume 14(1), pages 27–40, February 2006.

[30] K. Harras, K. Almeroth, and E. Belding-Royer. Delay tolerant mobile networks (dtmns): Controlled flooding schemes in sparse mobile networks. In *IFIP Networking*, May 2005.

[31] H. Hassanein and J. Luo. Reliable energy aware routing in wireless sensor networks. In *Proceedings of 2nd IEEE Workshop on Dependability and Security in Sensor Networks and Systems, DSSNS'06*, pages 54–64, April 2006.

[32] F. Hui. Experimental characterization of communications in vehicular ad hoc network. Technical report, University of California, Davis, 2005.

[33] B. Hull, V. Bychkovsky, Y. Zhang, K. Chen, M. Goraczko, A. Miu, E. Shih, H. Balakrishnan, and S. Madden. Cartel: a distributed mobile sensor computing system. In *Proceedings of the 4th international conference on embedded networked sensor systems, SenSys'06*, pages 125–138, New York, New York, USA, 2006. ACM Press.

[34] C.-C. Hung, H. Chan, and E. H.-K. Wu. Mobility pattern aware routing for heterogeneous vehicular networks. In *Proceedings of WCNC2008*, pages 2200–2205, 2008.

[35] M. Jerbi, R. Meraihi, S.-M. Senouchi, and Y. Ghamri-Doudane. Gytar: improved greedy traffic aware routing protocol for vehicular ad hoc networks in city environments. In *Proceedings of the 3rd International Workshop on Vehicular Ad Hoc Networks*, pages 88–89, 2006.

[36] J. G. Jetcheva, Y.-C. Hu, S. PalChaudhuri, A. K. Saha, and D. B. Johnson. Design and evaluation of a metropolitan area multitier wireless ad hoc network architecture. In *Proceedings of the 5th IEEE Workshop on Mobile Communications*, 2001.

[37] D. Jiang and L. Delgrossi. IEEE 802.11p: Towards an international standard for wireless access in vehicular environments. In *Vehicular Technology Conference*, pages 2036–2040, May 2008.

[38] D. B. Johnson and D. A. Maltz. Dynamic source routing in ad hoc wireless networks. In *Mobile Computing*, pages 153–181. Kluwer Academic Publishers, 1996.

[39] B. Karp and H. T. Kung. GPSR: greedy perimeter stateless routing for wireless networks. In *Proceedings of the Mobile Computing and Networking Conference*, pages 243–254, 2000.

[40] Y. Kim, J.-J. Lee, and A. Helmy. Impact of location inconsistencies on geographic routing in wireless networks. In *Proceedings of the 6th ACM International Conference on Modeling Analalysis and Simulation of Wireless and Mobile Systems, MSWiM'03*, pages 124–127, 2003.

[41] Y. Ko and N. Vaidya. Geocasting in mobile ad hoc networks: Location-based multicast algorithms. In *Proceedings of the 2nd IEEE Workshop on Mobile Computer Systems and Applications*, page 101, February 1999.

[42] V. Kone. Data dissemination in vehicular ad hoc networks, [online] http://www.cs.ucsb.edu/ṽinod/docs/vinod_vanet_readinglist.pdf.

[43] J. LeBrun, C. N. Chuah, D. Ghosal, and M. Zhang. Knowledge-based opportunistic forwarding in vehicular wireless ad hoc networks. In *Proceedings of IEEE Vehicular Technology Conference*, May 2005.

[44] U. Lee, J.-S. Park, E. Amir, and M. Gerla. Fleanet: A virtual market place on vehicular networks. In *Proceedings of 3rd Annual International Conference on Mobile and Ubiquitous Systems: Networking and Services*, pages 1–8, July 2006.

[45] U. Lee, J.-S. Park, J. Yeh, G. Pau, and M. Gerla. Code torrent: content distribution using network coding in vanet. In *Proceedings of the 1st international workshop on decentralized resource sharing in mobile computing and networking, MobiShare'06*, pages 1–5, New York, New York, USA, 2006. ACM.

[46] U. Lee, B. Zhou, M. Gerla, E. Magistretti, P. Bellavista, and A. Corradi. Mobeyes: smart mobs for urban monitoring with a vehicular sensor network. In *IEEE Wireless Communication*, volume 13, pages 52–57, October 2006.

[47] T. Leinmuller and E. Schoch. Greedy routing in highway scenarios: The impact of position faking nodes. In *Proceedings of the 3rd International Workshop on Intelligent Transportation, WIT'06*, Hamburg, Germany, March 2006.

[48] T. Leinmuller, E. Schoch, and C. Maihofer. Security requirements and solution concepts in vehicular ad hoc networks. In *Security requirements and solution concepts in vehicular ad hoc networks*, pages 84–91, January 2007.

[49] I. Leontiadis and C. Mascolo. Geopps: Opportunisitic geographical routing for vehicular networks. In *Proceedings of the IEEE Workshop on Autonomic and Opportunistic Communications*, pages 1–6, June 2007.

[50] J. Li and C. Chigan. Achieving robust message dissemination in vanet: Challenges and solutions.

[51] L. Lin, N. B. Shroff, and R. Srikant. Energy-aware routing in sensor networks: A large system approach. In *Ad Hoc Networks*, volume 5(6), pages 818–831, August 2007.

[52] G. Liu, B.-S. Lee, B.-C. Seet, F. Chuan-Heng, W. Kai-Juan, and L. Keok-Kee. A routing strategy for metropolis vehicular communications. In *Proceedings of ICOIN*, pages 134–143, 2004.

[53] C. Lochert, H. Hartenstein, J. Tian, H. Fussler, D. Hermann, and M. Mauve. A routing strategy for vehicular ad hoc networks in city environments. In *Proceedings of the IEEE Intelligent Vehicles Symposium*, pages 156–161, June 2003.

[54] C. Lochert, M. Mauve, H. Fubler, and H. Hartenstein. Geographic routing in city secnarios. In *ACM SIGMOBILE Mobile Computing and Communications Review*, pages 69–72, January 2005.

[55] M. Mauve, J. Widmer, and H. Hartenstein. A survey on position-based routing in mobile ad hoc networks. In *IEEE Network*, pages 30–39, 2001.

[56] J. Morillo-Pozo, O. Trullols, J. M. Barcelo, and J. Garcia-Vidal. A cooperative arq for delay-tolerant vehicular networks. In *Proceedings of the 28th International Conference on Distributed Computing Systems Workshops*, 2008.

[57] J. D. Morillo-Pozo and J. Garcia-Vidal. A low coordination overhead c-arq protocol with framed combining. In *Proceedings of the 18th Annual IEEE International Symposium on Personal, Indoor and Mobile Radio Communications, PIMRC'07*, 2007.

[58] T. Nadeem, S. Dashtiezhad, C. Liao, and L. Iftode. Trafficview: traffic data dissemination using car-to-car communication. In *Proceedings of the MC2R*, pages 6–19, 2004.

[59] V. Namboodiri and L. Gao. Prediction-based routing for vehicular ad hoc networks. In *IEEE Transactions on Vehicular Technology*, volume 56(4), pages 2332–2345, July 2007.

[60] A. Nandan, S. Das, G. Pau, M. Sanadidi, and M. Gerla. Cooperative downloading in vehicular ad-hoc wireless networks. In *Proceedings of IEEE/IFIP International Conference on Wireless On Demand Network Systems and Services*, pages 32–41, St. Moritz, Switzerland, January 2005.

[61] L. Pelusi, A. Passarella, and M. Conti. Opportunistic networking: Data forwarding in disconnected mobile ad hoc networks. In *IEEE Communications Magazine*, volume 44, November 2006.

[62] Y. Peng, Z. Abichar, and J. M. Chang. Roadside-aided routing (rar) in vehicular networks. In *Proceedings of ICC 2006*, volume 8, pages 3602–3607, June 2006.

[63] A. Pentland, R. Fletcher, and A. Hasson. Daknet: Rethinking connectivity in developing nations. In *IEEE Computer*, volume 37(1), pages 78–83, January 2004.

[64] A. K. Saha and D. B. Johnson. Modeling mobility for vehicular ad-hoc networks. In *Proceedings of the 1st ACM workshop on Vehicular ad hoc networks*, October 2004.

[65] R. A. Santos, R. M. Edwards, and N. L. Seed. Using the cluster-based location routing (cblr) algorithm for exchanging information on a motorway. In *Proceedings of the 4th International Workshop on Mobile and Wireless Communications Network*, pages 212–216, September 2002.

[66] E. Schoch, F. Kargl, T. Leinmüller, and M. Weber. Vulnerabilities of geocast message distribution. In *Proceedings of the 2nd IEEE Workshop on Automotive Networking and Applications, AutoNet'07, in conjunction with GlobeCom'07)*, pages 1–8, Washington, DC, USA, November 2007.

[67] J.-H. Song, V. W. S. Wong, and V. C. M. Leung. A framework of secure location service for position-based ad hoc routing. In *Proceedings of the 1st ACM International Workshop on Performance Evaluation of Wireless Ad Hoc, Sensor and Ubiquitous Networks*, pages 99–106, October 2004.

[68] A. Stampoulis and Z. Chai. A survey of security in vehicular networks.

[69] W. Sun, H. Yamaguchi, K. Yukimasa, and S. Kusumoto. Gvgrid: A qos routing protocol for vehicular ad hoc networks. In *Proceedings of IEEE IWQoS*, pages 130–139, June 2006.

[70] M. Tacca, P. Monti, and A. Fumagalli. Cooperative and non-cooperative arq protocols for microwave recharged sensor nodes. In *Proceedings of the EWSN*, February 2005.

[71] C. A. T. H. Tee and A. C. R. Lee. Survey of position based routing for inter vehicle communication systems. In *Proceedings of the 1st International Conference on Distributed Framework and Applications*, pages 174–182, October 2008.

[72] Y. Toor and P. Muhlethaler. Vehicle ad hoc networks: Applications and related technical issues. In *IEEE Communications Surveys*, volume 10(3), pages 74–88, 3rd Quarter 2008.

[73] S. Wang. Predicting the lifetime of repairable unicast routing paths in vehicle-formed mobile ad hoc networks on highways. In *Proceedings of the 15th IEEE International Symposium on Personal, Indoor and Mobile Radio Communications*, pages 2815–2819, September 2004.

[74] I. WG. IEEE 802.1f/final version. In *Recommended Practice for Multivendor Access Point Interoperability via an Inter-Access Point Protocol Across Distribution Systems Supporting 802.11 Operation*, January 2003.

[75] I. WG. IEEE 802.11p/d3.0, draft amendment for wireless access vehicular environments (wave). July 2007.

[76] H. Wu, R. Fujimoto, R. Guensler, and M. Hunger. Mddv: a mobility-centric data dissemination algorithm for vehicular networks. In *Proceedings of the 1st ACM International Workshop on Vehicular Ad Hoc Networks, VANET'04*, pages 47–56, October 2004.

[77] X. Zhang, J. Kursoe, B. Levine, D. Towsley, and H. Zhang. Study of a bus-based disruption tolerant network: Mobility modeling and impact on routing. In *Proceedings of the ACM Mobicom*, September 2007.

[78] Z. Zhang. Routing in intermittently connected mobile ad hoc networks and delay tolerant networks: Overview and challenges. In *IEEE Communications Surveys*, volume 8(1), pages 24–37, 1st Quarter 2006.

[79] B. Zhao and M. C. Valenti. Practical relay networks: A generalization of hybrid arq. In *Proceedings of the IEEE JSAC*, volume 23, January 2005.

[80] J. Zhao and G. Cao. Vadd: Vehicle-assisted data delivery in vehicular ad hoc networks. In *IEEE Transactions on Vehicular Technology*, volume 57, pages 1910–1922, May 2008.

[81] W. Zhao, Y. Chen, M. Ammar, M. D. Corner, B. N. Levine, and E. Zegura. Capacity enhancement using throwboxes in dtns. In *Proceedings of IEEE International Conference on Mobile Ad hoc and Sensor Systems, MASS'06*, October 2006.

Chapter 9

Delay Tolerant Networking (DTN) Protocols for Space Communications

Ruhai Wang, Xuan Wu, Tiaotiao Wang, and Tarik Taleb

9.1   Introduction 262
9.2   Space Communications and DTPs for Space/Interplanetary
       Internet 263
       9.2.1   Overview of Space Communications 263
       9.2.2   Performance Problems of TCP in Space 264
       9.2.3   Data Transport Protocols for Space/Interplanetary
       Internet 266
9.3   DTN, BP, and Convergence Layer Protocols 267
       9.3.1   Overview of DTN 267
       9.3.2   DTN for Space 268
       9.3.3   Bundle Protocol (BP) 269
       9.3.4   DTN Convergence Layer Protocol (CLP) 269
9.4   Recent Research and Experimental Activities on DTN for
       Space Communications 271
9.5   Summary 275
9.6   Open Issues 275
9.7   Acknowledgments 276

Delay/disruption tolerant networking (DTN) technology offers a new solution to highly stressed communications in space environments, especially those with long link delay and frequent link disruptions in deep space missions. It is becoming a recognized research area in computer networks and space communications. Extensive research work has been done on DTN for space communications in the past several years including numerous publications (Internet Engineering Task Force (IETF) Requests for Comments (RFCs) among them), DTN protocol implementations, and some experimental work done over simulation testbeds and in space, both low-Earth orbit and deep space. In this chapter, we provide an overview of the emerging DTN protocols for space communications with focuses on DTN architecture, BP, convergence layer protocols, and their application in space. An overview of recent research and experimental activities on DTN for space communications is also presented.

9.1 Introduction

The National Aeronautics and Space Administration (NASA) is moving towards packet-switched space communications using appropriate network architecture and protocols [1]–[3]. Transmission control protocol (TCP) [4] experiences severe performance degradation in space due to its inherent design and the vast differences between terrestrial and space environments [2] [3]. Delay/disruption tolerant networking (DTN) technology [5] [6] offers a new solution that uses a store-and-forward mechanism to combat long link delay and frequent link disruptions generally characterizing space communications. DTN communications use a bundle protocol (BP) [7], to construct a store-and-forward overlay network. The BP, situated at the application layer of the Internet model, utilizes conventional Internet protocols to send and receive data bundles. The DTN architecture requires a convergence layer protocol (CLP) [7] [8] in order to run the bundle protocol over the underlying transport layer protocols. CLPs currently available or under development for space internetworking include the TCP-based CLP (i.e., TCPCL) [8], user datagram protocol (UDP)-based CLP [9] [10], Saratoga [11], Licklider transmission protocol (LTP, also called a long-haul transmission protocol) [12]–[14], and LTP-Transport (LTP-T) [15]. The TCPCL protocol for the DTN system works together with the well-known TCP to assure reliable communication services between DTN nodes. In other words, the DTN nodes are capable of invoking various types of recovery mechanisms if they experience data losses because of link interruption or a TCP connection outage. LTP is intended to operate over point-to-point, long-haul, deep-space radio frequency links or similar links characterized by an extremely long transmission delay and/or frequent interruptions in connectivity [12]. Unlike TCPCL, LTP does not include any flow or congestion control signaling.

Although originally targeting challenged interplanetary Internet (IPN) [16], the DTN techniques and concepts have also been considered for use in other operational environments that are subject to link disruption and long link delay, such as terrestrial wireless networks, sensor-based networks, and tactical/military ad-hoc networks [17]–[26]. NASA considers DTN the most suitable technology to be employed in space internetworking and hopes to fly with it on space missions soon [27].

In this chapter, we provide an overview of the emerging DTN protocols for space communications with focuses on DTN architecture, BP, convergence layer protocols, and their application in space. We also review recent research and experimental activities on DTN for space communications.

The remainder of this chapter is organized as follows. In Section 9.2, we provide an overview of space communications, review the performance problems of TCP in the space environment, and then discuss the recently developed data transport protocols for space and interplanetary Internet. In Section 9.3, we introduce general DTN, DTN for space, BP, and DTN convergence layer protocols. In Section 9.4, we review recent research and experimental activities on DTN for space communications. A summary is provided in Section 9.5. Finally, open research issues are discussed in Section 9.6.

9.2 Space Communications and Data Transport Protocols for Space/ Interplanetary Internet

9.2.1 Overview of Space Communications

From early geo-stationary orbit (GEO) relay and broadcast communication satellites and the Advanced Communications Technology Satellite (ACTS) testing of the Internet technology, through the advanced tracking and data relay satellite system (TDRSS) and the deep space networks (DSN), to the current space Internet and Interplanetary Internet, space communications have been changing significantly with respect to both communication architecture and operating model. This movement from the traditional model of managed, mission-specific pairwise links to an automated network involving multiple nodes using relay and the Internet technologies [28] is of historical significance.

Planetary exploration missions have historically accomplished communication between data sources on mission planets and stations on Earth with dedicated direct communication circuits. Considering the limitations of direct connection time between any two moving planets, missions have recently begun to use relaying in space communication for effective data transmission [29]. Over 95% of the data received from the Mars Exploration Rovers has been returned to Earth via the Mars Global Surveyor and Mars Odyssey spacecraft [30]. This experience has demonstrated that using orbiters to relay data from the Mars surface can greatly increase data return. Similar benefits are expected around the moon and at other planets [27] [29].

In a typical space communication scenario, this kind of relay can be achieved in two ways: (1) through the planet-orbiting spacecraft; and (2)

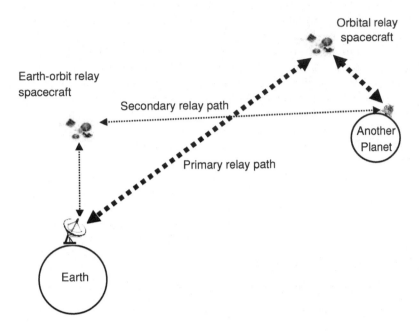

Figure 9.1: Relay infrastructure in interplanetary communications.

through the Earth-orbiting spacecraft such as the TDRSS [31] [32]. The first way is the primary method, while the second can be considered the backup choice. In Figure 9.1, we illustrate a typical architecture using relaying spacecraft to forward data from another planet to Earth. The data sources on the planetary surface can be roving robotics, crews, surface stations, or ascent/decent vehicles. Although the relay operation shown in Figure 9.1 is done through a single spacecraft, multiple spacecrafts or even constellations of satellites can be involved in relaying communications. As the number of relays and relay users increases, international cross-support for this relaying capability will increase mission robustness and data return, giving planetary landers multiple opportunities to forward data to Earth.

It is important to bring internetworking technology to space communications to enable as much autonomous operation as possible. The more autonomous we can make operations, the less human controlling and scheduling is involved and the greater the savings in reduced cost of operations [28].

9.2.2   Performance Problems of TCP in Space

As a reliable and widely-used transport protocol of the Internet protocol suites, TCP [4] performs very well in today's Internet operations. However, its performance in space communications is severely degraded by the inherent conflict between its design and the challenging characteristics of the very different space environment [2] [3].

Several issues of the space internetworking are inappropriately addressed by TCP, limiting its performance. One major issue is that TCP is designed to use window-based transmission control algorithms. With the sliding window flow-control mechanisms, TCP regulates the amount of data a source can send by adjusting the window size in response to the acknowledgement information from the destination; the timeliness of that information is key to the effectiveness of the technique.

Another major issue is that TCP cannot distinguish between data losses caused by network congestion and link errors.

The impact on performance of TCP caused by these issues is not very obvious in the terrestrial Internet. However, it becomes obvious and even serious due to the challenging space communication environment. Long and variable propagation delays, frequent and lengthy link outages, strong channel noise, and high channel-rate asymmetry in space all conspire to adversely degrade the performance of TCP.

Round-trip delays involved in space communications, especially those in deep space, are much longer than those of the terrestrial Internet. They can be up to 40 minutes long for the Martian channel and even much longer for a channel from Jupiter to Earth. The problem with long end-to-end RTTs in space communications is that they hurt TCP interactions between the space node and the ground Internet node. As a result, this limits the usefulness of TCP acknowledgment information from the remote destination node and influences the effectiveness of TCP transmission control.

Planetary bodies, asteroids, or spacecrafts moving in space all may periodically and randomly interrupt the communication link and cause frequent and long link outage between two communication endpoints [2]. This can cause a significantly large number of packet losses and retransmissions, resulting in severe performance degradation of TCP.

Space communications feature much noisier channels than terrestrial communications, resulting in very high bit error rates (BER) with 10^{-5} very common and even rates on the order of 10^{-1} in the deep space environment [2]. As also discussed in the literature, TCP makes an erroneous congestion decision which misinterprets any data loss as a congestion loss. In general, multiplicative-decrease in sending data rate is designed as a TCP response to data loss caused by network congestion. Consequently, the throughput of TCP is unnecessarily decreased when data loss is caused by bit error corruption instead of real network congestion. In fact, data loss due to a high channel BER in the space environment has a disproportionately negative effect on TCP performance.

Space communications channels are frequently asymmetric in terms of channel bandwidth: the bandwidth of the uplink, from the ground to the spacecraft, is generally much lower than the bandwidth of the downlink channel, from the spacecraft to the ground. The space channel asymmetry occurs mainly due to the scientific purpose of space missions and the cost that would be incurred by an increase in receiver power and antenna size as would be required for high bandwidth reception [3]. For most space missions, the data flowing from space to the ground over the downlink is substantially larger than the data flowing from the ground to space over the uplink. For this reason, the downlink designed for bulk data transfer has a broad bandwidth, while the uplink designed to send acknowledgement (ACK) messages or telecommand messages for commanding the spacecraft generally has a much narrower bandwidth. Channel rate asymmetry in space can severely affect the performance of TCP because TCP is ACK-clocked, relying on the timely feedback of ACKs through the uplink to make steady progress of data transmission. However, the slow uplink rate cannot not handle the transmission of returning ACKs effectively. This may result in frequent losses of ACKs and consequential performance degradation of the protocol. For a detailed discussion on how the network channel asymmetry affects the performance of TCP, see [33]. TCP generally cannot work effectively when the channel asymmetric ratio (defined as the ratio of ACK channel rate over data channel rate) is lower than $\frac{1}{50}$ [3][34][35]. The ratio of channel asymmetry can be as low as $\frac{1}{1000}$ in Earth-orbit communications and even significantly lower in the deep space environment.

The above performance issues of TCP in satellite and space communications have been fully investigated, and thus are not discussed in detail in this chapter. For a thorough literature review, see [2][3][34][35].

9.2.3   Data Transport Protocols for Space/Interplanetary Internet

While the feasibility of operating space communications using the Internet-type protocol has been a source of contention within NASA [1], a large number of transmission control protocols have been developed for space and other environments with similar communication conditions [2][3]. Based on the design and operating methodology, these protocols can be roughly classified into the following three categories:

- Category I: The protocols in this category involve changes only to the TCP protocol. The changes are generally on the congestion-control and error-control algorithms of TCP.

- Category II: The protocols generally involve modifications to the TCP protocol and/or network operation infrastructure and elements.

- Category III: This category involves new protocols that are designed to provide data transport functionality but operate at the application and link layers. DTN bundle BP [7] and LTP [12]–[14] produced by the DTN Research Group (DTNRG) of the Internet Research Task Force (IRTF) [36] are considered the representative protocols in this category.

For a comparative summary and classification of these protocols, their design techniques and performance evaluation, please refer to [37].

The current issues in network, link, and physical layers for space and Interplanetary Internet are not discussed here. Refer to [2] for a detailed discussion.

9.3   DTN, BP, and Convergence Layer Protocols

9.3.1   Overview of DTN

In comparison to conventional Internet architecture (i.e., TCP/IP architecture), DTN is a networking architecture designed to provide communications in highly stressed environments characterized by long or variable delays, intermittent loss of link connectivity, high error rates, and asymmetric data rates [38]. Originally developed for IPN, DTN was extended to terrestrial wireless networks and wireless sensor networks (WSNs) as those types of networks also suffer from link disruption and delay.

Application data units in DTN are carried in variable-length protocol data units (PDUs) called bundles that are intended to minimize the round-trip exchanges needed to complete a protocol transaction. In other words, if all the information required to complete a transaction is bundled, the number of exchanges between the message sender and the receiver can be reduced. This is significant to the protocol performance if the link propagation delay is extremely long, such as hours or longer (i.e., in deep space).

Extensive work has been recently done in DTN routing with most researchers studying routing algorithms. The DTN routing algorithms are mostly designed under the assumption that paths are not continuously connected end-to-end between message source and destination. Some recent studies have also integrated DTN concepts into mobile ad-hoc network (MANET) research which mainly focuses on routing in dense mobile ad-hoc networks where end-to-end connectivity is possible [39]. If DTN and MANET are combined, the routing of bundles will involve both simple forwarding and store-carry-and-forward operation as required.

DTN adopts a traditional custody transfer mechanism to deal with harsh communication environments [7]. This mechanism is optional. However, general description of DTN architecture assumes the "custody transfer" option is enabled. With this mechanism, DTN keeps track of a current custodian for each bundle. A DTN custodian retains in persistent memory a copy of each bundle it has sent, discarding that persistent copy only when an ACK is received from another node confirming that this node has received the bundle successfully and has accepted responsibility for forwarding it, as illustrated by the DTN operation architecture in Figure 9.2. By this means, no bundles are lost even if contact with the designated receiving node is interrupted due to a transient change in network topology. The adoption of custody transfer

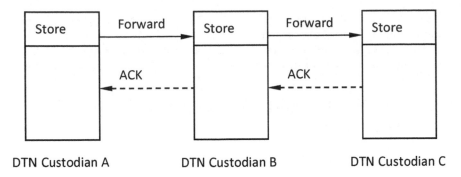

Figure 9.2: Custody transfer of DTN operation architecture [38].

and persistent storage at intermediate nodes makes DTN capable of delegating data retransmission responsibility to nodes other than the original source node; this feature is a significant departure from Internet architecture, in which all retransmission is end-to-end and the source node bears all responsibility for retransmission.

The DTNRG of IRTF [36] concerns itself with the whole range of potential uses for DTN, including "DTN for space."

9.3.2   DTN for Space

As the Internet protocols (mainly TCP/IP) are not effective across frequently disrupted links, long and variable delays, and high rates of data loss due to corruption, DTN technology is being considered for space communications. It is believed that integrating DTN into current space communication architecture will enable Internet-like user interaction to be maintained even in highly stressed space exploration environments.

In the Internet, connectivity is generally continuous and signal propagation latency is very small. This means that changes in network topology can be discovered dynamically and communicated to routers in time to revise computed routes before much traffic is misdirected. To some degree this is also true of terrestrial DTNs, making it plausible to use routing protocols such as Probabilistic Routing Protocol using History of Encounters and Transitivity (PROPHET) [26]. But these conditions do not hold in space communications, especially in deep space, so the kinds of routing protocols that work well in the Internet or even in some terrestrial DTNs are of little utility in space networking: by the time a node learned of the start of a communication opportunity the opportunity might already have ended. On the other hand, most changes in space network topology are intentional and scheduled rather than inadvertent, so a different sort of "routing protocol" may be used instead: a "contact plan" enumerating planned episodes of connectivity may be simply declared to routers well in advance of the scheduled changes in topology. Routes may be computed based on topology changes that are announced rather than discovered.

One other significant difference between Internet protocol implementations and DTN is the site of retention of data in transit. Because data may need to be retained for hours or even days in a relay satellite before it can be forwarded, bundles may best be stored in long term persistent mass memory [38] rather than dynamic memory. In contrast, an Internet router generally need not retain a received packet for more than a few milliseconds before transmitting it, transient retention in high-speed dynamic RAM is appropriate.

9.3.3 Bundle Protocol (BP)

As introduced, a store-and-forward message switching with the optional custody transfer is adopted by DTN to deal with challenging communication environments [7]. The store-and-forward mechanism of DTN moves a message or fragments of a message from one node to the next until it reaches the destination [38]. The adoption of custody transfer and persistent storage at intermediate nodes makes DTN capable of delegating data retransmission responsibility to nodes other than the original source node. To provide the store-and-forward message transmission service with the optional custody transfer, BP [7] is designed to operate as an overlay protocol on top of heterogeneous underlying CLP stacks, among which may be the Internet protocol stack itself.

The major capabilities of BP are the ability to cope with connectivity interruption and the ability to take advantage of scheduled, predicted, and opportunistic connectivity. BP can also be configured for retransmission on any failure of custody transfer, and it supports optional requests for end-to-end acknowledgment at the application layer.

In order to utilize an underlying convergence-layer protocol stack such as TCP/IP or UDP/IP, BP needs a convergence layer adapter (CLA) [8] deployed between the bundle layer and transport layer. (This implies that both the CLA and the BP function at the application layer in the Internet protocol stack.) The bundle protocol agent (BPA) [7] of a DTN node executes BP procedures, invoking the services of the CLA to do so. A CLA sends and receives bundles on behalf of the BPA, using the transport service of the underlying internetworking protocols.

9.3.4 DTN Convergence Layer Protocol (CLP)

Convergence-layer adapters for a variety of transport protocols have been developed to support data transmission in space DTN networks. As introduced, TCPCL, UDPCL and LTPCL are currently the most broadly supported CLAs. In Figure 9.3, a basic structure of the DTN protocol stack corresponding to the OSI stack is provided. As observed, both BP and the various CLAs operate at the DTN overlay layer above the transport layer.

Application Layer Presentation Layer Session Layer	DTN Application
	Bundle Protocol (BP)
	Convergence Layer (TCPCL, LTPCL, UDPCL, or Hybrid of them)
Transport Layer	TCP or UDP
Network Layer	IP
Link Layer	PPP, Ethernet, etc.
Physical Layer	RF, Optical, Wire, etc

OSI Stack   **DTN Protocol Stack**

Figure 9.3: Basic structure of DTN protocol stack in comparison to the canonical OSI stack layers.

The TCPCL protocol adapter uses TCP to provide reliable communication services between DTN nodes. When a bundle node establishes a TCP connection, it establishes a TCPCL connection at the same time for bundle communication. The TCPCL connects DTN nodes via a TCP channel. Over the established TCPCL connection, the sender can send bundles to the destination through the next node. While TCP itself is a reliable protocol, the TCPCL adaptation is additionally capable of invoking various types of recovery mechanisms if the transfer of a bundle experiences interruption because of a TCP connection outage. For an idle connection, a one-byte keep-alive message may optionally be sent at a defined interval. The protocol uses this to detect loss of connection in the absence of bundle traffic.

A simple UDP-based CLP has been implemented in the DTN-2 reference implementation [10]. This UDP-based CLP is intended for use over dedicated private links where congestion control is not required. Its design is based on a presumption that a bundle will always fit into a single UDP datagram, which is limited to around 64 Kbytes. In other words, this CLP is not able to support segmentation of large DTN bundles across multiple UDP packets. Another simple UDP convergence layer has been defined for unidirectional transport over both unicast and multicast networks [40].

Compared to these basic UDP-based CLP implementations, Saratoga [11], LTP [12], and LTP-T [15] are relatively complicated CLPs that support segmentation of large DTN bundles across multiple UDP packets or, in the case of LTP, packet structures for other link-layer protocols. These CLPs inherit core design ideas from the Consultative Committee for Space Data Systems (CCSDS) file delivery protocol (CFDP) [41]. While Saratoga was originally developed for efficient transfer of image data on board the IP-based satellites, it can now work in DTN as a CLP for exchanging bundles between DTN peer nodes. It is a lightweight UDP-based transfer protocol intended for use between point-to-point peers that have sporadic, intermittent connectivity using dedicated IP links [11]. In other words, Saratoga is not developed for use over shared paths, so it requires no congestion control mechanism. Saratoga is intended to run above UDP and UDP-Lite [42]. It implements checksums across each hop to protect data integrity.

Intended to serve as a reliable CLP underlying BP, the newly-developed LTP provides reliable communications between adjacent DTN nodes for interplanetary space. In contrast to Saratoga, LTP is designed to be able to run over both UDP and CCSDS link-layer protocols and to perform retransmission-based recovery of lost data with a selective repeat ARQ mechanism. LTP recognizes each block of data as having two parts: a "red-part," whose delivery must be assured by acknowledgment and retransmission, followed by a "green-part," whose delivery is attempted, but not assured. The length of either part may be zero; that is, any given block may be designated entirely red or entirely green. Thus, LTP can provide both TCP-like and UDP-like functionality concurrently in a single session. Unlike TCP, LTP includes no flow or congestion control signaling. Like CFDP, however, LTP implements optional mechanisms for accelerating retransmission at the cost of some additional signal traffic.

An extension of LTP, LTP-T is proposed as an end-to-end capable transport protocol for data transmissions over multi-hop space networks [15].

As an alternative transmission mechanism for deep-space data transfers, Deep-Space Transport Protocol (DS-TP) [43] has also been proposed. DS-TP is designed as a rate-based, reliable transport protocol whose advantage is its efficient and fast retransmission mechanisms.

For a detailed discussion of DTN, BP, CLPs and DS-TP, refer to [6]-[8][13][43].

9.4 Recent Research and Experimental Activities on DTN for Space Communications

DTN has become a recognized research area in computer networks and space communications [39]. Extensive research work has been done on DTN for space communications in the past several years, leading to numerous publications including several Internet Engineering Task Force (IETF) Request for Comments (RFCs): RFC 5050 Bundle Protocol Specification [7], RFC 5325 Licklider Transmission Protocol — Motivation [12], RFC 5326 Licklider Transmission Protocol — Specification [13], and RFC 5327 Licklider Transmission Protocol — Security Extensions [14]. While RFC 5050 introduced the means by which a DTN store-and-forward overlay is formed for custody-based message transmission, RFCs 5325–5327 describe how retransmission-based reliability can be ensured and secured for DTN over deep space links.

It has recently developed a CCSDS Green Book [44] for the application of DTN in space. This book discusses the DTN rationale, use scenarios, and networked architecture in the space internetworking environment ranging from LEO to deep space. It serves as a reference for space mission designers and mission operation personnel considering the needs of DTN services and capabilities to invoke [44]. The CCSDS is currently also developing a Draft Red Book of LTP [45].

Agreements are also being established on the standards that will enable interoperability and cross support operations. The Space Internetworking Strategy Group (SISG), which is composed of technical experts appointed by the Inter-agency Operations Advisory Group (IOAG) agencies, considers DTN to be the only mature candidate protocol available to handle long propagation delays, the frequent and lengthy network disruptions inherent in space missions involving multiple spacecraft [27]. The SISG concludes that DTN should be provided as the main end-to-end routing service in the future space networks and it is a high priority to mature DTN to full flight readiness for a wide variety of space missions by 2012. For the protocol implementation, NASA JPL's Interplanetary Overlay Network (ION) system for DTN in space [46], and the DTN reference implementation in C++, DTN-2 [10], are already openly available for evaluation.

Some experimental work has recently been done on DTN in the space environment.

Experiments with a United Kingdom disaster monitoring constellation (UK-DMC) satellite in low-Earth orbit have demonstrated data transfer from space using DTN BP [47]. In this experiment, an implementation of a subset of the bundle protocol on board the DMC satellite, utilizing Saratoga as the CLP, was the first successful use of bundles to transmit images from space to a ground data system in the presence of link disruption [47]. The experiment showed that providing a bundle integrity checksum can improve the reliability of BP. It also shows that network time synchronization is an important network configuration consideration.

As the first of a series of NASA experiments to mature DTN to flight readiness for a wide variety of missions, the Deep Impact Network Experiment (DINET) was conducted by JPL in October and November of 2008, using the EPOXI spacecraft located about 15 million miles away from Earth as the first DTN router in space [48]. The DINET project validated the use of DTN protocols (JPL's ION implementation) in deep-space Internet. During this month-long experiment, dozens of space images (about 14.5 MB) were transmitted from DTN nodes at JPL to the spacecraft and then automatically forwarded from the spacecraft back to the JPL nodes, hours or days later, without data loss or corruption. The experiment exercised the principal elements of DTN technology including automatic, contact-sensitive relay operations, custody transfer, rate control, and delay/disruption-tolerant retransmission [49] in the presence of significant light-time delay and both planned and unplanned network disruptions. The experimental results show that DTN BP and LTP work very effectively. They can tolerate signal propagation delay longer than 80 seconds and network outage on the order of days, and they effectively handle station handovers and transient failures in DSN uplink service [49]. ION had no failure in four weeks of continuous operation on VxWorks, Solaris, and Linux platforms, although several bugs were found in the contact graph routing implementation which resulted in some underutilization of network capacity. It was also found that the EPOXI spacecraft clock drifted more rapidly (around 1 second per day) than expected. However, time synchronization errors and one-way-light-time (OWLT) estimation errors of several seconds had no significant impact on DTN network operation [49]. In summary, the DINET project demonstrated that DTN is ready for operational use in space missions [48] and that the ION implementation of DTN is of sufficient quality that future space missions can use it at low risk.

NASA also began experimenting with DTN on the International Space Station (ISS) in July 2009, in a deployment led by the University of Colorado, Boulder (CU-Boulder) [50]. With the ION implementation of DTN installed on a BioServe payload known as Commercial Generic Bioprocessing Apparatus 5 (CGBA5), DTN bundles are transmitted from ISS to Marshall Space Flight Center (MSFC) in Huntsville, AL, and then forwarded to the payload operation control center at CU-Boulder. NASA and CU-Boulder are currently working to extend the DTN experiments on ISS so that other space agencies such as European Space Agency (ESA) and the Japanese Aerospace Exploration Agency can participate [50].

Some other projects are also under way to investigate DTN operation in space. In close collaboration with NASA JPL, extensive work has been done at Lamar University, Texas, to study DTN operation and performance in space communications, including establishment of two PC-based testbeds and experimental performance evaluation of various DTN implementations. One testbed is built to simulate a slow-speed point-to-point space link [51]–[53] and another to simulate high-speed Solar System Internetwork (SSI). With these two testbeds, extensive experiments have been done in evaluating the DTN protocols in cislunar and deep-space communication environments, including CFDP [52], DTN-2 [53][54], and ION [55][56].

The experimental results in [53] show that the TCPCL-based DTN protocol of DTN-2 implementation works effectively over less error-prone cislunar links with short delay and has performance advantage over LTP-based DTN. Along with the increase in link propagation delay and channel noise level, the performance of TCPCL is getting far behind the LTP-based DTN. The TCPCL-based DTN protocol can successfully handle an experimented short link disruption around $30s \sim 120$ s even accompanied by a very long link delay and a high channel BER, but its performance is poor in comparison to the LTPCL.

The investigation results in [54] indicate that the TCPCL-based DTN works effectively in handling a long link disruption experienced in data transmission accompanied by a cislunar link delay and a high BER. In cislunar communications, the performance of the DTN is most adversely affected by link disruption time in comparison to the effect of link delay and BER. In a presence of a very long link disruption of hours, the variations in goodput are nominal with respect to the change in cislunar link delay. It was also found that the goodput difference among the DTN transmissions at all three experimented BERs is trivial because the effect of a high BER is so small compared to a very long link disruption.

In [55], Wang *etal.* presented a rigorously quantified experimental study of the performance of the recently developed LTP in cislunar communications using a PC-based testbed. The performance of BP/LTPCL/UDP/IP in realistic file transfers was compared with two other DTN protocol stack options, BP/TCPCL/TCP/IP and BP/UDPCL/UDP/IP. With the help of a statistical method of t-test, it was found that LTPCL has a significant performance advantage over TCPCL for link delays longer than 4000 ms when BER is 10^{-6} or lower. For the transmissions over a very lossy channel with a BER of around 10^{-5}, LTPCL has a significant goodput advantage over TCPCL, with an advantage of around 3000 bytes/s for delays longer than 1500 ms. It was also found that LTPCL has a consistently significant goodput advantage over UDPCL, around 2500 $\sim$ 3000 bytes/s, at all levels of link delays and BERs. In [55], a summarized side-by-side comparison between TCPCL and LTPCL is also provided in a tabular form to illustrate their design differences.

The experiments at Lamar University also investigated possible hybrid of ION protocols, with LTPCL, TCPCL and UDPCL running over different hops of a typical three node, relay-type of interplanetary links [53][56]. It is found that the hybrid of TCPCL and LTPCL has significant goodput advantage over absolute LTPCL transmissions (i.e., LTPCL over all hops) and any other protocol combinations, and the configuration involving LTPCL on a longer hop (e.g., the hop from orbital relay spacecraft to Earth over the primary relay path in Fig. 9.1) is better able to tolerate long propagation delay and high BER than other configurations [56].

Considering the existence of channel-rate asymmetry in space communications environment, the team at Lamar University also investigated the effect of space channel asymmetry on DTN protocols. According to the study in [56], a hybrid protocol option, TCPCL—LTPCL, shows significant performance advantage over TCPCL as well as LTPCL over all the experimented asymmetric channel ratios (varying between 25/1 800/1). In addition, the study results showed that aggregation of multiple BP bundles within a single LTP block mostly brings significant goodput advantage to LTPCL, especially over a less noisy cislunar channel with a BER around 10^{-6}. The team also found that the variation of channel ratios does not affect the performance of LTPCL significantly but it affects the performance significantly when the cislunar channel noise level is as high as a BER of 10^{-5}.

The ESA has established a DTN/IP Space-Ground testbed and is developing protocol implementations for evaluation, mainly targeting Martian communications [57].

9.5 Summary

In this chapter, we provide an overview of the emerging DTN protocols for space communications. We discuss the operation of space communications, the performance problems of TCP in space, and available data transport protocols for space and Interplanetary Internet. The chapter focuses on a discussion of DTN technologies including the general DTN, DTN for space, BP, and convergence layer protocols. As an essential part of the chapter, the latest development and experimental activities on DTN for space communications are also reviewed. Based on the available simulation and experimental results, it can be concluded that DTN BP and LTP protocols work very effectively with space missions in the presence of significantly long link delay and frequent and lengthy network outage. While international organizations are working on DTN standardization, flight experiments have already demonstrated DTN readiness for operational use in space missions. DTN for space communications is gradually approaching maturity, despite some issues that need to be addressed.

9.6 Open Issues

Although some file transfer testing using DTN has been done successfully for space communications in both testbed [53]–[56] and real-world deployment [47]–[49], these exercises have all been experimental activities. While the DINET experiment demonstrated DTN readiness in space missions, DTN for space is still far from being incorporated into mission-critical long-running applications. At this stage, several issues must be addressed prior to pervasive deployment, especially in regard to routing, security, and clock synchronization [39][47][49].

Although contact graph routing was shown to be generally effective in the DINET experiment (despite some implementation bugs), no dynamic routing protocol has yet been standardized for DTN in space.

Security will soon become an important issue in space missions. Clock synchronization was shown as a problem in both LEO UK-DMC [47] and deep-space DINET projects [48].

The DTN architecture remains a topic of some controversy. In [58], the authors assert that the DTN Bundle Protocol is not well-suited to many operational scenarios that it was intended to support, especially with respect to error detection and transmission reliability. While evaluation of BP and supporting routing protocols is recognized as critical for DTN development [59], evaluation and optimization of LTP is likewise critical for deployment of DTN in space flight. This suggests that additional experimental and the-oretical investigations need to be done to better understand the operation of space DTN, and practical remediation strategies may be needed for enhanced performance.

According to our experimental results, the TCP-based DTN-2 shows better performance over less error-prone links (such as terrestrial links) while the LTP protocol has performance advantages over lossy space links. With this conclusion, we suggest a hybrid of DTN-2 and LTP protocols should be tested to evaluate their performance in an integrated terrestrial and space networking scenario.

According to our research results [60], the traffic shaping mechanism of a rate-based transmission is much more effective than the bursty flow of window-based transmission over a long-delay, lossy space communication channel. There exist new protocols such as DS-TP and Saratoga for deep space com-munications to control data transmissions in a rate-based manner. We suggest these new protocols should be evaluated to investigate the performance advan-tage of a rate-based transmission mechanism in the deep-space environment.

9.7 Acknowledgments

The authors would like to thank the guest editors, Drs. Athanasios Vasilakos, Yan Zhang, Thrasyvoulos Spyropoulos, and the anonymous reviewers for their constructive comments and suggestions.

The research work described in this chapter was performed in part at the JPL, California Institute of Technology, under a contract with the NASA, and was supported in part with the National Natural Science Foundation of China (NSFC) under Grant 61032003.

The authors would also like to thank Scott C. Burleigh, the globally well known expert and scholar in IPN/DTN and the primary developer of ION and CFDP, at JPL, for significant discussion and help in improving the quality of the chapter. This chapter could not be completed without his help.

Bibliography

[1] J. Jackson, "The interplanetary Internet," *IEEE Spectrum*, vol. 42, no. 8, August 2005, pp. 31–35.

[2] I. F. Akyildiz, O. B. Akan, C. Chen, J. Fang, and W. Su, "InterPlanetary Internet: State-of-the-art and research challenges," *Computer Networks Journal (Elsevier)*, vol. 43, no. 2, pp. 75–113, October 2003.

[3] R. Durst, G. Miller, and E. Travis, "TCP extensions for space communi-cation," *ACM/Kluwer WINET Journal*, vol. 3, no. 5, October 1997, pp. 389–403.

[4] J. Postel, "Transmission control protocolDDarpa Internet programDProtocol specification," *IETF Request for Comments RFC 793*, September 1981.

[5] S. Burleigh, A. Hooke, L. Torgerson, K. Fall, V. Cerf, R. Durst, K. Scott, and H. Weiss, "Delay-tolerant networking: an approach to interplanetary Internet," *IEEE Communications Magazine*, vol. 41, no. 6, June 2003, pp. 128–136.

[6] K. Fall, "A delay-tolerant network architecture for challenged internet," in *Proceedings of SIGCOMM'03*, Karlsruhe, Germany, August 2003.

[7] S. Burleigh and K. Scott, "Bundle protocol specification," *IETF Request for Comments RFC 5050*, November 2007.

[8] M. Demmer and J. Ott, "Delay tolerant networking TCP convergence layer protocol," *Internet Draft draft-irtf-dtnrg-tcp-clayer-01.txt*, February 2008.

[9] H. Kruse and S. Ostermann, "UDP Convergence Layers for the DTN Bundle and LTP Protocols," IETF DTNRG IRTF Research Group, draft-irtf-dtnrg-udp-clayer-00.txt (Work in Progress), November 2008.

[10] DTN Reference Implementation, October 2007 release, [Online]: http://www.dtnrg.org/wiki/Code.

[11] L. Wood, W. M. Eddy, W. Ivancic, J. McKim, and C. Jackson, "Saratoga: A delay-tolerant networking convergence layer with efficient link utilization," In *Proceedings of the International Workshop on Space and Satellite Communications (IWSSC)*, Salzburg, Austria, September 2007.

[12] S. Burleigh, M. Ramadas, and S. Farrell, "Licklider Transmission ProtocolD Motivation," *IETF Request for Comments RFC 5325*, September 2008.

[13] M. Ramadas, S. Burleigh and S. Farrell, "Licklider Transmission ProtocolDSpecification," *IETF Request for Comments RFC 5326*, September 2008.

[14] S. Farrell, M. Ramadas, and S. Burleigh, "Licklider Transmission ProtocolDSecurity Extensions," *IETF Request for Comments RFC 5327*, September 2008.

[15] S. Farrell and V. Cahill, "LTP-T: A Generic Delay Tolerant Transport Protocol," *Technical Report TCD-CS-2005-69*, Department of Computer Science, Trinity College Dublin, College Green, Dublin, Ireland, December 7, 2005.

[16] V. Cerf, S. Burleigh, A. Hooke, L. Torgerson, R. Durst, K. Scott, K. Fall, E. Travis, and H. Weiss, "Interplanetary Internet (IPN): Architectural definition," *IRTF*, May 2001.

[17] S. Jain, K. Fall, and R. Patra, "Routing in a delay tolerant network," in *Proceedings of ACM SIGCOMM'04*, August–September 2004.

[18] W. Zhao, M. Ammar, and E. Zegura, "Controlling the mobility of multiple data transport ferries in a delay-tolerant network," in *Proceedings of IEEE INFOCOM'05*, Miami, Florida, March 2005.

[19] S. Jain, M. Demmer, R. Patra, and K. Fall, "Using redundancy to cope with failures in a delay tolerant network," in *Proceedings of ACM SIGCOMM'05*, Philadelphia, Pennsylvania, August 2005.

[20] W. Zhao, M. Ammar, and E. Zegura, "Multicasting in delay tolerant networks: Semantic models and routing algorithms," in *Proceedings of ACM SIGCOMM'05 Workshop*, Philadelphia, Pennsylvania, August 2005.

[21] Y. Wang, S. Jain, M. Martonosi, and K. Fall, "Erasure-coding based routing for opportunistic networks," in *Proceedings of ACM SIGCOMM05 Workshop*, Philadelphia, Pennsylvania, August 2005.

[22] Z. Zhang, "Routing in intermittently connected mobile ad hoc networks and delay tolerant networks: overview and challenges," *IEEE Communications Survey and Tutorials*, vol. 8, no. 1, 1st Quarter 2006, pp. 24–37.

[23] X. Zhang, J. Kurose, B. Levine, D. Towsley, and H. Zhang, "Study of bus-based disruption-tolerant network: Mobility modeling and impact on routing," in *Proceedings of ACM MobiCom'07*, Montreal, Quebec, Canada, September 2007.

[24] E. Jones, L. Li, J. Schmidtke, and P. Ward, "Practical routing in delay-tolerant networks," *IEEE Transactions on Mobile Computing*, vol. 6, no. 8, August 2007, pp. 943–959.

[25] T. Spyropoulos, K. Psounis, and C. S. Raghavendra, "Efficient Routing in Intermittently Connected Mobile Networks: The Multiple-Copy Case," *IEEE/ACM Transactions on Networking*, vol. 16, no. 1, February 2008, pp. 77–90.

[26] A. Lindgren, A. Doria, E. Davies, and S. Grasic, "Probabilistic Routing Protocol for Intermittently Connected Networks," DTN Research Group InternetDraft, draft-irtf-dtnrg-prophet-02, March 2009.

[27] The Space Internetworking Strategy Group (SISG), "Recommendations on a strategy for space internetworking," Report of the Interagency Operations Advisory Group, July 28, 2008.

[28] K. Bhasin and J. Hayden, "Communication and Navigation Networks in Space System of Systems," *Chapter 15*, System of Systems Engineering, Wiley & Sons, New York, 2009, pp. 348–384.

[29] Consultative Committee for Space Data Systems, "Cislunar space internetworking architecture," *Draft Informational Report, CCSDS 730.1-G-0*, Draft Green Book, National Aeronautics and Space Administration, Washington, DC, USA, December 27, 2006.

[30] National Aeronautics and Space Administration (NASA), "NASA Space DTN Readiness Program," *Space Operations Mission Directorate SOMD Technology ProgramDProgram Plan: 2007–2010*, 2007.

[31] National Aeronautics and Space Administration (NASA) Tracking and Data Relay Satellite System (TDRSS) Overview.

[32] National Aeronautics and Space Administration, "Space Network Users Guide (SNUG)," *Rev. 8*, Goddard Space Flight Center, Greenbelt, Maryland, June 2002.

[33] H. Balakrishnan and V. Padmanabhan, "How network asymmetry affects TCP," *IEEE Communications Magazine*, vol. 39, no. 4, April 2001, pp. 60–67.

[34] E. Criscuolo, K. Hogie, and R. Parise, "Transport protocols and applications for Internet use in space," in *Proceedings of IEEE Aerospace Conference*, vol. 2, Big Sky, Montana, March 2001, pp. 951–962.

[35] K. Hogie, E. Criscuolo, and R. Parise, "Using standard Internet protocols and applications in space," *Computer Networks Journal*, vol. 47, no. 5, April 2005, pp. 603–650.

[36] Internet Research Task Force (IRTF) Delay Tolerant Networking Research Group (DTNRG).

[37] R. Wang, T. Taleb, A. Jamalipour, and B. Sun, "Protocols for reliable data transport in space Internet," *IEEE Communications Surveys and Tutorials*, vol. 11, no. 2, Second Quarter 2009, pp. 21–32.

[38] F. Warthman, "Delay-Tolerant Networks (DTNs): A tutorial," Wartham Associates, 2003.

[39] K. Fall and S. Farrell, "DTN: An Architectural Retrospective," *IEEE Journal on Selected Areas in Communications*, vol. 26, no. 5, June 2008, pp. 828–836.

[40] D. Kutscher, K. Loos, and J. Greifenberg, "Uni-DTN: A DTN convergence layer protocol for unidirectional transport," Internet Draft, draft-kutscher-dtnrg-uni-clayer-00.txt, April 2007.

[41] Consultative Committee for Space Data Systems, "CCSDS File Delivery Protocol (CFDP)Part 1: Introduction and overview," *Informational Report, CCSDS 720.1-G-3*, Green Book, National Aeronautics and Space Administration, Washington, DC, USA, April 2007.

[42] L. Larzon, M. Degermark, S. Pink, L. Jonsson, and G. Fairhurst, "The Lightweight User Datagram Protocol (UDP-Lite)," *IETF Request for Comments RFC 3828*, July 2004.

[43] I. Psaras, G. Papastergiou, V. Tsaoussidis, and N. Peccia, "DS-TP: Deep-Space Transport Protocol," in *Proceedings of IEEE Aerospace Conference 2008*, Big Sky, Montana, March 2008, pp. 1–13.

[44] Consultative Committee for Space Data Systems, "Rationale, scenarios, and requirements for DTN in space," Informational Report, CCSDS 734.0-G-1, Green Book, National Aeronautics and Space Administration, Washington, USA, August 2010.

[45] Consultative Committee for Space Data Systems, "Licklider Transmission Protocol (LTP) for CCSDS," Draft Recommended Standard, CCSDS 734.1-R-1, Draft Red Book, National Aeronautics and Space Administration, Washington, USA, February 2011.

[46] Interplanetary Overlay Network (ION), November 2009 release.

[47] W. Ivancic, W. Eddy, L. Wood, D. Stewart, C. Jackson, J. Northam, A. da S. Curiel, "Delay/Disruption-Tolerant Network Testing Using a LEO Satellite," in *Proceedings of the 8th Annual NASA Earth Science Technology Conference (ESTC)*, University of Maryland, June 2008.

[48] J. Wyatt, S. Burleigh, R. Jones, L. Torgerson, and S. Wissler, "Disruption tolerant networking flight validation experiment on NASAs EPOXI," in *Proceedings of the First International Conference on Advances in Satellite and Space Communications (SPACOMM) 2009*, Colmar, France, July 2009, pp. 187–196.

[49] V. Cerf, S. Burleigh, R. Jones, J. Wyatt, and A. Hooke, "First Deep Space Node on the Interplanetary Internet: the Deep Impact Networking Experiment (DINET)," presented at *Ground System Architectures Workshop 2009*, Torrance Marriott South Bay, Torrance, California, March 2009.

[50] L. Clare, "Delay/Disruption tolerant networking for space," presented at *Space-Enabled Global Communications & Electronic Systems Industry Update*, Cisco Systems, Irvine, California, August 2009.

[51] S. Horan and R. Wang, "Design of a space channel simulator using virtual instrumentation software," *IEEE Transactions on Instrumentation and Measurement*, vol. 51, no. 5, October 2002, pp. 912–916.

[52] R. Wang, B. Shrestha, X. Wu, T. Wang, A. Ayyagari, E. Tade, S. Horan, and J. Hou, "Unreliable CCSDS File Delivery Protocol (CFDP) over cislunar communication links," *IEEE Transactions on Aerospace and Electronic Systems*, vol. 46, no. 1, January 2010, pp. 147–169.

[53] R. Wang, X. Wu, T. Wang, X. Liu, and L. Zhou, "TCP Convergence Layer-based operation of DTN for long-delay cislunar communications," *IEEE Systems Journal*, vol. 4, No. 3, September 2010, pp. 385–395.

[54] R. Wang, X. Wu, Q. Zhang, T. Taleb, Z. Zhang, and J. Hou, "Experimental evaluation of TCP-based DTN for cislunar communications in presence of long link disruption," special issue on opportunistic and delay tolerant networks of *EURASIP Journal on Wireless Communications and Networking*, November 2010.

[55] R. Wang, S. Burleigh, P. Parik, C-J Lin, and B. Sun, "Licklider Transmission Protocol (LTP)-based DTN for cislunar communications," *IEEE/ACM Transactions on Networking*, vol. 19, No. 2, April 2011, pp. 359–368.

[56] R. Wang, Z. Wei, V. Dave, B. Ren, Q. Zhang, J. Hou, and L. Zhou, "Which DTN CLP is best for long-delay cislunar communications with channel-rate asymmetry?", *IEEE Wireless Communications* (in press).

[57] Extending Internet into Space: ESA/ESOC DTN/IP Testbed Implementation and Evaluation: COMNET Research Group, Democritus University of Thrace, Greece, November 2009 release.

[58] L. Wood, W. Eddy, and P. Holliday, "A Bundle of Problems," In *Proceedings of IEEE Aerospace Conference* 2009, Big Sky, Montana, March 2009, pp. 1–17.

[59] B. Walker, M. Tsuru, A. Caro, A, Keranen, J. Ott, T. Karkkainen, S. Yamamura, and A. Nagata, "Panel: The state of DTN evaluation," In *Proceedings of the ACM MobiCom Workshop on Challenged Networks* (CHANTS2010), Chicago, IL, September 2010, pp. 29–30.

[60] R. Wang, B. Gutha, and Paradesh Kumar Rapet, "Window-based and rate-based transmission control mechanisms over space-Internet links," *IEEE Transactions on Aerospace and Electronic Systems*, vol. 44, No. 1, January 2008, pp. 157–170.

Chapter 10

DTN and Satellite Communications

Carlo Caini and Rosario Firrincieli

10.1	Introduction	284
10.2	Bundle Layer DTN Architecture Overview	286
	10.2.1 End-to-End Reliability and Custody Transfer Option	287
	10.2.2 DTN Concept Development	288
10.3	Satellite Channel Impairments and Possible Solutions	288
	10.3.1 End-to-End TCP Enhancements and Other Transport Protocols for Satellite and Space Communications	289
	10.3.2 Performance Enhancing Proxies (or "Protocol Accelerators")	290
	10.3.3 The Delay/Disruption Tolerant Networking Alternative	292
10.4	Scenario 1: Satellite Systems with End-to-End Continuous Connectivity	294
	10.4.1 Testbed and Tools Used in Performance Evaluations	294
	10.4.1.1 The TATPA Testbed	294
	10.4.1.2 The DTNperf_2 tool	295
	10.4.2 Performance Evaluation	296
	10.4.2.1 Performance in the Presence of Congestion	296
	10.4.2.2 Performance in the Presence of Link Losses	297
	10.4.2.3 Performance in the Presence of Congestion and Link Losses	298
	10.4.2.4 Sending Window Performance Improvement	298

10.5   Scenario 2: Satellite Systems with Random Intermittent
       End-to-End Connectivity 300
       10.5.1   TCP Resilience to Disruption 301
       10.5.2   Analysis of DTN Behavior in the Presence of
               Disruption 302
               10.5.2.1   Hop-by-hop DTN Micro-Analysis: DTN
                       "Link Class" Retransmission Timers
                       and Backoff Policy 303
               10.5.2.2   End-to-End DTN Micro-Analysis: Bundle
                       Transmission Process from DTN Sender
                       to DTN Receiver 306
       10.5.3   Performance Evaluation in a Realistic Case Study:
               Railway Satellite Communications 308
               10.5.3.1   Performance in the Absence of Competing
                       Traffic 309
               10.5.3.2   Performance in the Presence of Competing
                       Traffic 310
10.6   Scenarios 3 and 4 312
       10.6.1   Scenario 3: Satellite Systems with Scheduled
               Intermittent End-to-End-Connectivity 312
       10.6.2   Scenario 4: Satellite Systems without End-to-End
               Connectivity 312
10.7   Ongoing Research 313
       10.7.1   NASA JPL October 2008 Experiments 313
       10.7.2   DTN Application to a Military Heterogeneous
               Network 314
       10.7.3   DTN Application to the UK-DMC LEO System 314
10.8   Conclusions 315

10.1   Introduction

Satellite communications are an interesting and promising application field for Delay and Disruption Tolerant Networking (DTN). Although primarily conceived for deep space communications and sensor networks, DTN was immediately recognized as applicable to satellite environments, in particular to cope with the intermittent channels typical of LEO (Low Earth Orbit) constellation satellite systems. However, DTN applicability to the satellite field is not limited to LEOs, but extends to other systems, like GEOs (Geostationary Earth Orbit), where performance of reliable transport protocols, such as TCP, is impaired by the characteristics of the satellite channel itself. Even in the most favorable case of continuous end-to-end connectivity, long delays inherent in GEO systems, together with the possible presence of a high Packet Error Rate (PER) due to the wireless channel, severely affect performance. The most common countermeasure is the insertion of intermediate agents, called PEPs (Performance Enhancing Proxies), which generally offer good performance but violate the end-to-end semantics of transport protocols. Here too DTNs can have an important role.

The possible applications of the DTN concept in satellite communications are various and can be conveniently classified on the basis of the presence or absence of end-to-end connectivity:

- continuous end-to-end connectivity; end-to-end connectivity is usually present, this may be the case with GEO satellite systems with terrestrial fixed terminals;

- random intermittent end-to-end connectivity; end-to-end connectivity may be present but channel disruptions are frequent, and difficult, or impossible, to predict; for example, a GEO satellite connecting mobile terminals on means of transport, where satellite link is disrupted by tunnels or other obstructions;

- scheduled intermittent end-to-end connectivity; end-to-end connectivity is assured at regular intervals; for example, a LEO satellite for Earth observation, which can connect to its gateway stations only at intermittent but predictable intervals due to its orbital motion;

- no end-to-end connectivity; there is never end-to-end connectivity between end-points; for example, a single LEO satellite acting as a "mule" between a terrestrial sensor network and a remote satellite gateway station, which are never in the satellite coverage area at the same time.

These scenarios will be investigated in next sections. However, here let us briefly anticipate the reasons why DTN can be more advantageous with respect to other approaches, like end-to-end transport protocols or PEPs. To this end, let us re-consider the four above scenarios, bearing in mind that the more challenging a scenario, the more suitable the DTN concept.

The first scenario (continuous end-to-end connectivity) is the most interesting to investigate, as in it the advantage of the DTN approach is far from obvious. DTN alternatives, and in particular PEPs, offer good performance and are in fact widely and successfully applied. However, as it will be shown later, DTN can be helpful in this case too, as it can offer the same level of performance as PEPs, with the additional advantage of avoiding any violation of end-to-end semantics.

In the second scenario end-to-end connectivity is still present but disruptions are frequent. Here too the usual solutions, namely, end-to-end transport protocols and PEPs, are not prevented from working. Yet, the presence of disruptions actually makes DTN more appealing than in the previous case. Performance assessment, however, is as demanding as before, because the DTN advantage is highly dependent on disruption characteristics (e.g. frequency, length) and on the DTN implementation used.

The third scenario (intermittent end-to-end connectivity) is still more suited to DTN, as end-to-end transfers are now possible, but only at scheduled times and for a limited period. This latter restriction limits the total amount of data (or "contact volume") that can be transferred at each availability interval ("contact time"). Files exceeding the contact volume must be divided into multiple parts. This is a task that the DTN bundle protocol can perform automatically thanks to its proactive and reactive fragmentation features.

The last scenario is the most challenging from a communications point of view and therefore benefits the most from DTN. The absence of end-to-end connectivity prevents the establishment of TCP, or TCP-like, connections and even unreliable UDP transfers are impossible, due to the lack of a continuous path between end-nodes. Here the only possible approach is the application of DTN "store-and forward" techniques. Data must first be transferred to intermediate nodes, then, when possible, transferred to the receiver. This task can easily be accomplished by DTN, unlike PEPs or end-to-end transport protocols.

In the rest of the chapter we will present a brief overview of bundle layer DTN architecture and a summary of satellite impairments and possible solutions. Then, advantages and disadvantages of DTN will be discussed in detail for the first two scenarios, which are the most interesting to investigate, as here the DTN benefit is more uncertain. To assess this, comparison will be made with present solutions, supported, whenever possible, by data collected through testbeds working on real DTN implementations. The other two scenarios will also be considered, but in less detail. Finally, a survey of ongoing research involving DTN and satellites will conclude the chapter.

10.2   Bundle Layer DTN Architecture Overview

Bundle layer DTN architecture [1], [2], [3] aims to support communication in challenged environments by adding the new "bundle layer" between transport and application layers. The bundle protocol can interface with different transport protocols through "convergence layer adapters." This new architecture is particularly suitable as an overlay on top of a heterogeneous network consisting of many homogeneous segments, such as wired Internet, satellite links, wireless LANs, etc. By installing the DTN bundle protocol on end-points and on nodes at the border of homogeneous segments, the end-to-end transport protocol features are redefined (Figure 10.1). In fact, end-to-end data transfer across the heterogeneous network is now provided by the bundle protocol, which exchanges large data packets, called "bundles," between DTN nodes through a store-and-forward relay. End-to-end transport protocol features are now confined inside one DTN hop, which may coincide, as in Figure 10.1, with a homogeneous network segment (A, B, and C). The advantage is that in this architecture different transport protocols (but also different network stacks) can be used in different network segments. In this way it is possible to use

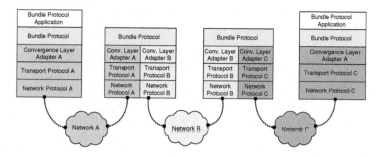

Figure 10.1: Bundle protocol DTN architecture and protocol stack.

protocols (or versions of the same protocol specifically designed) to tackle the specific characteristics of different network segments (e.g., a satellite channel or a sensor network) [4], [5]. Note that from this point of view, DTN architecture resembles and extends the TCP connection splitting technique used in PEPs [6]. Similarities and differences between DTN and PEPs will be further discussed below.

10.2.1 End-to-End Reliability and Custody Transfer Option

End-to-end reliability at bundle layer can be provided by selecting a reliable transport protocol, like TCP, at DTN nodes, and/or through two optional mechanisms: end-to-end acknowledgments and custody transfer. Applications are left free to exploit the former, which is just a signaling process (like parcel tracking by a courier), while the latter implies persistent storage (e.g., on hard disks) and bundle retransmission capabilities at some selected DTN nodes. Quoting RFC 4838 [2]: 'Transmission of bundles with the custody transfer requested option specified generally involves moving the responsibility for reliable delivery of bundles among different DTN nodes in the network. For unicast delivery, this will typically involve moving bundles "closer" (in terms of some routing metric) to their ultimate destination(s), and retransmitting when necessary. The nodes receiving these bundles along the way (and agreeing to accept the reliable delivery responsibility) are called "custodians." The movement of a bundle (and its delivery responsibility) from one node to another is called a "custody transfer." Note that in the case of link disruption, DTN bundles continue to be safely stored at DTN custodians in local databases, until the next hop becomes reachable again. Bundles are deleted from custodian databases once the transfer of their custody is explicitly acknowledged by another custodian or by the destination node, or their lifetime expires. As shown later, this feature can offer significant advantages in the presence of discontinuous or disruptive channels. The interested reader is referred to [7] for an exhaustive discussion of the many implications of the custody transfer option.

10.2.2   DTN Concept Development

DTN was first conceived within the Inter-Planetary Network Research Group (IPNRG) of the Internet Research Task Force (IRTF), to address the problems of space communications. Later, its scope was enlarged to encompass all challenged networks, whether spatial or terrestrial. To this end, in 2002 the IRTF Delay Tolerant Networks Research Group (DTNRG) was established to promote DTN. An interesting survey of DTN motivation and development, written by the two DTNRG leaders, can be found in [1], while for further information on DTN origin and development see [8]. The DTNRG website [9] is the most authoritative source of DTN documentation and software. API DTN bundle protocol reference implementation (DTN2) is freely available from [10]. Other DTN applications are also available, including DTNperf_2 [11], a DTN performance evaluation tool co-developed by the authors and used in the numerical evaluations presented in this paper (the latest release can be downloaded from [12]). Finally note that, although the bundle layer is the most important DTN architecture, it is not the sole possible option, as documented on the DTNRG Web site.

10.3   Satellite Channel Impairments and Possible Solutions

Satellite systems offer several advantages with respect to terrestrial counterparts, such as fast deployment and effective coverage of wide areas. Hence, they can play an important part in reducing the digital divide, by offering fast and ubiquitous access to many important services, including Internet. However, satellite links pose some significant problems, not only at physical but also at transport layer [13]. In particular, the TCP protocol [14], [15] is severely affected by the following impairments: long RTTs, possible presence of segment losses not due to congestion, and possible link disruption (especially with mobile satellite terminals). Here we will focus on the first two. First, the RTT may be greatly increased by the long propagation delay of the satellite leg (especially in GEO systems, where the RTT is about 600 ms). This greatly reduces TCP ability to open the congestion window (cwnd), with severe impact on: start-up performance (i.e. the satellite TCP connection is unable to exploit the bandwidth made available by lower layers); fairness (the satellite connection is severely penalized by competing short RTT wired traffic); and ability to recover fast from cwnd reductions caused by segment losses.

Second, the satellite channel, like all wireless channels, may be affected by packet losses, as packets that fail error control checks due to physical channel errors are discarded. The impact of this on performance depends on the robustness of link design and on the severity of propagation impairments. As it is very difficult to disambiguate the origin of segment losses (whether due to congestion or link errors), they are all ascribed by TCP to congestion; in order to reduce it, the standard TCP [16] roughly halves the cwnd, which in turn results in a halving of the segment transmission rate. This conservative choice is justified in wired networks, where bit errors are infrequent. However, in more error-prone environments, like wireless networks, a significant PER may cause many unnecessary, and actually harmful, cwnd halvings. In satellite communications the consequences of losses not due to congestion are further worsened by long RTTs, which prevent fast cwnd reopening after recovery.

All these impairments require proper countermeasures [13]. For clarity we distinguish between three possible approaches. The first, the end-to-end approach, uses enhanced versions of TCP or alternative transport protocols. The second relies on the introduction of PEPs [6], which are widely adopted commercially, as they provide good performance and are also "transparent" to end users. In this case, we focus on solutions that implement the "TCP splitting" technique. Last, but of particular interest here, comes the DTN approach, which will be treated separately. Note that the first approach, dealing with transport protocols, is not necessarily an alternative to the other two, which, by contrast, introduce enhancements in the network architecture. In particular, both PEPs and DTN can benefit from the introduction of a transport protocol optimized against satellite impairments.

10.3.1 End-to-End TCP Enhancements and Other Transport Protocols for Satellite and Space Communications

In recent years, the number of proposed TCP congestion control variants has constantly increased, encouraged by the widely accepted opinion that Van Jacobson's (VJ) original algorithm [15] is not well suited to both high-speed networks and wireless environments. Among the many TCP enhancements compatible with NewReno [16], here we mention TCP Hybla, because it was conceived with the primary aim of counteracting long RTTs typical of satellite connections. Hybla [17] consists of a set of enhancements that includes a modified congestion control algorithm, the mandatory adoption of the SACK [18] and timestamp options, the introduction of "packet pacing," and the use of Hoe's algorithm for initial bandwidth estimation. In particular, Hybla congestion control differs from VJ's in that the additive cwnd increase is not fixed, but parameterized to the minimum RTT of a reference connection in order to speed up cwnd growth rate. This feature is particularly beneficial in heterogeneous networks, as it allows long RTT satellite connections to compete effectively with wired cross traffic. However, it also provides performance improvements at connection start-up and in the presence of PER. Comparative evaluations of different TCP enhancements in satellite environments can be found in [19], [20].

When the satellite link can be isolated from the rest of the network (e.g., in PEP and DTN architectures; see below), the possibility of employing transport protocol other than TCP should also be considered. Among the many proposals, here we mention SCPS-TP (Space Communications Protocol Standard-Transport Protocols) [21], LTP (Licklider Transmission Protocol) [22], [23], [24], Saratoga [25], and NORM (Nack-Oriented Reliable Multicast) [26]. SCPS-TP is actually more a suite of TCP extensions than an entirely new protocol. It was designed and standardized by CCSDS (Consultative Committee for Space Data Systems, a committee formed by the major space agencies of the world) to allow communication over potentially unreliable space data transmission paths, but it can also be applied in other challenging environments. The main difference from previous TCP enhancements is that some SCPS-TP features are not compatible with standard TCP nodes, like SNAKs (Selective Negative ACK). For this reason SCPS-TP is widely adopted in distributed PEPs where the satellite link is isolated at both ends. LTP is a point-to-point protocol designed to provide retransmission-based reliability over deep-space links, whose design was based on experience with the CCSDS File Delivery Protocol (CFDP). LTP is worth mentioning here because in the DTN bundling protocol stack it can serve as a reliable convergence layer for deep-space point-to-point links. Saratoga is a transfer protocol built on UDP and UDP-Lite, originally developed to efficiently transfer remote-sensing images from a LEO constellation [25]. It can be used as a file transfer protocol to move files or streaming data between peers that may have only intermittent connectivity. Like LTP, Saratoga too can be used in the bundle protocol stack as a convergence layer [27]. Finally, NORM was designed for end-to-end reliable multicast, but can also be used as a unicast transport protocol in environments with long delays and high PER. It relies on selective negative acknowledgments and provides for the use of packet-level forward error correcting codes.

10.3.2 Performance Enhancing Proxies (or "Protocol Accelerators")

A different approach, based on a network architecture modification, is followed by "protocol accelerators" or PEPs [6], [28] and in particular by PEPs based on the TCP splitting technique. They aim to isolate the satellite segment from the rest of the network by inserting one (integrated PEP) or two (distributed PEP) intermediate agents. In the distributed PEPs, the most widely adopted architecture, the TCP connection established between end hosts is split into three connections by the insertion of two intermediate nodes (Figure 10.2). Thanks to PEP, the impairments introduced by the satellite leg (long RTTs and random losses) are removed from both the first and last connections, which become short-RTT and basically error free, and thus can continue to use a standard TCP variant (PEP is transparent to end users). By contrast, for the satellite connection it is more advantageous to use a TCP variant

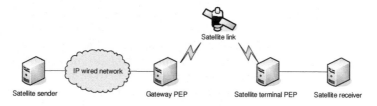

Figure 10.2: Distributed PEP architecture layout.

specifically designed for satellite environments, or even a different transport protocol (e.g., SCPS-TP or proprietary solutions).

The integrated PEP differs from the previous scheme because of the presence of just one intermediate node, which splits the original end-to-end connection into only two connections (Figure 10.3). The differences between the two are subtle. One advantage of distributed PEP is complete freedom of choice in selecting the transport protocol between the two PEP devices. In integrated PEP, this choice is limited to TCP variants compatible with a standard TCP receiver, which is however true for most TCP variants currently available on real operating systems, including Hybla. If this optimized TCP is adopted on the intermediate PEP only, performance improvement can be obtained only in the forward link (i.e. from the server to the satellite client), which is of course the most important one, at least for surfing the Web or downloading files. On the other hand, the integrated solution is easier to update and maintain, as well as less expensive, as a PEP device is not required at the receiver side. As an example of integrated PEP, we cite PEPsal, an Open Source (GNU license) TCP splitting implementation for Linux [29], [30]. PEPsal allows the user to select the version to be adopted on the satellite connection from among all the TCP versions implemented in the Linux kernel, including Hybla. Note that PEPsal does not require any modification to either end host, an advantage in common with distributed architectures.

Despite the very good performance, PEPs present some problems. First, the splitting technique violates end-to-end TCP semantics: the sender receives acknowledgments from an intermediate agent instead of the final receiver. Sec-

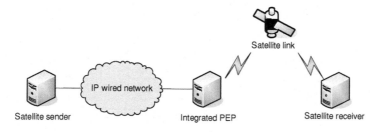

Figure 10.3: Integrated PEP architecture layout.

ond, the intermediate agent needs to access the TCP header of the packets in order to send back ACKs and perform all its optimization procedures, but this is incompatible with the use of IPsec technique, which, by encrypting the IP payload, would make the TCP header unavailable to intermediate agents. If IPsec is active, PEPs will allow encrypted packets to transit, without splitting the TCP connection and therefore without achieving any performance improvement.

10.3.3 The Delay/Disruption Tolerant Networking Alternative

Bundle layer DTN architecture [2], [3], although not specifically conceived for satellite communications, offers a promising alternative to previous approaches, and can be seen as a direct extension of the PEP TCP splitting concept. To illustrate this fundamental point, let us again consider the integrated PEP in Figure 10.3, and assume that PEPsal is used on the intermediate node. The corresponding protocol stack is given in Figure 10.4. The original end-to-end satellite connection is split into two TCP connections, a wired one, from the satellite sender to the PEP, and a satellite one, from the PEP to the satellite receiver. On PEPsal Hybla is selected, while end-nodes continue to use a standard TCP. If we focus on the forward link on the wired and the satellite connections, two TCP versions are in use: TCP A (standard) and TCP B (Hybla), respectively.

Alternatively, the DTN bundle protocol can be installed on three nodes, the two end-nodes and the intermediate node (instead of the integrated PEP), as shown in Figure 10.5. As we have two DTN hops, we can still distinguish between a wired and a satellite connection, where two different transport protocols can be used, Transport A and Transport B (Figure 10.6). Maximum correspondence between these PEP and DTN architectures is reached when TCP A and TCP B are adopted as Transport A and B.

Despite the evident similarity, there are still significant differences between the two architectures. First, from a formal point of view, DTN avoids any infringement of transport layer end-to-end semantics, thanks to the implicit

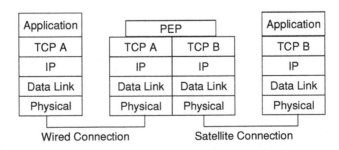

Figure 10.4: PEPsal protocol stack (integrated PEP architecture).

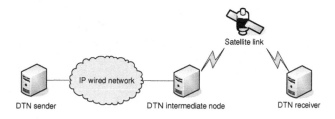

Figure 10.5: DTN architecture reference layout.

Application			Application
Bundle Layer			
Transport A	Transport A	Transport B	Transport B
IP	IP	IP	IP
Data Link	Data Link	Data Link	Data Link
Physical	Physical	Physical	Physical

Wired Connection   Satellite Connection

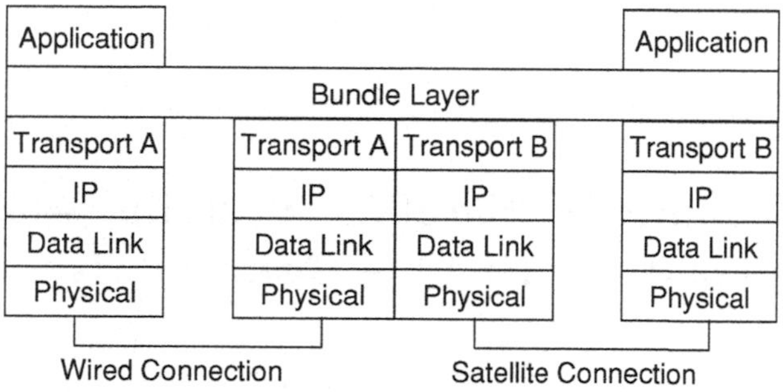

Figure 10.6: DTN protocol stack (DTN architecture reference layout).

redefinition of end-to-end connectivity at bundle layer. Second, data transfer between DTN nodes can start even in the absence of continuous end-to-end connectivity, which would be impossible for PEPs. Third, resilience to disruptions may be greatly increased by exploiting bundle protocol reliability features, like custody transfer. On the negative side, DTN software must be installed on end hosts and on every intermediate node that wants to offer DTN services (i.e., DTN architecture is not transparent to end hosts). Moreover, the bundle protocol necessarily introduces a transmission overhead, whose impact on performance must be evaluated.

Before concluding, let us extend our comparison to more complex heterogeneous networks, consisting of many homogeneous segments (e.g. wired, wireless, satellite, etc.). While PEPs are designed to cope with the specific impairments of a particular segment (isolated from the rest of the network by means of the effective but also formally questionable TCP splitting technique), DTN architecture intrinsically enables the use of a different optimized transport protocol on each homogeneous segment. In view of this, the authors believe that DTN can be considered an effective, as well as formally elegant, extension of the TCP splitting concept.

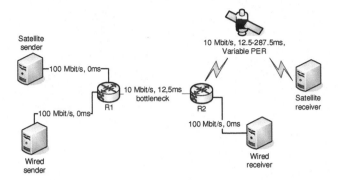

Figure 10.7: Layout of the TATPA testbed used for performance evaluation.

10.4 Scenario 1: Satellite Systems with End-to-End Continuous Connectivity

This scenario, characterized by the continuous presence of an end-to-end path, is the most uncertain with respect to the DTN approach, as both end-to-end transport protocols and PEPs are not only possible, but are also currently the usual solution. However, we will show that DTN can offer the same performance as PEPs, without infringing on end-to-end semantics. Before examining a selection of performance evaluations presented in [4], let us briefly describe the characteristics of the testbed and tools used.

10.4.1 Testbed and Tools Used in Performance Evaluations

Performance evaluations presented in this chapter were carried out using the TATPA testbed (Testbed on Advanced Transport Protocols and Architectures) [31]. An essential role was also played by the DTNperf_2 tool.

10.4.1.1 The TATPA Testbed

This testbed aims to reproduce the essential characteristics of heterogeneous networks that include satellite links (see Figure 10.7). TATPA consists of a set of Linux PCs, patched with the Multi TCP package [32] to add "packet pacing" and Hoe's algorithm to the Hybla congestion control, which is already included in the official Linux kernel. Moreover, Multi TCP allows users to collect logs of TCP internal variables otherwise not directly available (like cwnd and RTO). PEPsal can be optionally enabled on the router R2. The DTN2 reference implementation [10] of the bundle protocol is installed on three nodes: satellite sender, satellite receiver, and R2, which acts in DTN experiments as a DTN intermediate node. For the sake of fairness in comparison with other approaches, we adopted TCP in both DTN hops, reproducing the same layout presented in Figure 10.5 and using the same protocol stack given in Figure 10.6 (NewReno on the wired connection and Hybla on the satellite one). In the testbed we can distinguish between satellite connections (from satellite sender to satellite receiver), composed of both wired and satellite legs, and wired connections (from wired sender to wired receiver), composed of entirely wired paths. This allows us to emulate not only pure satellite environments, but also more complex heterogeneous networks. All the connections share the R1-R2 link, whose bandwidth has been deliberately limited to 10 Mbit/s in order to study congestion effects. The router R1 at the bottleneck input, where all the congestion events are confined, uses a RED (Random Early Detection) queue policy. The satellite channel emulator is based on the NistNet package [33], which allows the user to set the PER and the additional delay due to satellite link. On the satellite emulator a tool specifically developed to emulate random and deterministic disruptions is also installed.

DTN performance was evaluated using DTNperf_2 (described below). Linux TCP defaults were used when not otherwise specified, with a Maximum Segment Size of 1448 Bytes. The DTN bundle size was set to 1 MB (i.e., a value larger than the Bandwidth Delay Product, or BDP, of the satellite channel) and the DTN custody transfer option was enabled. The DTN routing configuration used in our experiments is given in Table 10.1. Although a discussion about DTN configuration parameters would be beyond the scope of this section, note, however, that with the configuration used (i.e. ONDEMAND) two (bidirectional) TCP connections for each DTN node pair are opened by the bundle protocol. We will see later how it is possible to exploit this effect.

Table 10.1: DTN Routing Configuration Adopted

Satellite Sender (SS)	R2 (Gateway)	Satellite Receiver (SR)
link add Gateway IP_R2:4556 ONDEMAND tcp	link add Sender IP_SS:4556 ONDEMAND tcp	link add Gateway IP_R2:4556 ONDEMAND tcp
	link add Receiver IP_SR:4556 ONDEMAND tcp	
route add SR.dtn Gateway	route add SS.dtn Sender	route add SS.dtn Gateway
route add G.dtn Gateway	route add SR.dtn Receiver	route add G.dtn Gateway

10.4.1.2   The DTNperf_2 tool

To assess DTN performance we used the Open Source (GNU license) DTNperf_2 tool, a client-server application specifically designed to evaluate goodput (average number of bits acknowledged per second) and collect logs in DTN architectures. This tool is an enhanced version of DTNperf, which was named after the famous Iperf tool [34] (used in TATPA to evaluate end-to-end TCP and PEP performance). DTNperf is included in the official suite released by DTNRG and the DTNperf_2 enhanced version (used in our tests) can be downloaded from [12], which host the latest ("bleeding edge") versions of DTN2 code. Among the most important new DTNperf_2 features, we cite the introduction of a bundle sending window, W, to overcome the throughput limitation of one bundle per RTT of the first version. This window allows W bundles to be in flight (sent but not yet acknowledged) at the same time. By default, bundle acknowledgments are given by "status delivered" bundle reports (sent by the DTN destination node). However, by enabling the SONC (Send ON Custody) option, also "custody accepted" reports sent by the first DTN intermediate node can be interpreted as bundle acknowledgments, which may be useful in coping with disruptive channels. Another important feature widely used in our tests is the possibility of logging all status reports exchanged during a bundle flow. The interested reader is referred to [11] for further information.

10.4.2   Performance Evaluation

In this section various satellite environments are considered, all characterized by continuous end-to-end path availability. In particular, DTN performance is compared with end-to-end Hybla and PEPsal, but also, for the sake of completeness, with end-to-end NewReno and SACK (henceforth jointly denoted as NewReno/SACK, because of their performance equivalence). With regard to this, we would mention a minor modification in the DTN TCP convergence layer adapter (we disabled the use of segment sizes larger than MTU), introduced in order to assure full compatibility with adopted TCP variants. The DTNperf_2 sending window, W, is set to one when not diversely specified. Most of the DTN results presented below are taken from [4], but new data, referring to the benefits of $W > 1$ will also be given. Results are always averaged over multiple runs.

10.4.2.1   Performance in the Presence of Congestion

Because of the RTT-unfairness of TCP, long RTT satellite connections are severely penalized by the presence of competing short RTT wired connections. To illustrate the problem, Figure 10.8 shows the goodput of a satellite connection, supposed error free ($PER = 0\%$) and with a variable RTT, in the presence of 5 wired connections (NewReno/SACK, $RTT = 25ms$) on the R1-R2 bottleneck. The goodput of these background connections (not reported) is always close to the maximum fair share (i.e. the bottleneck bandwidth divided by the number of competing connections, 1.66 Mbit/s here). As expected, with TCP NewReno/SACK satellite performance shows rapid degradation with in-

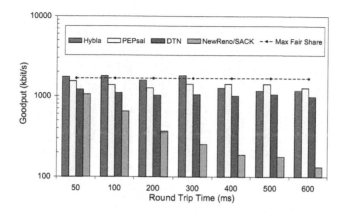

Figure 10.8: Performance comparison in the presence of congestion on the R1-R2 bottleneck: goodput of a satellite connection vs. its RTT; 5 background wired TCP connections ($RTT = 25ms$, NewReno). From [4].

creasing RTT, with totally unsatisfactory performance for the RTTs typical of GEO satellites (about 600 ms). By contrast, end-to-end Hybla, being specifically designed to counteract the RTT-unfairness problem, offers much better performance. PEPsal is able to obtain the same results as Hybla. If we consider DTN performance, similar observations can be made. DTN, as PEPsal, splits the original connection into a pair of connections (although at the bundle layer), achieving the same advantages provided by the splitting approach. Despite the limitations introduced by conservatively setting the DTNperf_2 sending window, W, to one, DTN performance is only marginally inferior to that obtained by PEPsal or end-to-end Hybla, especially for long RTTs. The choice of a bundle size larger than the satellite link BDP contributes to this result and also helps in making the DTN overhead impact negligible.

10.4.2.2 Performance in the Presence of Link Losses

In error-prone environments, the inability of TCP to distinguish the origin of packet losses may severely interfere with the congestion control mechanism, especially in the presence of long RTTs. This behavior is clear if we examine the performance of NewReno/SACK given in Figure 10.9, which refers to a single satellite connection without any background traffic. Having eliminated any form of congestion, all the losses are due to the non ideal satellite channel ($PER = 1\%$). Here too the improvement obtained by end-to-end Hybla and PEPsal, whose performance is basically the same, is remarkable. If we consider DTN, performance is still much higher than with end-to-end NewReno/SACK, but we can observe a slight drop in performance with respect to end-to-end Hybla and PEPsal. This reduction, however, depends on having set W to one, as it will be shown below.

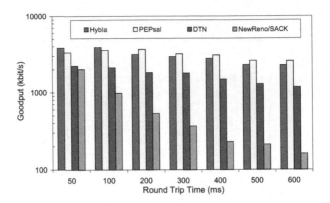

Figure 10.9: Performance comparison in the presence of random losses on the satellite channel ($PER = 1\%$): goodput of a satellite connection vs. its RTT; no background TCP connections. From [4].

10.4.2.3   Performance in the Presence of Congestion and Link Losses

The simultaneous presence of congestion and PER leads to the scenario considered in Figure 10.10. By comparing it with Figure 10.8 (congestion-only case), we can observe a predictable performance worsening, which leaves, however, the qualitative terms of comparison between the different possible approaches substantially unchanged.

10.4.2.4   Sending Window Performance Improvement

Let us introduce the problem addressed by the DTNperf_2 sending window by considering the example reported in Figure 10.11, which refers to the same scenario as Figure 10.9 ($PER = 1\%$ on the satellite leg, no congestion). Three time-sequences (obtained from DTNperf_2 logs) are plotted: "status sent," which denotes the beginning of the bundle transfer from the satellite sender; "R2 status received," the R2 acknowledgment of bundle reception; "R2 status forwarded," and the R2 signaling of the end of the forwarding process. Note that, in the sending window absence (which is equivalent to $W = 1$), a new bundle cannot be sent by the satellite sender before the acknowledgment of the previous one. This results in an R2 idle time (1180 ms, see Figure 10.11), during which R2 has no bundle to send, waiting for the complete reception of the next bundle from the satellite sender, thus causing underutilization of the satellite channel. This problem can be avoided by setting $W > 1$, i.e. by allowing more than one bundle in flight, in order to continuously feed R2. This, moreover, actually allows the DTN intermediate node R2 to exploit the two TCP connections opened between R2 and the satellite receiver when the DTN configuration presented in Table 10.1 is used. In other words, $W > 1$

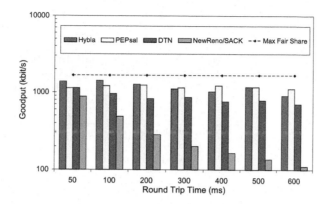

Figure 10.10: Performance comparison in the presence of both congestion on the R1-R2 bottleneck and random losses ($PER = 1\%$) on the satellite channel: goodput of a satellite connection vs. its RTT; 5 background wired TCP connections ($RTT = 25ms$, NewReno). From [4].

allows R2 to send two bundles in parallel on both the opened connections, with an additional performance advantage. To assess the two effects separately, it is possible to open just one TCP connection on each DTN hop, by adopting the alternative DTN configuration given in Table 10.2.

Results achievable in the same scenario as before ($PER = 1\%$ on the satellite leg, no congestion) by increasing W are presented in Figure 10.12. for both one or two TCP connections opened. Note that when just one TCP connection is opened, $W = 2$ is enough to achieve most of the gain. In the

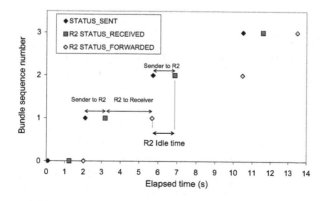

Figure 10.11: Bundle time-sequences without DTNperf_2 sending window: R2 idle time. Random losses on the satellite channel ($PER = 1\%$), no background TCP connections. From [4].

Table 10.2: Alternative DTN Routing Configuration (Single TCP Connections on DTN Hops)

Satellite Sender (SS)	R2 (Gateway)	Satellite Receiver (SR)
link add Gateway IP_R2:4556 ALWAYSON tcp	link add Receiver IP_SR:4556 ALWAYSON tcp	link add Gateway IP_R2:4556 OPPORTUNISTIC tcp remote_eid=G.dtn
	link add Sender IP_SS:4556 OPPORTUNISTIC tcp remote_eid=SS.dtn	
route add SR.dtn Gateway	route add SS.dtn Sender	route add SS.dtn Gateway
route add G.dtn Gateway	route add SR.dtn Receiver	route add G.dtn Gateway

case of two TCP connections opened, a gain close to saturation requires $W \geq 4$, because the two parallel TCP connections between R2 and the satellite receiver need twice the number of bundles as before. The advantage of $W > 1$ also is found in the case of congestion on the R1-R2 bottleneck. Note, however, that in our experiments the second parallel TCP connection between the sender and R2 (if present) is not actually used and $W > 1$ benefits are therefore limited to continuous feeding of the bottleneck. As a final remark, let us point out that even in the conservative case of a single TCP connection, for $W > 2$ the limited DTN performance gap with PEPsal performance (the best non-DTN solution) is made null (see Figure 10.12). This leads to the fundamental conclusion that, in case of continuous channel availability, there is no penalization with respect to the best non-DTN solution in using DTN.

10.5   Scenario 2: Satellite Systems with Random Intermittent End-to-End Connectivity

In this scenario, there is still end-to-end connectivity but disruptions are frequent and difficult to predict. Although end-to-end transport protocols and PEPs are not prevented from working, here the advantages of DTN are more evident than before. As a realistic case study, at the end of this section we will consider a railway scenario, where train passengers are connected to the Internet through a GEO satellite link and disruptions are caused by tunnels. However, first we provide the reader with a review of TCP and DTN bundle protocol behavior in the case of disruption. Although this part (namely, Sections 10.5.1 and 10.5.2, partially derived from [35]) is necessary to explain the superior performance of DTN, it is also quite complex and necessarily

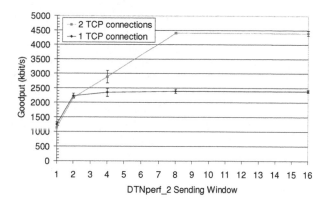

Figure 10.12: Goodput as a function of the DTNperf_2 sending window W. Random losses on the satellite channel ($PER = 1\%$), no background TCP connections.

detailed. It can be skipped by the reader mainly interested in the experiment results (Section 10.5.3, partially derived from [36]).

10.5.1   TCP Resilience to Disruption

Although not conceived to cope with disruptive channels, TCP itself is relatively robust. This robustness, however, largely depends on TCP implementation (RFC 2988 [37] allows great freedom) and on the setting of some retransmission parameters. In particular, an established TCP connection can be closed, in the absence of any network feedback, either when a timer expires, or after a given number of unsuccessful retransmission retries; this choice is left to the implementer. In Linux, the latter alternative is pursued and the default maximum number of retries is 15. This figure can be easily modified by the user (TCP_RETR2 sysctl parameter), which makes actual disruption resilience highly configurable. Moreover, while the original retransmission mechanism of TCP doubled the retransmission timeout interval (RTO) at each retry (exponential backoff algorithm), RFC 2988 [37] allows users to limit the maximum RTO. In Linux this is set to 120 s in the kernel code; this means that, by contrast with TCP_RETR2, this value can be changed only by recompiling the kernel. Note that, having fixed the maximum number of retries, putting a ceiling on RTO produces two effects: first, it sets a limit to the maximum restart delay (i.e. the time between link disruption end and transmission restart), which is equal to the RTO itself, as channel availability is probed at RTO intervals; second, it can reduce the Maximum Tolerable Disruption Length (MTDL), i.e. the maximum disruption time after which the connection is closed by TCP.

Basically the same retransmission algorithm is used by TCP to establish a new connection, but the maximum number of SYN segment retransmissions is governed by another variable ($TCP\_SYN\_RETRIES$ sysctl, default 5). PEPs that use TCP at transport layer inherit TCP performance. However, it should be noted that satellite providers using PEPs may close (at application layer) idle TCP connections after a relatively short timeout, to free memory resources.

In conclusion, TCP disruption resilience is not nil, but it is largely dependent on implementations and settings; just to provide some quantitative insight, MTDL is a little less than 20 minutes with Linux defaults. Moreover, its tuning is far from obvious, and, in the case of PEPs, may go against satellite providers' wishes. As shown below, DTN definitively provides superior resilience performance, with an MTDL longer than 24 hours as default.

10.5.2   Analysis of DTN Behavior in the Presence of Disruption

The following analysis considers bundle protocol DTN architecture with TCP as convergence layer adapter. We call it "micro-analysis," in contrast to the "macroscopic" performance evaluations presented later. As always in scientific experiments, micro-analysis is necessary for the researcher to fully understand the mechanisms that lead to macro results (e.g. atomic theory presented by John Dalton in 1803–1810 is necessary to explain the conservation mass law formulated by Antoine Lavoisier in 1789). More modestly, the micro-analysis presented here aims to fully explain the DTN mechanisms that make DTN disruption resilience better than other solutions. That is to say, the aim of this subsection is to explain why performance can be better, before showing how much better it can be (Section 10.5.3).

The micro-analysis consists of two parts: first it considers a link between two consecutive DTN nodes (hop-by-hop analysis) and then a typical bundle flow from a DTN sender to a DTN receiver (end-to-end analysis). The first part highlights bundle protocol interaction with TCP, such as closing disrupted TCP connections and subsequent reconnection attempts; it is therefore closely related to bundle protocol specific implementation (e.g., DTN2, ION, etc. [9]); it was made by direct inspection of the DTN2 code (rel. 2.6.0) [10] and by using packet level performance evaluation tools, like Tcpdump [38] and Wireshark [39]. The second part deals with bundle protocol internal mechanisms in the presence of disruption, such as bundle re-transmissions, custody transfer, bundle fragmentation and reassembling, possible deletion and final bundle delivery; as these features are specified in RFC [2], [3], they should be common to all implementations; analysis is carried out here by making extensive use of DTNperf_2 logs.

10.5.2.1 Hop-by-hop DTN Micro-Analysis: DTN "Link Class" Retransmission Timers and Backoff Policy

We will examine a disruptive link between two consecutive DTN nodes that adopt TCP at transport layer. All default TCP timers and retransmission mechanisms described above are retained, and bundle protocol algorithms applied on top. We focus on DTN retransmission timers specified in the DTN2 "link class" (upper part of the convergence layer adapter). Note that these timers are not exclusive to the TCP convergence layer, which makes the analysis more general. The algorithm followed by the DTN2 implementation is summarized in Figure 10.13. Note that other DTN bundle protocol implementations [9] may differ.

When the bundle protocol wants to send a bundle, it checks the status of the link towards the next DTN node by asking the lower layer (e.g., TCP) for socket availability. If the link is active the bundle is passed to the lower layer for transmission. However, let us first consider the opposite case (right-hand flow in Figure 10.13). If the socket is unavailable (e.g., due to a link disruption), DTN application sets a timer (DTN variable *data_timeout_*, default 30 s), and waits for the socket to be available again. If this does not happen before timer expiry, the socket is closed (the underlying TCP agent sends an FIN segment to the other peer), and after a retry interval (*min_retry_interval_* in Link.cc [10], initial value = 5 s) it tries to open a new socket, adopting the standard TCP procedure based on SYN retransmissions. If the other peer is still unreachable after the maximum number of SYN retries, the DTN application doubles the retry interval and again asks TCP to open a new connection. This DTN backoff algorithm fixes the maximum retry interval to 600 s; moreover, the maximum amount of time during which the DTN application will try to open a new connection is set to 24 h. If, after 24 hours, the other peer is still unreachable, the DTN application stops trying to use this link and delegates the discovery algorithm to check link availability. As the same backoff procedure is used in the case of a disrupted transmission (see below), this extends the MTDL of the DTN architecture beyond 24 hours, a figure in no way comparable to the few minutes of TCP with default settings.

Now, let us consider what happens when the socket is found available (left-hand flow in Figure 10.13). The bundle, waiting in a DTN queue, starts to be transferred to the transport layer (here TCP). If no disruption occurs, the bundle is successfully transmitted to the next DTN node. Otherwise, the socket becomes unavailable and the *data_timeout_* timer is set as before. At its expiry DTN forces link closure (an FIN segment is sent) and starts a new socket opening, which follows the algorithm already described. However, in order not to retransmit successfully received data, the current bundle is split into two "fragments" as a result of connection closing ("reactive fragmentation" [3]). The first fragment consists of bundle data already sent, the second of its complement. The algorithm continues trying to send this second fragment and terminates when all bundle data are sent; note that consecutive

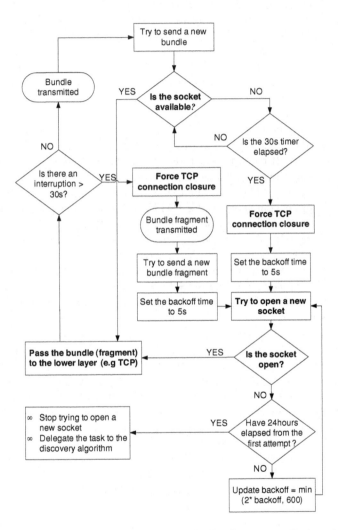

Figure 10.13: Flow chart of the DTN2 "link class" retransmission timers and backoff policy (hop-by-hop analysis). From [35].

fragmentations are possible.

A numerical example can help clarify this algorithm. Let us consider a GEO satellite connection with a channel disruption of 1200 s (20 min.): tests are carried out using TATPA (Figure 10.7), with DTN configuration as before (Figure 10.5, Figure 10.6, Table 10.1) and all TCP agents with default Linux retransmission timer parameters. DTN behavior is evaluated by examining the satellite TCP connection flow, which is the only one affected by disruptions. Moreover, although two TCP sockets may be concurrently active (one opened by the DTN intermediate node, R2, the other by the satellite receiver) we limit analysis to the TCP sockets opened by R2, which are the only ones actually used to transfer bundles.

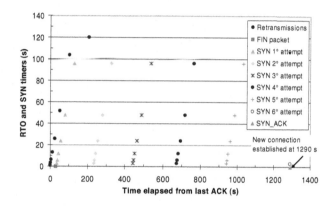

Figure 10.14: An example of DTN hop-by-hop analysis. Flow of retransmissions, FIN and SYN segments; disruption length = 20 min; initial RTO = 0.82 s. DTN2 reference implementation. From [35].

Figure 10.14 shows all TCP segments exchanged between R2 and the satellite receiver after reception of the last ACK segment (time axis origin). After some retransmissions of the last unacknowledged TCP segment performed at increasing intervals (RTO exponential backoff), the bundle protocol triggers a FIN segment at 30 s to force the TCP connection closure. Then we have 5 consecutive unsuccessful socket opening attempts (spaced by the DTN backoff increasing interval), each consisting of 6 SYN segment transmissions (the first plus *TCP_SYN_RETRIES* retransmissions). As for retransmissions, the y-axis shows the new RTO values set by TCP backoff at transmission times given on the x-axis. After the first five unsuccessful attempts, the first SYN segment of the sixth attempt finds the channel available again. The transmission therefore restarts with a new TCP connection after the corresponding SYNACK segment reception (1290 s after disruption start). The restart delay, during which the link is active but not exploited, is therefore 90 s, which is more than satisfactory if compared with 1200 s. As a final remark, note that end-to-end TCP would have closed the connection just before interruption end, as MTDL with Linux default is slightly shorter than the disruption considered.

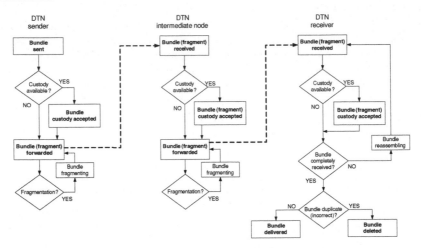

Figure 10.15: Flow chart of the bundle transmission process from the DTN sender to the DTN receiver (end-to-end analysis). From [35].

10.5.2.2 End-to-End DTN Micro-Analysis: Bundle Transmission Process from DTN Sender to DTN Receiver

The next step is the transmission of one bundle from DTN sender to DTN receiver (Figure 10.15), which should be common to all implementations. Here, the state of a bundle is described by the corresponding "status report" administrative bundle (in bold), which can optionally be sent to the "report-to" node (e.g., the DTN sender). A status "sent" indicates that the application has passed the bundle content to the bundle layer. We assume that custody transfer is required; if the node can accept this request ("custody accepted"), the bundle is written on a local database, waiting for transmission. The bundle can be fragmented (reactive fragmentation [2]) if the transport socket becomes unavailable because of link disruption. Fragments are then treated as ordinary bundles by subsequent nodes (until reassembly). The transmission of each bundle (or fragment) is acknowledged by "forwarded," its reception by "received." The intermediate node may or may not accept custody. The process continues as before (possibly through other DTN intermediate nodes not shown in the figure) until the receiver is reached. Here bundle fragments are reassembled and the bundle is eventually either delivered to the application or deleted ("delivered" or "deleted," respectively). The deletion of a bundle may be the result of duplicate reception. The deleted status report is the only mandatory one in the case of custody transfer request. Figure 10.15 is simplified as it does not consider the possibility of bundle reassembly at intermediate nodes or proactive fragmentation (i.e. fragmentation performed before bundle transmission) [2]. Finally, let us remember that custody transfer is carried out between DTN nodes through the exchange of "custody signals" independent of the (optional) corresponding status reports.

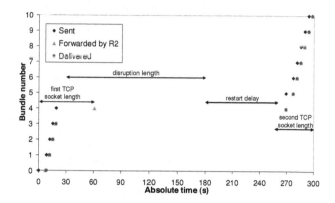

Figure 10.16: An example of end-to-end bundle transmission process. A 300 s transmission perturbed by a 150 s channel interruption. From [35].

To illustrate this end-to-end process, we consider a disruption of 150 s in the same GEO satellite environment as before. The total transfer length is 300 s, with disruption starting after 30 s from transfer launch. Each bundle "history" is given in Figure 10.16. through its three most significant status reports, namely "sent," "forwarded" (by R2), and "delivered." Note that, in contrast to Figure 10.13, here the time on the x-axis is computed from transfer launch, and not from interruption start. Looking at the graph, we see that the first four bundles (bundle#0–bundle#3) are regularly sent by the satellite sender and then delivered to the satellite receiver after a few seconds. By contrast, the transmission of bundle#4 is interrupted by disruption start; then, at 60 s (i.e. after *data_timeout_* seconds from disruption start), the DTN bundle protocol forces TCP connection closure and a first "forwarded" is sent from R2 to the satellite sender. From DTNperf_2 logs, we can discover that this bundle was only partially transmitted to the satellite receiver (meaning that reactive bundle fragmentation happened), and then, by examining tcpdump traces, that transmission was re-established at the first SYN of the second re-opening attempt (at 263 s). After connection reopening, the second fragment of bundle#4 is transmitted, generating a second "forwarded" for bundle#4 (sent by R2) and a "delivered" (by satellite receiver), both at 270 s. Then, bundle transmission continues regularly till the end of the data transfer. The figure also shows the duration of the first TCP socket terminated by the disruption, disruption length, restart delay (about 83 s) and the duration of the second TCP socket opened after disruption end.

Finally, for the sake of completeness, in Table 10.3 the complete history (all available status reports from DTNperf_2 logs) for bundle#4 is given. The first column shows the time at which the "report-to" node (here the DTN sender) receives the corresponding status report (including those sent by itself). The second, third and fourth columns contain report type, sender, and bundle ID, respectively. Finally, the last column contains the fragment byte offset; an offset greater than zero indicates a bundle fragment whose first payload byte is equal to the offset.

Table 10.3: Bundles Process: Status Reports for Bundle #4

Time (s)	Status Report	Report Sender	Bundle Number	Fragment Offset
19.042	STATUS SENT	SS	4	0
19.050	STATUS CUSTODY ACCEPTED	SS	4	0
20.104	STATUS RECEIVED	R2	4	0
20.119	STATUS FORWARDED	SS	4	0
20.120	STATUS CUSTODY ACCEPTED	R2	4	0
60.620	STATUS FORWARDED	R2	4	0
265.353	STATUS RECEIVED	SR	4	0
265.930	STATUS CUSTODY ACCEPTED	SR	4	0
269.472	STATUS FORWARDED	R2	4	4002
269.475	STATUS RECEIVED	SR	4	4002
269.499	STATUS CUSTODY ACCEPTED	SR	4	4002
269.500	STATUS DELIVERED	SR	4	0

10.5.3 Performance Evaluation in a Realistic Case Study: Railway Satellite Communications

Having reviewed both TCP and DTN behavior in the presence of disruption, now we can come back to the railway scenario cited at the beginning of this section to provide comparative performance evaluation of DTN [36]. Train passengers are connected to an Internet server through a GEO satellite link and disruptions are caused by railway tunnels. For goodput evaluation we assume continuous file transfer from the server to a client on board of the train. Results are always averaged over multiple runs and 90% confidence intervals are given.

Disruptions are emulated on the basis of real railway tunnel traces (tunnel lengths and positions), with reference to the Bologna–Florence "Direttissima," one of the main Italian railway lines. This 96 km long line was chosen because of its interesting mix of tunnels of different lengths alternating with open segments. The total length of open segments is 59 km, meaning that if we assume a constant train speed, we have clear sky propagation for about 61% of total trip time. At 120 km/h (typical of passenger trains on this line), tunnels cause

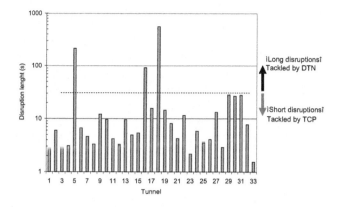

Figure 10.17: Bologna–Florence railway line; length of channel disruptions caused by tunnels (train speed = 120 km/h). From [36].

disruption durations given in Figure 10.17 (signal reacquisition times at tunnel ends can be safely neglected, being orders of magnitude lower, e.g., about 100 ms for a DVB-RCS system). Note that the most protracted disruption caused by the longest tunnel (18.5 km) lasts less than the TCP Maximum Tolerable Disruption Length (MTDL) which we found above to be a little less than 20 min. in Linux with default settings. This allows performance comparison between TCP, PEPsal, and DTN, as all can cope with this disruption. In fact, disruptions longer than the TCP MTDL (achievable by considering longer tunnels or simply a lower train speed) would have resulted in premature closing of TCP connections (both end-to-end and PEPsal). In this case, the comparison would have been easier, but also less significant, due to the lack of any DTN competitor (we recall that DTN can cope with disruptions longer than 24 hours, considering DTN2 implementation with default settings).

Tests were carried out, as usual, through the TATPA testbed, with TCP, PEPsal, and DTN configurations as before. The layout is the same as in Figure 10.7, where the satellite sender represents the web server and the satellite client is on board of the train. Tunnels are emulated by means of a specific application installed on the satellite channel emulator. The router R2 acts as satellite gateway and background wired traffic can be generated by the wired pair to create congestion on the R1-R2 bottleneck, when desired. Note that in the experiments the DTNperf_2 sending window W is set to 8 to maximize R2 utilization (we checked that parallel TCP connections, if opened, are not actually exploited in the experiments considered here).

10.5.3.1 Performance in the Absence of Competing Traffic

We consider first the case of a single GEO satellite connection ($RTT = 600ms$), from the satellite sender to the satellite receiver, in the absence of

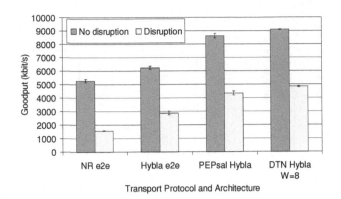

Figure 10.18: Impact of channel disruptions on various techniques; goodput of a single satellite connection ($RTT = 600ms$) in the absence of competing traffic (no background TCP connections).

any concurrent traffic, which means that the R1-R2 bottleneck bandwidth (10 Mbit/s) is fully available. To isolate the effects of disruptions, satellite PER has been considered negligible. Comparative evaluations are presented in Figure 10.18. The first series, introduced by way of reference, shows the goodput (data successfully transferred divided by the total transfer time) of a long term connection in the absence of any disruption. By contrast, disruptions caused by tunnels are considered in the second series, which refers to a connection that lasts for the entire train journey, from Bologna to Florence, at 120 km/h. The figure clearly shows the superiority of all the techniques considered over end-to-end NewReno, especially in the presence of disruptions. DTN always achieves the best result, with an 11% improvement on the best non-DTN approach (PEPsal Hybla) in the case of disruptions.

10.5.3.2 Performance in the Presence of Competing Traffic

Let us now add the effect of congestion on the R1-R2 bottleneck: 5 competing wired TCP connections ($RTT = 25ms$, NewReno), generated by the wired sender and directed to the wired receiver, are now present (see Figure 10.7). The R1-R2 bottleneck bandwidth fair share is about 1.6 Mbit/s. Figure 10.19 shows that end-to-end NewReno is severely penalized by the long RTT of the satellite connection even in the absence of disruptions. In end-to-end Hybla, this is compensated by its countermeasures against the RTT unfairness problem, and also, in PEPsal and DTN, by the isolation of the satellite channel. We considered two alternative DTN solutions ($W = 8$ and $W = 8$ SONC). In the first case (SONC disabled), the DTNperf_2 window is updated after the reception of status "delivered" reports; in the second (SONC enabled), after the reception of "custody accepted" reports sent by the first intermediate node (R2 here). Reasons for the introduction of the SONC option are fully discussed in [36]. Here let us just observe that when a bottleneck is present before a faster but disrupted leg, it is possible to exploit the DTN custody transfer mechanism to temporarily store bundles that cannot be sent to destination in DTN intermediate nodes. In other words, the DTN intermediate node can serve as a sort of bundle buffer. In our case, by enabling the SONC option the first DTN hop (the bottleneck) can be used continuously, independently of disruption length on the second hop, provided that R2 continues to accept custody of incoming bundles. At disruption end, bundles stored in R2 are quickly transmitted (thanks to the 10 Mbit/s speed) to the DTN receiver. In this way, it is largely possible to "mask" the effects of channel disruptions and actually to achieve significant (25%) goodput improvement (from about 800 kbit/s, common to PEPsal and DTN, to 1 Mbit/s achieved by DTN with SONC enabled; see Figure 10.19)

Before concluding, we must mention a practical problem we encountered with the SONC option enabled. While in theory R2 should stop accepting bundles when its local database is nearly full, in practice this sort of congestion control has not yet been correctly implemented in DTN2. At the time of experiments (DTN2 release 2.6.0), the only possible way to overcome this problem seems to be to select a database size large enough to accommodate all the bundles sent to R2 during the longest satellite disruption. Luckily, the corresponding 120 MB required are no real constraint.

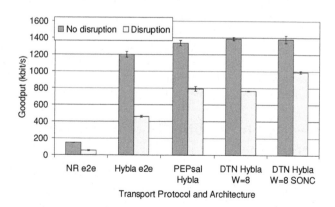

Figure 10.19: Impact of channel disruptions on various techniques; goodput of a single satellite connection ($RTT = 600ms$) in the presence of 5 background wired TCP connections ($RTT = 25ms$, NewReno).

10.6　Scenarios 3 and 4

This section deals with the last two scenarios. The treatment is briefer, as in this case DTN superiority over other solutions (if available) is clearer. Examples of satellite applications that can be included in these scenarios will be presented in the next "Ongoing research" section. Other examples can be found in [40], where a wide variety of possible DTN applications (most inter-planetary, some involving satellites or Mars orbiters), is given.

10.6.1　Scenario 3: Satellite Systems with Scheduled Intermittent End-to-End-Connectivity

This scenario applies to satellite systems that provide scheduled intermittent connectivity, such as communications between satellite ground stations and LEO systems. Due to their low orbit (160–2000 km from Earth), LEO satellites have a short revolution period (about ninety minutes at 520 km altitude) to counteract Earth's gravity [41]. This implies that, unlike GEOs, a LEO satellite passes above satellite ground stations on a scheduled basis and for a given short period, thus providing scheduled, intermittent (and also periodic) connectivity. The short transmission time window and the limited channel bandwidth pose a limit on "contact volume" [2], i.e., the total amount of data that can be transferred at each link availability interval. If very large files (e.g., images of Earth) could not be transferred during a single pass, it would be necessary to divide them into multiple segments to be transmitted during consecutive passes. In this case DTN can fruitfully use the "proactive fragmentation" feature [2] of the bundle protocol, which automatically subdivides large bundles into multiple fragments of a predetermined size whenever contact volume between two successive DTN nodes is known in advance. In the following section devoted to ongoing research, a practical case of large file transfer on an intermittent channel will be presented and discussed.

10.6.2　Scenario 4: Satellite Systems without End-to-End Connectivity

The last scenario is the most favorable to DTN applications, as the total absence of end-to-end connectivity prevents the establishment of TCP (or TCP-like) connections. In the absence of feedback from the receiver end-node, the "three way handshake," which implies the exchange of SYN and SYNACK segments between end-nodes [14], cannot be completed and thus the connection opened. Moreover, the lack of end-to-end connectivity also makes UDP transfers impossible. In conclusion, the only possible approach is a "store-and-forward" technique with data storage at intermediate nodes. Relatively large amounts of data (not just a few IP packets) need to be transferred first to an intermediate node, which stores them temporarily, then to the successive intermediate node, or final destination, when possible. This is a task the DTN bundle protocol was actually conceived for.

This scenario applies to communications through a "data mule," usually a terrestrial or a maritime "vehicle" traveling around data sensors [42]. It also applies to some deep space communications, for example to two satellites orbiting around different planets (e.g., Earth and Mars), or to LEO systems with data sources and sinks never concurrently in visibility.

10.7 Ongoing Research

In this section some of the current research relating to the use of DTN with satellite systems is presented: first, NASA experiments on interplanetary communications; second, DTN use in a military network with satellites characterized by random disruptions; finally, an application of DTN technology to LEO systems characterized by intermittent periodic connectivity between satellites and ground stations.

10.7.1 NASA JPL October 2008 Experiments

In October 2008 NASA JPL carried out a series of tests [43] to prove the feasibility of DTN architecture for deep space communications [44]. Researchers used DTN to transmit space images to and from the NASA science spacecraft EPOXY, located more than 32 million kilometers from Earth. This was the first deep space DTN transmission and, quoting Adrian Hooke, team leader and manager of space-networking architecture, it was: "a first step in creating a totally new space communications capability, an interplanetary Internet." Although the EPOXY spacecraft is not a satellite, it was used during the experiments as a Mars data-relay orbiter (it was one of the ten DTN nodes in the experiments, the other nine were actually on the ground and simulated other network nodes). These experiments had a great impact on the public at large, as they were covered by all the main newspapers throughout the world, on the Internet, and also for the first time, on a You-Tube video clip (by V. Cerf himself, one of fathers of the Internet and also a DTN pioneer) [45]. They made a huge contribution towards making DTN known outside the restricted circle of DTN researchers. From a more technical point of view, it is worth mentioning that during the experiments the DTN bundle protocol was operated on top of the LTP convergence layer protocol. A NASA demonstration using new DTN software loaded on board the International Space Station was performed in the summer of 2009. Other activities on DTN are planned [46].

10.7.2 DTN Application to a Military Heterogeneous Network

A DTN approach to improve performance in a military network scenario is described in [5]. It is characterized by a high level of heterogeneity, as it can include satellite links, airborne relays, and line-of-sight radios. Moreover, disruptions and disconnections are frequent due to node dynamics, and high PER can be present even when the links are not completely disrupted. To cope with this extremely challenging network, the use of the DTN bundle protocol architecture is suggested, in conjunction with NORM (Nack-Oriented Reliable Multicast) transport protocol [26] at convergence layer. Simulations were carried out to compare performance of end-to-end TCP, end-to-end NORM, DTN plus TCP, and DTN plus NORM. Results show that DTN gives better performance in scenarios that include satellite links. The DTNperf_2 tool was used to evaluate performance.

10.7.3 DTN Application to the UK-DMC LEO System

An interesting application of DTN technology to LEO systems characterized by scheduled intermittent connectivity is given in [47]. The United Kingdom-Disaster Monitoring Constellation (UK-DMC) is a multi-satellite LEO sensor network able to capture and store onboard images for subsequent download to ground stations. Image downloading can happen only when the satellite and a ground station are in visibility (i.e., when the satellite passes over a ground station). For the DMC systems, this means that a contact of 5 to 14 minutes is established every 100 minutes (thanks to the presence of multiple ground stations). By fully exploiting the 8.134 Mbps downlink speed, the maximum contact volume ranges from about 300 MB to about 850 MB depending on the connection length. Note, however, that this figure is an upper limit, because in practice the speed offered by the physical layer is difficult to exploit, due to upper layer inefficiencies. The experiments used 160 Mbyte images, which allowed file transfer to be completed in a single pass, provided that the link was efficiently exploited. To this end, the Saratoga protocol, developed by Surrey Satellite Technology Ltd (SSTL), was used at transport layer in the DMC architecture. The experiments aimed to compare the existing Saratoga-based system with bundle layer DTN architecture, with Saratoga at convergence layer. Note that both possible uses of Saratoga have recently been promoted by the authors of the experiments as IETF Internet Drafts, [25], [27]. Results show that images were correctly received even in the case of proactive bundle fragmentation. Some bundle protocol issues, mainly related to bundle checksum and clock synchronization, were also identified during the experiments. Despite these minor issues, these experiments proved that the addition of a common bundle protocol overlay can facilitate automated routing of sensor data and increase integration between terrestrial Internet and satellite observation networks.

10.8   Conclusions

The aim of this chapter was to investigate the feasibility and the possible advantages of DTN in satellite communications. After a brief review of the main characteristics of DTN bundle protocol and satellite impairments, four different application scenarios were considered.

In the case of continuous end-to-end connectivity without disruptions, the least favorable to DTN, it was shown that DTN can offer the same performance as the best current solutions, including PEPs. This is mainly due to the opportunity DTN architecture offers to employ different transport protocols on various network segments, and, in particular, on satellite links. This feature actually extends the PEP TCP splitting technique to large heterogeneous networks.

When end-to-end connectivity is usually present, but short and long disruptions are frequent, DTN becomes more advantageous than end-to-end TCP or PEPs. Hop-by-hop bundle retransmission mechanisms, working atop the usual TCP backoff retransmission timers, actually extend the maximum tolerable disruption length from a few minutes to more than 24 hours. Moreover, reactive bundle fragmentation and custody option optimize channel utilization and transfer reliability.

Moving to the last scenarios, characterized by intermittent or totally absent end-to-end connectivity, DTN becomes essentially the only possible solution (excluding manual "handling" of data transfers). In these scenarios the custody transfer option assumes a particularly relevant role and proactive fragmentation must be added to all bundle protocol features mentioned above.

The variety of recent application examples presented in the last section shows that DTN, although still in its infancy, has a great future in the satellite communications field too. This is the authors' opinion and also what they hope.

Acknowledgments

The authors would like to thank all those who contributed to the DTN experiments realization. In particular, Piero Cornice and Marco Livini for developing the DTNpef_2 tool; Daniele Lacamera for Hybla and PEPsal implementations; and Stefano Tamagnini for the TATPA design and implementation. A special thanks to reviewers for their helpful remarks and valuable suggestions.

Bibliography

[1] K. Fall and S. Farrell, "DTN: an architectural retrospective," *IEEE Journal on Selected Areas in Communications*, vol. 26, no. 5, pp. 828–836, June 2008.

[2] V. Cerf , A. Hooke, L. Torgerson, R. Durst, K. Scott, K. Fall, and H. Weiss, "Delay-Tolerant Networking Architecture," IETF RFC 4838, Apr. 2007.

[3] K. Scott and S. Burleigh, "Bundle Protocol Specification," IETF RFC 5050, Nov. 2007.

[4] C. Caini, P. Cornice, R. Firrincieli, and D. Lacamera, "A DTN Approach to Satellite Communications," *IEEE Journal on Selected Areas in Communications*, special issue on Delay and Disruption Tolerant Wireless Communication, vol. 26, no. 5, pp. 820–827, Jun. 2008.

[5] C. Rigano, K. Scott, J. Bush, R. Edell, S. Parikh, and R. Wade, "Mitigating naval network instabilities with disruption tolerant networking," in Proc. IEEE MILCOM 2008, San Diego, Nov. 2008, pp. 1–7.

[6] J. Border, M. Kojo, J. Griner, G. Montenegro, and Z. Shelby, "Performance Enhancing Proxies Intended to Mitigate Link-Related Degradations," IETF RFC 3135, June 2001.

[7] K. Fall, W. Hong, and S. Madden, "Custody Transfer for Reliable Delivery in Delay Tolerant Networks," Technical Report IRB-TR-03-030, Intel Research, Berkeley, California, July 2003, pp. 1–6. Available at DTNRG Web site.

[8] A. McMahon and S. Farrell, "Delay- and Disruption-Tolerant Networking," *IEEE Internet Computing*, vol. 13, no. 6, pp. 82–87, Nov./Dec. 2009.

[9] DTNRG Web site: http://www.dtnrg.org/wiki.

[10] DTN reference implementation and related projects: http://sourceforge.net/projects/dtn/.

[11] C. Caini, P. Cornice, R. Firrincieli, and M. Livini, "DTNperf_2: a Performance Evaluation Tool for Delay/Disruption Tolerant Networking" in Proc. of E-DTN 2009, St. Petersburg, Russia, October 2009, pp. 1-8.

[12] DTNperf_2 source code: DTN2 Mercurial repository, http://dtn.hg.sourceforge.net/hgweb/dtn/DTN2.

[13] Y. Hu and V.O.H. Li, "Satellite-based internet: a tutorial," *IEEE Commun. Mag.*, vol. 39, no. 3, pp. 164–171, Mar. 2001.

[14] W. R. Stevens, "TCP/IP Illustrated, Volume 1," Addison-Wesley, Reading, Massachusetts, Nov. 1994.

[15] M. Allman, V. Paxon, E. Blanton, "TCP Congestion Control", IETF RFC 5681, Apr. 1999.

[16] S. Floyd, T. Henderson, and A. Gurtov, "The NewReno Modification to TCP's Fast Recovery Algorithm," IETF RFC 3782, 2004.

[17] C. Caini and R. Firrincieli, "TCP Hybla: a TCP Enhancement for Heterogeneous Networks," *Int. J. Satell. Commun. Network*, vol. 22, pp. 547–566, Sept.-Oct. 2004.

[18] M. Mathis and J. Mahdavi, "TCP Selective Acknowledgment Options," IETF RFC 2018, Oct. 1996.

[19] C. Caini, R. Firrincieli, and D. Lacamera, "Comparative Performance Evaluation of TCP Variants on Satellite Environments," in Proc. of IEEE ICC 2009, Dresden, Germany, June 2009, pp. 1–5.

[20] C. Caini, R. Firrincieli, D. Lacamera T. De Cola, M. Marchese, C. Marcondes, M. Y. Sanadidi, and M. Gerla, "Analysis of TCP Live Experiments on a Real GEO Satellite Testbed," Performance Evaluation, Elsevier, vol. 66, no. 6, pp. 287–300, Jun. 2009.

[21] SCPS Web site: http://www.scps.org/scps/.

[22] S. Burleigh, M. Ramadas, and S. Farrell, "Licklider Transmission Protocol Motivation," IETF RFC 5325, Sept. 2008.

[23] M. Ramadas, S. Burleigh, and S. Farrell, "Licklider Transmission Protocol Specification," IETF RFC 5326, Sept. 2008.

[24] S. Farrell, M. Ramadas, and S. Burleigh, "Licklider Transmission Protocol Security Extensions," IETF RFC 5327, Sept. 2008.

[25] L. Wood, J. McKim, W. Eddy, W. Ivancic, and C. Jackson, "Saratoga: A Scalable File Transfer Protocol," IETF Internet draft, work in progress, http://tools.ietf.org/html/draft-wood-tsvwg-saratoga.

[26] B. Adamson, et al., "Negative-acknowledgment (NACK)-Oriented Reliable Multicast (NORM) Protocol," IETF RFC 3940, Nov. 2004.

[27] L. Wood, J. McKim, W. Eddy, W. Ivancic, and C. Jackson, "Using Saratoga with a Bundle Agent as a Convergence Layer for Delay-Tolerant Networking," IETF Internet draft, work in progress, http://tools.ietf.org/id/draft-wood-dtnrg-saratoga.

[28] ETSI TR 102 676: "Satellite Earth Stations and Systems (SES); Broadband Satellite Multimedia (BSM): Performance Enhancing Proxies (PEPs)."

[29] C. Caini, R. Firrincieli, and D. Lacamera, "PEPsal: a Performance Enhancing Proxy for TCP Satellite Connections," *IEEE Aerospace and Electronic Systems Magazine*, vol. 22, issue 8, pp. b-9, b-16, Aug. 2007.

[30] PEPsal: http://sourceforge.net/projects/pepsal/.

[31] TATPA Web site: http://tatpa.deis.unibo.it.

[32] MultiTCP package: http://sourceforge.net/projects/multitcp.

[33] NistNet Web site: http://snad.ncsl.nist.gov/nistnet/.

[34] Iperf wiki page: http://en.wikipedia.org/wiki/Iperf.

[35] C. Caini, R. Firrincieli, and M. Livini, "DTN Bundle Layer over TCP: Retransmission Algorithms in the Presence of Channel Disruptions," to appear in *Journal of Communications (JCM)*, Academy Publishers, 2010.

[36] C. Caini, P. Cornice, R. Firrincieli, D. Lacamera, and M. Livini, "TCP, PEP and DTN Performance on Disruptive Satellite Channels," in Proc. of IEEE IWSSC 2009, Siena, Italy, Sept. 2009, pp. 371–375.

[37] V. Paxson and M. Allman, "Computing TCP's Retransmission Timer," IETF RFC 2988, Nov. 2000.

[38] Tcpdump Web site: http://www.tcpdump.org/.

[39] Wireshark Web site: http://www.wireshark.org/.

[40] W. Ivancic, "Delay/Disruption Tolerant Networking - Network Management Requirements," work in progress, http://tools.ietf.org/html/draft-ivancic-dtnrg-network-management-reqs.

[41] http://en.wikipedia.org/wiki/Low_earth_orbit.

[42] S. Farrell and V. Cahill, "Delay- and Disruption-Tolerant Networking," Artech House, 2006.

[43] J. L. Torgerson, L. Clare, S. Y. (Cindy) Wang, and J. Schoolcraft, "The Deep Impact Network Experiment Operations Center," in Proc. IEEE Aerospace Conference 2009, March 2009, pp. 1–12.

[44] S. Burleigh, A. Hooke, L. Torgerson, K. Fall, V. Cerf, B. Durst, K. Scott, and H. Weiss, "Delay-tolerant networking: an approach to interplanetary Internet," *IEEE Communications Magazine*, vol. 41, issue 6, pp. 128–136, June 2003.

[45] http://www.youtube.com/watch?v=lQzjUvn_hWY.

[46] https://www.spacecomm.nasa.gov/spacecomm/programs/technology/DTN/default.cfm.

[47] W. Ivancic, W. M. Eddy, D. Stewart, L. Wood, J. Northam, and C. Jackson, "Experience with delay-tolerant networking from orbit," *Int. J. Satellite Commun. And Networking*, vol. 28, no. 5–6, pp. 335–351, Sep.–Dec. 2010.

Index

A

Absolute utility criterion,
  forwarding, 39
Access control, VANET, 247–251
  cooperative ARQ, 247, 249
  802.11p MAC, 249–251
Accuracy, map
  eMAP, 206, 208
  gMAP
    evaluation metrics, 210–211
    open issues, 215, 216
    regioning performance, 214
Acknowledgments
  bundle transmission, 9, 304
  DTN overview, 267
  open issues, VANET, 252
  SNAKs, 290
  TATPA testbed, 296
  TCP problems in space, 265,
    266
  TCP splitting and, 292
Addressing, VANET
  characteristics, 229
  open issues, 253
  research, 225
Address space, ITR overlay
  network, 176, 177, 178
Ad Hoc Networks, x, 160
  mobile; See MANETs
  vehicular; See Vehicular Ad
    Hoc Networks
Ad-hoc On demand Distance
  Vector (AODV), 19, 33,
  47, 49, 71, 171, 175
  MADPastry, 181

MANET forwarding schemes,
  235
Ad hoc routing protocols, 19
Ad-hoc Storage Overlay System
  (ASOS), 131
  comparison among systems,
    181
  MPP, 171, 173
Advanced Communications
  Technology Satellite
  (ACTS), 263
Advertisements
  incentives, 105
  P2PSI, 174
  priority messages, 61
  RPP, 128, 129, 131, 132
    diffusion of, 136–137
    interactions with, 142
    location information, 134
    service container, 139, 140,
      141, 142
    throwbox nodes, 143
  social networks, 44
  SPAWN, 233–234
Age of last encounter parameter,
  43
Aggregate signatures, SMART
  scheme efficiency, 115
Aggregation
  isocluster, 194
  WSN, 191, 193–194
Agg-SMART, 117
Almost-connected networks, 49–50
Almost connected networks, 49–50
AN; See Assist nodes, mWSN
  gMAP

Ant-colony based Routing
        Algorithm (ARA), 173,
        174, 181
AODV; *See* Ad-hoc On demand
        Distance Vector
Application agent, defined, 11
Application data units
    DTN, 267
    store-and-forward message
        switching, 5–6
Application layer, data transport
        protocols for space, 266
Application requirements
    RPP system, 150
    taxonomy of DTNs, 55–56
Applications with priorities,
        routing case studies,
        61–63
ARA; *See* Ant-colony based
        Routing Algorithm
Architecture, 4–9
    custody transfer, 7–9
    fragmentation and
        reassembly, 6–7
    MPP systems, 170–175, 180,
        181
    overlay, 4–5
    routing and forwarding, 6
    RPP, 151
    store and forward message
        switching, 5–6
ARQs; *See* Automatic repeat
        requests
ASOS; *See* Ad hoc storage overlay
        system
Assist nodes (AN), mWSN
        gMAP, 197
    data collection, 204, 205
    data collection performance,
        213
    open issues, 215
Association procedures, Wave
        BSS, 250
Asymmetric communications, 165,
        265–266

Asymmetric data rates, DTN
        requirements, 4
Attacks, SMART scheme
    prevention of, 109–111
    types of, 107–108
Authentication
    bundle service, 9
    VANET open issues, 251
    Wave BSS, 250
Authentication, authorization,
        and accounting (AAA),
        27
Authorization, security issues, 27
Automatic repeat requests
        (ARQs), 165
    cooperative (C-ARQ), 247,
        249
    selective, 271
Automotive diagnostics, CarTel
        applications, 232

B
Backoff policy, DTN
        microanalysis, 303–306
Ballistic nodes, 40
Bandwidth, node resources, 55
Bandwidth Delay Product (BDP),
        TATPA testbed, 295
Bandwidth-limited environments,
        ORION and, 171
Bank, SMART scheme, 106
Base layer, SMART scheme, 104,
        110
Basic Service Set (BSS), IEEE
        802.11p, 249–250
Battery/battery lifetime
    DTN classification, 34
    mWSN model, 197
    node resources, 55
    residual battery information,
        93–94
Beaconing, VANETs, 238
Best effort priority service, 61
Big-M-formulations, 202

Binary spray and wait (SWB),
SMART simulation, 119
BioServe, 273
Bit error rates (BER), TCP
problems in space, 265
Bloom filtering, 174
BP/LTP/UDP/IP, 273
BP/TCPCL/TCP/IP, 273
Breakpoints
mWSN gMAP open issues,
215
mWSN path planning,
200–201, 209
Broadcast, 244
BSP tree, 206
Bundle blocks, formats, 12, 13
Bundle endpoint, defined, 11
Bundle layer DTN architecture, 4,
286–288, 292
Bundle node, defined, 10
Bundle protocol (BP), 9–19
data transport protocols for
space, 266
definitions, 10–11
DTN, 273–274
format, 11–14
end-point IDs, 12
formats of bundle blocks,
14
self-delimiting numeric
values, 11–12
JAVA-BP, 273
LTP and, 272
processing, 14–19
creation at source, 15, 16
first hop processing and
forwarding, 17
reception by destination,
18–19
second-hop processing and
forwarding, 18
transmission by source, 15,
16
satellite communications,
290, 312-313

service, 9–11
space communications, 269,
270
TATPA testbed, 294
UK-DMC LEO system, 314
Bundle protocol agent (BPA), 11
Bundle reports, TATPA testbed,
296
Bundles, 14–19, 267
creation
SMART scheme, 113
at source, 15, 16
defined, 10
forwarding
notification, 9
SMART scheme, 113–114
processing and forwarding
first-hop, 17
second-hop, 18
reception by destination,
18–19
transmission
end-to-end DTN
microanalysis, 306–308,
by source, 15, 16
Bundle time sequences, sending
window performance
improvement, 299
Bus-based DTN, 246

C
Caching, 175
Prophet-based information
retrieval, 174
WBSS, 251
Capability, DTN classification, 34
CarTel, 232–233
CCSDS, 270, 271–272, 290
CCSDS File Delivery Protocol
(CFDP), 273, 290
CCSDS link-layer protocols, LTP
and, 271
Cell misclassification percentage
(CMP)
evaluation metrics, 210–211

regioning performance, 213,
214
Certificate Authorities (CA),
public key revocation in
DTNs, 123
Certificate Revocation List
(CRL), 123
CFDP (CCSDS File Delivery
Protocol), 273, 290
Challenges, 2–4
Channel asymmetry, TCP
problems in space,
265–266
Channel impairments, satellite
communications,
288–293
delay/disruption tolerant
networking alternative,
292–293
end-to-end TCP
enhancements and other
optimized protocols,
289–290
performance enhancing
proxies (protocol
accelerators), 290–292
Charge model, 111
Charging, SMART scheme, 114
Chatty application protocols,
VANET characteristics,
231
Chloropleth map, 210, 213
Chord, 163
Civil infrastructure monitoring,
CarTel applications, 232
Class of Service (CoS), SMART
scheme, 104
Clock synchronization, space
communications, 273,
274
Cluster based forwarding/routing,
VANETs, 241, 244
Clusters
MADPastry, 175–176

peer organization in P2P
hybrid systems, 165
CMP; See Cell misclassification
percentage
Code, routing module
applicability, 57
Coding
intermittently connected
networks, routing on,
40–42
opportunistic routing
components, 34
routing primitives, 47, 48
Coding-based forwarding,
VANETs, 241, 247, 248
Coin, SMART scheme, 104
Colorado University, 273
Commercial Generic
Bioprocessing Apparatus
5 (CGBA5), 273
Communications
interplanetary, 264; See also
Space communications
mWSN model, 197–198
RPP design space, 149
SMART scheme
cryptographic overhead,
117
U-Hopper, 153
VANET open issues, 252
Completeness, mWSN gMAP,
210, 211, 212, 214, 215
Complexity of message, mWSN
gMAP, 214, 215
Comprehensibility, eMAP, 206,
208
Computation cost/overhead,
SMART scheme, 105,
115, 117, 118
Concatenation, layer, 105, 110
CONCORDE, 208
Congestion
data transport protocols for
space, 266
MPP systems, 169

open issues, 25–26
satellite communications
performance evaluation,
296–297, 298, 299, 300
TATPA testbed, 295
TCP problems in space, 265
VANET characteristics, 228
and VANET
communications, 224
Congestion control
LTP and, 271
TATPA testbed, 294
Connection lifetime, VANET
characteristics, 228
Connectivity
DTN classification, 34
DTN requirements, 3
end-to-end; *See* End-to-end
connectivity in satellite
communications
intermittently connected
networks
almost-connected networks,
50
connectivity islands, 51–52
sparse networks, 50–51
routing module applicability,
57
taxonomy of DTNs, 49–52
VANET, 224–225, 228
Connectivity islands, 51–52
Consultative Committee for Space
Data Systems (CCSDS),
270, 271–272, 290
Contact graph routing, 272, 274
Content Addressable Network
(CAN), 163
Content based
forwarding/forwarding,
VANETs, 238, 246
Content distribution network
(CDR), 240
Content manager, RPP, 142
functional blocks, 138, 139
user nodes, 143

Context-based routing, VANETs,
241
Continuous end-to-end
connectivity, satellite
communications, 285
Controlled replication
intermittently connected
networks, 37–38
routing case studies, 61–62
Controlled utility-based
replication, 38–39
Control systems, LTP and, 271
Convergence layer adaptor (CLA),
4, 11, 269
Convergence layer protocol (CLP)
DTN for space, 269
LTP and, 290
space communications,
269–271
Convex-Hull formulations, 202
Cooperation group, IEEE
802.11p, 249–251
Cooperative ARQ (C-ARQ),
VANET access control,
247, 249
Cooperative behavior,
destination-independent
(DI) utility, 46
Cooperative downloading,
VANET applications,
233–234
Cooperator nodes, 249
Coordinate-based prediction
services, 148
Coordinator, VANET routing,
238–239
Copies, number of
restricted epidemic routing
(RER) tradeoff function,
86–87, 88
SMART scenarios, 120–122
SMART scheme simulation,
121–122
Copy-limited replication, 37
CORSIM, 227

Coupon Collector's Problem, 247
CPLEX, 208
Creation at source, bundle
        protocol processing, 15,
        16
Credit-based incentive scheme;
        See SMART scheme
Credit-based schemes, 103
Credit creation, 105–106
Credit forgery attacks, 107
Credit value, 104
Cross-layered architecture
        Bundle protocol, 9–19
        MADPastry, 175
        MPP systems, 170, 173, 180,
        181
Cross-support, DTN
        standardization for space
        networking, 272
Cryptography, 26, 27
        SMART scheme, 117–118
        VANET open issues, 252
CSMA/CA-based MAC layer,
        mWSN model, 197
Cure-ack, 252
Currency, virtual, 103
Custody, open issues, 25–26
Custody-based retransmission,
        bundle service, 10
Custody transfer
        architecture, 7–9, 267–268,
        287–288
        bundle service, 9
        end-to-end DTN
        microanalysis, 306–307
        TATPA testbed, 295

D
DakNet, 231–232, 241
Data collection
        gMAP, 196–197
        gMAP construction
        performance, 209–214
        comparison among
        systems, 214

data collection
        performance, 211–213
        evaluation metrics, 210–211
        regioning performance,
        213–214
        simulation settings,
        209–210
        mobility-aided, 195
        VANET applications,
        232–233
Data completeness, gMAP,
        comparison among
        systems, 214, 215
Data gathering, U-Hopper, 152
Data lifetime, mWSN gMAP,
        open issues, 215–216
Data loss; *See* Packet drop/packet
        loss
Data mule, 195, 197
        CarTel applications, 233
        data collection techniques,
        195
        epidemic routing, 79
        mWSN gMAP
        comparison among
        systems, 214, 215
        open issues, 216
        satellite communications, 313
Data rate
        DTN requirements, 4
        TCP problems in space, 265
Data retrieval, RPP, 134–136
Data storage; *See* Storage
Data structures
        eMAP construction, 204
        P2P systems, structured, 162
Data units, DTN, 267
Decentralized P2P systems, 164
Dedicated Short Range
        Communications
        (DSRC), 249–251
Deep Impact Networking
        (DINET), 166, 272
Deep-Space Transport Protocol
        (DS-TP), 271, 275

Delaunay triangulation, 206
Delayed C-ARQ (DC-ARQ), 249
Delays
    DTN requirements, 4
    end-to-end
        unpredictable, 165
        WSN, 191
    routing case studies
        applications with priorities,
            62
        pocket switched networks,
            58, 59
    RPP application
        requirements, 150
    RTTs; *See* Round trip times
    TCP problems in space, 288
    VANET performance metrics,
        235
Delay-tolerant monitoring; *See*
        Mobility-assisted WSN,
        delay-tolerant
        monitoring
Delay-Tolerant Research Group,
        IRTF, 288
Delegation
    custody transfer model, 25
    data retransmission, 89, 268
    DTN2 link class transmission
        times, 303, 304
    DTN backoff algorithm, 302
    R-P2P
        interaction controller, 140
        P2P architectures, 160
Delegation forwarding, 42
Delivery
    CFDP protocols, 290
    ERP confidence level, 79, 80
    ERP network lifetime, 74, 75,
        76
    mobility-assisted, 195
    RER tradeoff function, 88
    routing schemes, 22
    VANET characteristics, 229
Delivery delay; *See* Delays
Delivery priority; *See* Priority

Delivery ratio
    routing case studies,
        applications with
        priorities, 62
    RPP application
        requirements, 150
    SMART scenarios, 120–122
Denial-of-Service (DoS) attacks,
        103
Density, VANET characteristics,
        228
Design
    routing, 32, 34, 56, 57
    RPP system, 149
    SMART scheme, 106–108
Destination-dependent (DD)
        utility, 42–45
    age of last encounter, 43
    history of last encounter,
        43–44
    patterns of locations visited,
        44
    social networks, 44–45
    traditional routing table
        entry, 45
Destination-independent (DI)
        utility, 45–46
Deterministic routing, 33
DHT; *See also* Distributed hash
        tables
Diffusion of advertisements, RPP,
        136–137
Diffusion of queries, RPP, 134–136
Dimensioning, RPP system,
        148–150
    application requirements, 150
    design space, 149
DINET, 272–273, 274
Direct Transmission/Delivery
        protocol, 106, 195
Disaster monitoring satellite, 272,
        274, 314
Disconnected islands, hybrid
        protocols for, 54

Dissemination/discovery process,
      162, 163, 169, 182
Dissemination length, VANET
      forwarding schemes, 245
Distance information, prediction
      services, 148
Distributed hash tables (DHT),
      132
   alternatives and
      modifications, 147
   opportunistic routing, 179
   structured MPP overlay
      networks, 175
   structured P2P systems, 162,
      163
   throwbox implementation,
      153
   throwbox nodes, 144, 146–147
   virtual ring routing, 178–179,
      180
Distributed PEP architecture,
      290–291
Distributed source coding (DSC),
      40, 47, 48
Distributed storage system (DSS),
      132, 138, 139
DSDV, 33, 235
DSR; *See* Dynamic source routing
DS-TP (Deep-Space Transport
      Protocol), 271, 275
DTN-2, 273, 275
DTN/IO Space-Ground testbed,
      273
DTNperf_2 tool, 295–296, 297,
      298–300, 310
   bundle transmission, 308
   military applications, 313
DTN plus TCP, military
      applications, 314
DTN Research Group (DTNRG),
      266
DTNR IRTF, 268
Dynamic source routing (DSR),
      19, 49
   enhanced (EDSR), 172, 181

MANET forwarding schemes,
      235
VANET, 225–226

E
Earth orbiting spacecraft, 264;
      See also Geostationary
      orbit satellites; Low
      earth orbit system
EDSR (Enhanced Dynamic
      Source Routing), 172,
      181
Effectiveness
   SMART scheme design goals,
      108
Efficiency
   mWSN gMAP
      data collection
      performance, 210, 211,
      213
   open issues, 215
   SMART scheme
      design goals, 108
      enhancement of, 114–117
      fragment authentication,
      115–116
      overhead reduction, 115
802.11p MAC, VANET access
      control, 249–251
eMap construction, 204, 206–208
Emerging new application,
      VANET, 253
Encounter-Based Routing (EBR),
      39
Encryption, 26, 27
Endorsed layer, 104, 110
Endpoint, bundle, 11
Endpoint IDs, 12
End-to-end connectivity in ad hoc
      networks, 160–161
End-to-end connectivity in
      satellite communications
      applications based on, 285
   channel impairments and
      possible solutions,

289–290
continuous, 294–300
    performance evaluation,
        296–300
    testbed and tools used in
        performance evaluations,
        294–296
DTN applications, 292–293
enhancements optimized for
    space communications,
    289–290
performance enhancing
    proxies (protocol
    accelerators), 290–292
random intermittent, 300-311
    analysis of behavior in
        presence of disruption,
        302–308
    performance evaluation in
        realistic case study of
        railway communications,
        308–311
    TCP resilience to
        disruption, 301–302
scheduled intermittent, 312,
    314
systems without, 312–314
End-to-end delays
    unpredictable, 165
    WSN, 191
End-to-end DTN microanalysis,
    306–308
End-to-end Hybla, 297
End-to-end latency, mWSN
    gMAP, 214, 215
End-to-end routing
    DTNs, 71–72
    Internet architecture
        assumptions, 3
    LTP-T and, 271
    space communications, 272
Energy, eMAP construction, 208
Energy-aware forwarding,
    MANETs, 235–236

Energy-aware routing protocols,
    70–95
    epidemic routing, 70
    epidemic routing protocol,
        72, 74–76
        network lifetime, 74–76
        properties, 74
    epidemic routing protocol
        analytic model, 77–82
        solution of, 79–82
        statistics of encounter
            times, 77
        transitional epidemic
            routing model, 77–79
    maximizing lifetime, 93–95
        comparison of lifetime
            performance, 94–95
        residual-battery
            information, 93–94
        residual-energy (RE) scheme,
            91–93
        LC scheme with
            residual-energy
            information, 92–93
    restricted epidemic routing
        (RER) protocol, 92–96
        exclusion scheme, 82–83
        limited number of copies
            (LC) scheme, 84–86
        limited time (LT) scheme,
            83–84
    restricted epidemic routing
        (RER) tradeoff function,
        86–91
        evaluation of tradeoff
            function, 88, 89
        network lifetime, 88, 90–91,
            92
        number of copies and time
            delay, 86–87, 88
    sparse networks, 70–74
    VANET, 234
Energy consumption, gMAP,
    comparison among
    systems, 214

Energy density, eMAP
    construction, 204, 205,
    206
Energy information
    mWSN, data collection, 204
    mWSN gMAP, data
        collection performance,
        211
Energy map (eMap)
    aggregation-based
        approaches, 194
    WSN, 191
Energy messages, collection
    efficiency, 210
Energy source, mWSN model, 197
Enforced routing, 33
Enhanced Dynamic Source
    Routing (EDSR), 172,
    181
Environmental factors, VANET,
    224, 229
Environmental monitoring,
    CarTel applications, 232
Epidemic routing protocol (ERP),
    21, 48, 70, 72, 74–76, 166
    analytic model, 77–82
    solution of, 79–82
    statistics of encounter
        times, 77
    transitional epidemic
        routing model, 77–79
    network lifetime, 74–76
    pocket switched networks, 58,
        59
    properties, 74
    routing case studies, 61–62
    routing in sparse networks,
        72–73
    routing module applicability,
        57
    VANET, 244
EPOXI spacecraft, 272–273, 311
Erasure Coding, 47
ERP; *See* Epidemic routing
    protocol

Error control, data transport
    protocols for space, 266
Error rates
    DTN requirements, 4
    TCP problems in space, 265
eScan, 192, 193–194, 214
European Space Agency, 273
Exclusion (EX) scheme
    comparison among restriction
        methods, 88, 89
    lifetime of, 90
    RER tradeoff function, 86,
        87, 88
    residual battery information,
        94
    restricted epidemic routing
        (RER) protocol, 82–83
Exotic media networks, ix–x
EX scheme; *See* Exclusion scheme

F
Fading, VANET characteristics,
    229
False position information,
    VANET open issues,
    251–252
Fault tolerance
    data delivery techniques, 195
    P2P architectures, 164, 165
Ferrying protocols, 166, 195, 197,
    214, 215
File Delivery Protocols, CFDP,
    290
File sharing applications, P2P
    systems, 163, 164, 172
First Contact protocol, 106
First hop processing and
    forwarding, 17, 103
First-in-First-out (FIFO)
    algorithm, MPP, 173
Flat peer hierarchy, 164–165
FleetNet project, 238
Flexible cities, gMAP
    construction, 191
Flooding, 104, 172, 195

Flooding-like strategy ORION, 171

Flow control
  LTP and, 271
  TCP problems in space, 265

Focus, 57; *See also* Spray and focus

Format, bundle protocol, 11–14
  end-point IDs, 12
  formats of bundle blocks, 14
  self-delimiting numeric values, 11–12

Forwarding; *See also* Routing; *specific routing schemes*
  bundle protocol processing
    first hop, 17
    second hop, 18
  bundle service, 9
  end-to-end DTN microanalysis, 306
  ODR, 179, 180
  routing
    intermittently connected networks, 39–40
    as opportunistic forwarding, 35–36
    opportunistic routing components, 34
  routing primitives, 47, 48
  RPP design space, 149
  selfishness issue, 103
  SMART scheme, 114; *See also* SMART scheme
  VANETs, 229, 236–247, 248
    MANET limitations, 235–236
    opportunistic, 239–247, 248
    position-based, 236–238
    trajectory-based, 238–239

Forwarding copy number impact, SMART scheme simulation, 121–122

Forwarding node set, 110–111

Forward message switching, 5–6

Fragmentation
  architecture, 6–7
  end-to-end DTN microanalysis, 306–308
  security issues, 26–27

Fragment authentication, SMART scheme, 115–116

Frame scheduling, IEEE 802.11p, 251

Freenet, 164

Free-riders, 174

FRESH, 48

G

Gateway, VANET routing, 242, 243

Generality, design goals, SMART scheme, 108

Generation control, 41

Geocast vulnerability, VANET open issues, 252

Geographic Source Routing (GSR), VANETs, 238

Geo-Imaging, CarTel applications, 233

Geostationary orbit (GEO) satellites, 263, 284, 288, 297

Global Maps (gMAPs)
  comparison with existing approaches, 196–197
  mobility-assisted monitoring with, 198–208
    eMap construction, 204, 206–208
    overview of approach, 198
    path planning of ANs, 199–203
    scenario classification, 198–199
  performance evaluation, 208–215
    comparison among systems, 214
    comparison to related work, 214–215

data collection and map
    construction, 209–214
data collection
    performance, 211–213
evaluation metrics, 210–211
path planning, 208–209
regioning performance,
    213–214
simulation settings,
    209–210
system model, 197–198
WSN, 191
Gnutella/Gnutella-like, 164, 171,
    181, 182
Goodput, 308, 310, 311
Gossip, 233–234
Graph routing, 272, 274
Greedy Perimeter Coordinator
    Routing (GPCR), 238
Greedy Perimeter Stateless
    Routing (GPSR), 238
Greedy replication
    intermittently connected
        networks, 36–37
    routing primitives, 47, 48
Greedy Spraying, 60
Greedy Traffic Aware Routing
    (GyTAR), 244

H
Handoff, WBSS, 251
Hash function, 162–163
Hash tables
    distributed (DHT), 144,
        146–147, 162
    structured MPP overlay
        networks, 175
Hash tree, Merkle, 115–116
Healing, 252
Heterogeneous mobility, 53–54
Heterogeneous node capabilities,
    55
Hierarchy
    P2P hybrid systems, 165
    peer, 164–165

History-based routing, 43–44
    epidemic, 48
    routing primitives, 47, 48
    VANETs, 241, 245–246
History of DTN, 2
History of last encounter, 43–44
History of past encounters, 43–44,
    47
Hoe's algorithm, 294
Homogeneous mobility, 53–54
Hop-by-hop DTN microanalysis,
    302–303
Hop-by-hop transfer, 4
HubCode, 241, 247, 248
Hybla, 291, 296
    features of, 289
    PEPsal, 292
    performance comparison
        with congestion, 297
        with congestion and link
            losses, 299
        with link losses, 297, 298
        railway satellite
            communication, 307, 308
    TATPA testbed, 294, 295
Hybrid optimal control theory,
    195–196
Hybrid protocols
    for disconnected islands, 54
    P2P architectures,
        unstructured, 164, 165
Hyper Text Transfer Protocol
    (HTTP), MPP and, 172

I
Ideal lifetime, ERP network, 76
Identifier
    ITR overlay network, 176
    peer, 162
IEEE 802.11 interfaces, 128
IEEE 802.11p MAC, 249–251
Incentive effectiveness, 119–120
Incentive scheme, credit based,
    116; See also SMART
    scheme

Incorrect location information,
VANET open issues,
251–252
Indirect routing, MPP
MADPastry, 176
tree-based (ITR), 175
comparison among
systems, 181
peer-to-peer overlay
networks for DTNs,
176–178
Information retrieval,
Prophet-based, 174–175
Infrastructure
CarTel applications, 232
VANET characteristics,
227–228
Infrastructure-based forwarding,
VANETs, 240–242, 243
Infrastructureless forwarding,
VANETs, 241, 242,
244–247
INLR, 194, 214
In-network aggregation, WSN,
191
Integrated architecture, 170, 171
MPP systems, 180, 181
PEPs (performance
enhancing processes),
291–292
Integrated path planning
algorithm, mWSN, 203
Intelligent caching, 174, 175
Interaction controller
RPP functional blocks, 138,
139
RPP user nodes, 140–143
interaction with
throwboxes, 142–143
interaction with user
nodes, 142
Interaction diagram for sequence
of operations, 145, 146
Inter-agency Operations Advisory
Group (IOAG), 272

Intermittent communications,
VANET, 224
Intermittently connected networks
bundle service, 10
defined, 165
DTN requirements, 3
routing, 34–42
message coding, 40–42
message forwarding, 39–40
message replication, 36–39
routing as opportunistic
forwarding, 35–36
RPP, 131
satellite communications,
284, 286
VANET, 224–225
International Space Station, 273
Internet Engineering Task Force
(IETF), 262, 271
Internet model, 10, 49
Internet Research Task Force
(IRTF), 266, 268, 288
Interoperability, space
networking, 272
Interplanetary Internet (IPN),
262, 263
Inter-Planetary Network Research
Group (IPNRG), 288
In-time delivery ratio, VANET,
234
ION system, 272, 273
IPOPT, 208
Island Hopping, 48
Isobar, 193–194, 214
Isocluster aggregation, 194
Isoline, 194, 210, 213, 214
Isoline map, 192
Iso-Map, 192, 194, 214
ITR (indirect tree-based routing),
175, 176–178, 181

J
J2ME, 151, 152, 153
Jamming, 252

Japanese Aerospace Exploration
    Agency, 273
JAVA-BP, 273
JavaMIDlet, 151–153
JSR, 153

K
Kademlia, 163
KazaA, 165
k-d tree, 206
Key management, public key
    revocation, 123
Kiosknet, 166

L
Landmark keys, MADPastry, 176
Landmarks, prediction of
    distance, 148
Latency
    mWSN
        comparison among
            systems, 215
        data collection
            performance, 213
        SMART scenarios, 120–122
        VANET characteristics, 229
Layer concatenation, 105, 110
Layered architecture
    DTNCLP, 269, 270
    MPP systems, 180, 181
        cross-layer architecture,
            172
        cross-layered, 170
        integrated, 170, 171
    RPP, 133
Layered coin, 104
Layered coin aggregation, 115
Layered coin batch verification,
    115
Layer injection attack, 107,
    109–111
Layer removal attack, 107–108
l bits, ITR overlay network, 176
LC scheme; *See* Limited number
    of copies scheme

Least Recently Used (LRU)
    algorithm, 173
LEO; *See* Low earth orbit system
Levy Walks, 40
L-hop neighborhood query
    spraying (LNS), 174–175
Licklider (long-haul) transmission
    protocol (LTP), 266,
    270–271, 290
    LTPCL hybrid, 273
    LTP-DTN2 hybrid, 275
    LTP-transport, 262, 270, 271
Lifetime, network; *See* Network
    lifetime
Lifetime maximization,
    energy-aware routing
        protocols, 93–95
    comparison of lifetime
        performance, 94–95
    residual-battery information,
        93–94
Limited number of copies (LC)
    scheme
    comparison among restriction
        methods, 88, 89
    lifetime of, 91, 92
    RER tradeoff function, 86,
        87, 88
    residual battery information,
        94
    with residual energy
        information, 92–93
    restricted epidemic routing
        (RER) protocol, 84–86
Limited time (LT) scheme
    comparison among restriction
        methods, 88, 89
    lifetime of, 90–91
    RER protocol, 83–84
    RER tradeoff function, 86,
        87, 88
    residual battery information,
        94
Linear network coding, 40

Link class retransmission timers,
DTN microanalysis,
303–306
Link-layer protocols, 271
Link losses, satellite, 297, 298, 299
Linux
ION performance, 272
MTDL, 309
TATPA, 294
TCP disruption resilience in,
302
TCP splitting, 291
Load balancing, mWSN gMAP,
216
Location
distance prediction services,
148
MPP systems, 182
VANET characteristics, 228
VANET open issues, 251–252
Location criteria, RPP throwbox
nodes, 147–148
Location-dependent identifiers,
ITR, 176, 178
Locations visited patterns, 44
Long haul transmission; *See*
Licklider (long-haul)
transmission protocol
Low earth orbit (LEO) system,
284, 312–314
satellite communications, 290
UK-DMC, 272, 314
LTP; *See* Licklider (long-haul)
transmission protocol
LT scheme; *See* Limited time
scheme

M
MAC; *See* Medium access control
MADPastry; *See* Mobile ad-hoc
Pastry
Management, P2P architectures,
164, 165
MANETs (mobile ad hoc
networks), 47, 49, 267

data collection techniques,
195, 197
as DTN special case, 166, 168
identifier queries, 178
MPP system applicability,
181
routing, 32–33
opportunistic, 239
in sparse networks, 70–73
VRR, 178
routing issues, 24
routing module applicability,
57
VANET applications, 234
and VANET forwarding,
235–236
VANET simulations, 227
Map-based monitoring; *See*
Mobility-assisted WSN,
delay-tolerant
monitoring
Markov Chain Model, 77, 78, 84,
85
Mars Global Surveyor, 263
Mars Odyssey, 263
Martian communications, 273
Max Fair Share, 299
Maximum Tolerable Disruption
Length (MTDL),
301–303, 306, 309
MaxProp, 22–23, 48, 241, 246
M-DART (Multi-path Dynamic
Address RouTing), 176,
181
MDDV, 241, 245
Medium access control (MAC)
mWSN data collection, 204
mWSN energy information
incompleteness, 211
VANETs, 242
cooperative ARQ, 247, 249
enhancements, 225
IEEE 802.11p, 249–251
Medium access control (MAC)
layer

mWSN model, 197
RPP network interface, 143
sparse networks, 51
VANET routing, 242
Meets and visits (MV), 241, 246
Memento, 193
Meng, 214
Merkle hash tree, 115–116
Message coding; *See* Coding,
      message
Message complexity, mWSN
      gMAP, 214, 215
Message content/priority,
      application requirements,
      56
Message fault tolerance, 195
Message ferrying protocols, 166,
      195, 197, 214, 215
Message forwarding; *See*
      Forwarding
Message replication; *See*
      Replication of message
Message switching, 5–6
Metropolitan networks with
      heterogeneous nodes,
      59–61
Middleware; *See* RPP
Military applications, 165–166,
      311
Minimum delivery probability
      (MDP) lifetime, 76, 88,
      90, 93–94, 95
Mixed-integer linear programming
      (MILP), 201, 202–203,
      208, 209
Mixed-integer nonlinear
      programming (MINLP),
      195–196, 201–202
Mixing time, mobility, 53
MKH-SMART, 117
Mobile access points (MAPs),
      DakNet, 231–232
Mobile Ad Hoc Networks; *See*
      MANETs

Mobile ad-hoc Pastry
      (MADPastry), 175–176,
      181
Mobile peer-to-peer control
      protocol (MPCP), 172
Mobile peer-to-peer (MPP)
      systems, 160–182
   comparison among, 181
   comparison among systems,
      181
   delay tolerant networks,
      165–168
      MANETs as special case
         of, 166, 168
      store-carry-forward
         paradigm, 166, 167
   overlay networks, 161–165
      structured, 162–163
      unstructured, 164–165
   overlay networks for DTNs,
      168–182
      challenges, 169–170, 171
      structured, 175–179, 180
      unstructured, 170–175
Mobility
   classes of, structured versus
      unstructured, 197
   destination-independent (DI)
      utility, 45
   DTN classification, 34
   routing,
      destination-independent
         (DI) utility, 45
      routing module applicability,
         57
      routing primitives, 47, 48
   taxonomy of DTNs, 52–54
      amount of, 52–53
      structure of mobility
         model, 53–54
   VANET characteristics, 228
Mobility-assisted data collection
      techniques, 195
Mobility-assisted delay-tolerant
      monitoring techniques,

194–196
Mobility-assisted WSN,
    delay-tolerant
    monitoring, 190–216
  evaluation, 208–215
    comparison to related
    work, 214–215
    data collection and map
    construction
    performance, 209–214
    path planning
    performance, 208–209
  gMAP, using global maps,
    198–208
    eMap construction, 204,
    206–208
    overview of approach, 198
    path planning of ANs,
    199–203
    scenario classification,
    198–199
  state-of-the-art, 192–197
    gMAP contributions,
    196–197
    monitoring techniques,
    194–196
    node-centric delay-critical
    monitoring techniques,
    193
    region-centric delay-critical
    monitoring techniques,
    193–194
  system model, 197–198
Mobility model
  routing module applicability,
    57
  VANET, 225–226
Mobility patterns
  RPP design space, 149
  VANET, 224, 228
Mobyspace, 48
Momento, 192
Monitoring; *See* Mobility-assisted
    WSN, delay-tolerant
    monitoring

Morpheus, 165
MoVe, 241, 244
MPP; *See* Mobile peer-to-peer
    systems
MTDL (maximum tolerable
    disruption length),
    301–303, 306, 309
Mule, data; *See* Data mule
Multi-copy forwarding, 104, 106,
    119
Multi-hop networks, 168, 169,
    191; *See also* Mobile
    peer-to-peer systems
Multi-path Dynamic Address
    RouTing (M-DART),
    176, 181
Multi TCP, TATPA testbed, 294
MV (Meets and Visits) routing,
    48, 246

N
Nack-Oriented Reliable Multicast
    (NORM), 290, 314
Naming scheme, DTN
    architecture, 4
Napster, 164
National Aeronautics and Space
    Administration (NASA),
    262, 272–273
  NASA JPL DINET, 165–166
  NASA JPL experiments,
    satellite communications,
    313
  NASA JPL ION system, 272
N-dimensional model, throwbox,
    148
Neighborhoods, TSP with
    (TSPN), 196
Network characteristics, routing,
    32
Network coding, 34, 40–41
Network infrastructure
  data transport protocols for
    space, 266
  RPP design space, 149

Network interface, RPP, 138, 139,
        143–144
Network lifetime
    epidemic routing protocol,
        70, 71, 74–76
    restricted epidemic routing
        (RER) tradeoff function,
        88, 90–91
Network load, SMART
        performance, 121
Network topology, MPP systems,
        182
NewReno, 289, 295, 296, 310, 311
NewReno/SACK, 296–297, 298,
        299
NLP (nonlinear problem), 208,
        209
NLP-solver, 208, 209
Node availability, VANET
        characteristics, 229
Node-centric monitoring, 193, 196
Node clusters, MADPastry,
        175–176
Node resources
    destination-independent (DI)
        utility, 46
    taxonomy of DTNs, 55–56
Nodes
    heterogeneous, 59–61
    SMART scheme, 111–112
Node speed, mobility, 52
Nodular tontine attack, 107–108,
        109–111
Noise, 265
Noncooperative behavior, VANET
        open issues, 252
Nonlinear problem (NLP), 208,
        209
Nonlinear programming,
        mixed-integer (MINLP),
        195–196, 201–202
NORM (Nack-Oriented Reliable
        Multicast), 290, 314
NoTalk, 244

Notification process, bundle
        service, 9
Number of copies, RER tradeoff
        function, 86–87, 88
Numeric values, self-delimiting,
        14–19

O

Octree, 206
ODR (Opportunistic DHT-based
        Routing), 175, 179, 180,
        181
Offline Security Manager (OSM),
        106, 112
Ohio University, 272
OLSR, 33, 47, 52, 71
One-way-light-time (OWLT), 273
Opportunistic connectivity,
        bundle service, 10
Opportunistic DHT-based
        Routing (ODR), 175,
        179, 180, 181
Opportunistic forwarding, 102
    rewarding models, 107
    routing as, 35–36
    RPP system dimensioning,
        148–151
    VANETs, 239–247, 248
        cluster based, 244
        coding-based, 247, 248
        history based, 245–246
        infrastructure-based,
            240–242, 243
        infrastructureless, 242,
            244–247
        predictive, 244
        priority/content based, 246
        topology based, 245
Opportunistic frame scheduling,
        251
Opportunistic network coding, 41
Opportunistic Networking
        Environment (ONE)
        simulator, 118–119

Opportunistic routing, 32, 33–34;
See also Routing
  examples, 47, 48
  routing as opportunistic
    forwarding, 35–36
  VANET, 234
Opportunistic Routing with
  Window-Aware
  Replication (OR WAR),
  42, 48
Optimal assist nodes, 216
Optimal control theory, 195–196
Optimization-based path
  planning, mWSN, 203
Optimized Routing Independent
  Overlay Network
  (ORION), 170, 171, 172,
  181
Optimized SMART, 117–122
ORION, 170, 171, 172, 181
OR-WAR (Opportunistic Routing
  with Window-Aware
  Replication), 42, 48
OSM; See Offline Security
  Manager
Overhead
  mWSN gMAP, comparison
    among systems, 214
  P2P architectures
    ITR address space overlay
      management, 178
    ORION and, 171
    unstructured, 163, 164
    SMART scenarios, 120–122
    SMART scheme, 105
    SMART scheme efficiency,
      115
  VANET performance metrics,
    235
Overlaps, mWSN
  gMAP
    data collection
      performance, 211
    open issues, 215
    path planning, 201–203

Overlay
  architecture, 4–5
  MPP; See Mobile
    peer-to-peer systems;
    Peer-to-peer overlay
    networks for DTNs
  structured P2P systems, 163
  throwbox, 148
Overlay identifiers, MADPastry,
  176

P
Pacing, TATPA testbed, 294
Packet delivery; See Delivery
Packet drop/packet loss
  incentive effectiveness, 119
  Internet architecture
    assumptions, 3
  satellite communications,
    288–289, 297
  space communications, 166,
    265
Packet Error Rate (PER), 284,
  298, 314
Packet flow, controlling on basis
  of energy information, 73
Packet pacing, TATPA testbed,
  294
Pairing technique, SMART
  scheme, 109
Parking space, VANET
  applications, 234
Partitions/partitioning
  eMAP construction, 206
  routing, 49
  VANET characteristics, 228
Passive cure, 252
Passive distributed indexing
  (PDI), 131
Past encounters parameters,
  43–45, 47
Pastry, 163
Pastry-like routing, MPP, 181
Path length, VANET, 228

Path planning of assist nodes,
        mWSN
    integrated, 203
    mobility-aided, 195–196
    open issues, 215, 216
    performance, 208–209
    semi-structured scenarios,
        199, 200–203
    structured scenarios, 198,
        199–200
PATH project, 232
Patterns of locations visited, 44
Pause time and frequency,
        mobility, 52–53
Peer id, 162
Peers
    defined, 160
    pocket switched networks, 58
Peer-to-Peer file sharing system
        based on Swarm
        Intelligence (P2PSI),
        171, 173–174, 181, 182
Peer-to-peer overlay networks for
        DTNs, 168–182
    challenges, 169–170, 171
    mobile, 168
    MPP systems, 161–165
    routing; See Routing
    structured, 162–163, 175–179,
        180
        ITP, 176–178
        MADPastry, 175–176
        ODR, 179, 180
        virtual ring routing,
            178–179
    unstructured, 164–165,
        170–175
        MPP, 172
        ORION, 171
        P2P swarm intelligence,
            173–174
        Prophet-based information
            retrieval, 174
PEPs; See Performance enhancing
        proxies

PEPsal, 291, 296, 297
    performance comparison, 298,
        299, 300, 309, 310
    railway satellite
        communication, 308–310
PER (Packet Error Rate), 284,
        298, 314
Performance, 3
Performance enhancing proxies
        (PEPs) (protocol
        accelerators), 284–285,
        287, 289, 290–292, 296
Performance evaluation
    gMAPs, 208–215
        comparison to related
            work, 214–215
        data collection and map
            construction, 209–214
        path planning, 208–209
    mobility, 52–54
    MPP systems, 181, 182
    routing case studies
        applications with priorities,
            61–63
        metropolitan networks
            with heterogeneous
            nodes, 59–61
        pocket switched networks,
            58, 59
    satellite communications with
        continuous end-to-end
        connectivity, 294–300
    satellite communications with
        random intermittent
        end-to-end connectivity,
        307, 309–310, 312
        with congestion, 296–297
        with congestion and link
            losses, 298, 299
        with link losses, 297
        sending window
            performance
            improvement, 298–300
    SMART scheme, 117–123

cryptographic overhead,
117–118
simulation, 118–122
TCP, in space, 264–266
VANET routing protocol
metrics, 234–235
Performance problems, space
communications,
264–266
Periodicity, VANET forwarding
schemes, 245
Persistent paths, 165, 166, 172,
176
Persistent storage; *See* Storage,
persistent
Phase transition, 49
Pheromones, P2PSI, 174
PHY/MAC, 249
Physical layer enhancements,
VANET research, 225
Pocket switched networks, 58, 59,
102
Polygon aggregation, 193–194
Position-based
forwarding/routing,
VANETs, 234, 236–238
Position information; *See also*
Location
VANET characteristics, 228
VANET open issues, 251–252
Power management framework,
message ferrying, 195
Predicability, VANET
characteristics, 228
Prediction-based routing,
VANETs, 239, 241
Predictive Directional Greedy
Routing (PDGR),
VANETs, 239
Predictive forwarding, VANETs,
244
PREP, 48
Priority
bundle service, 9
routing

application requirements,
56, 57
case studies, 61–63
Priority/content based
forwarding, VANETs,
246
Privacy, VANET, 225, 229, 251,
253
Proactive protocols
MANET forwarding schemes,
235
VANETs, 236
Probabilistic Routing Protocol
using History of
Encounters and
Transitivity
(PROPHET); *See*
PRoPHET
Probability-limited replication, 38
Processing
bundle protocol, 14–19
creation at source, 15, 16
first hop processing and
forwarding, 17
reception by destination,
18–19
second-hop processing and
forwarding, 18
transmission by source, 15,
16
DTN classification, 34
Propagation delay, TCP problems
in space, 288
Propagation impairment, satellite
communications,
288–289
PRoPHET, 22, 48, 268
Prophet-based information
retrieval, 174, 181
Protocol accelerators, 284–285,
290–292
Protocol data units, 267
Pseudocode, regioning algorithm,
206, 207
Public key revocation, 123

Pull-based forwarding, 241
Push-based forwarding, 241

Q
Query, U-Hopper, 151, 153
Query diffusion, RPP, 134–136
Query dissemination, MPP,
        174–175
Queue policy, TATPA testbed,
        295

R
Railway satellite communications,
        308–311
Random caching, 174
Random Early Detection (RED),
        295
Random intermittent end-to-end
        connectivity, satellite
        communications, 285,
        300–311
Random landmarking,
        MADPastry, 175–176
Random network coding, 40–41
Random networks, 49–50
Random Waypoint mobility
        model, 61
RAPID, 23, 42, 48
Reactive fragmentation, 306–307,
        315
Reactive protocols, MANET
        forwarding schemes, 235
Receipt, bundle service, 9
Reception by destination, 18–19
Record Management Store,
        U-Hopper, 152
Redundant node deployment,
        WSN, 191
Regional percentile accuracy
        (RPA), 211, 213
Region-centric delay-critical
        monitoring techniques,
        193–194
Region-centric (map-based)
        monitoring, WSNs, 192

Regioning
    eMAP construction, 206, 207,
        208
    evaluation metrics, 210–211
    gMap construction, 213–214
Relative utility criterion,
        forwarding, 39
Relay infrastructure,
        interplanetary
        communications, 264
Reliability
    application requirements, 56
    incentive schemes, 104
    routing module applicability,
        57
    satellite communications,
        bundle layer DTN
        architecture overview,
        285
Remuneration, SMART scheme,
        104
Replication-based routing
        protocols, 21–24
    epidemic routing, 21
    MaxProp protocol, 22–23
    PRoPHET protocol, 22
    RAPID protocol, 23
    spray and wait protocol,
        23–24
Replication of message
    intermittently connected
        networks
        controlled replication,
        37–38
        greedy replication, 36–37
        utility-based replication,
        38–39
    routing
        case studies, 61–62
        intermittently connected
            networks, 36–39
        opportunistic routing
            components, 34
        routing module
            applicability, 57

routing primitives, 47, 48
Replication Time Limit (RTL),
    83, 86
Reputation-based schemes, 103,
    105, 123
REQ beacon
    collection efficiency, 210
    mWSN data collection, 204
RER; *See* Restricted epidemic
    routing, tradeoff function
RE scheme; *See* Residual-energy
    scheme
Research
    DTN standardization for
        space networking,
        271 272
    satellite communications,
        312–313
    space communications,
        271–273
Residual-battery information,
    93–94
Residual-energy (RE) scheme,
    91–93
Resource Allocation Protocol for
    Intentional DTN routing
    (RAPID); *See* RAPID
Resources
    ITR queries, 178
    MPP systems, 180, 181
    routing module applicability,
        57
    routing primitives, 47, 48
    RPP application
        requirements, 150
Restricted epidemic routing
    (RER) protocol, 92–96
    exclusion scheme, 82–83
    limited number of copies
        (LC) scheme, 84–86
    limited time (LT) scheme,
        83–84
    tradeoff function, 86–91
        evaluation of tradeoff
            function, 88, 89

network lifetime, 88, 90–91
    number of copies and time
        delay, 86–87, 88
Retransmission
    bundle service, 10
    delegation of, 268
    Internet architecture
        assumptions, 3
    satellite communications,
        301–303
Retransmission timers, DTN
    microanalysis, 302–303
Return receipt, bundle service, 9
Reverse Peer to Peer systems
    (RPP), 128–154
    application scenario, 129
    architecture, novelty of, 132
    functional blocks, 138–148
        throwbox nodes, 139,
            144–148
        user nodes, 139–144, 145
    implementation, 151–153
        throwboxes, 153
        U-Hopper, 151–153
    related works, 131–132
    system dimensioning,
        148–150
    system elements, 132–137
        advertisements, diffusion
            of, 136–137
        query diffusion and data
            retrieval, 134–136
RFC 5050 Bundle Protocol
    Specification, 271
RFC 5325 Licklider Transmission
    Protocol C Motivation,
    271
Roadside-Aided Routing (RAR),
    VANETs, 240
Roadside infrastructure, VANET
    characteristics, 227–228
Round trip times (RTTs)
    RTT unfairness problem, 297,
        310
    satellite communications, 289

performance evaluation,
    296–299
protocol accelerators, 290
TCP problems in space, 288
Route stretch effect, MPP
    systems, 169
Routing, 19–24; *See also*
    Forwarding
architecture, 6
classification of, 20–21
credit-based incentive
    scheme; *See* SMART
    scheme
DTNs, 71–73
energy aware; *See*
    Energy-aware routing
    protocols
MPP
    comparison among
        systems, 181
    store-carry-forward
        paradigm, 166
MPP overlay networks,
    structured, 175–179, 180,
    181
    ITR, 176–178
    MADPastry, 176
    ODR, 179, 180
    virtual ring, 178–179
MPP overlay networks,
    unstructured, 170–175,
    180, 181
    ASOS, 173
    MPP, 172
    ORION, 171
    P2PSI, 173–174
    PRoPHET protocol,
        174–175
open issues, 24–25
as opportunistic forwarding,
    35–36
path planning, 195
replication-based, 21–24
    epidemic routing, 21
    MaxProp protocol, 22–23

PRoPHET protocol, 22
RAPID protocol, 23
spray and wait protocol,
    23–24
routing considerations, 20
SMART simulation, 119
space and satellite
    communications,
    268–269
    end-to-end DTN
        microanalysis, 306–308
    overview of, 267
sparse networks
    epidemic routing, 72–73
    spray and wait, 73
    store-carry-forward
        paradigm, 71–72
TATPA testbed, 295
VANETs, 224–225, 229, 234
    MANET protocols and
        their limitations,
        235–236
    routing protocols, 236–247
    simulations, 227
Routing, taxonomy and design,
    31–63
case studies, 57–63
    applications with priorities,
        61–63
    metropolitan networks
        with heterogeneous
        nodes, 59–61
    pocket switched networks,
        58, 59
design guidelines, 56, 57
intermittently connected
    networks, 34–42
    message coding, 40–42
    message forwarding, 39–40
    message replication, 36–39
    routing as opportunistic
        forwarding, 35–36
taxonomy of DTNs, 49–52
    application requirements,
        55–56

connectivity, 49–52
node resources, 55–56
utility functions, 42–47, 48
additional considerations,
47
destination-dependent
(DD) utility, 42–45
destination-independent
(DI) utility, 45–46
examples of routing
protocols, 47, 48
Routing modules, 32
Routing tables, 47, 48, 179
RPA; *See* Regional percentile
accuracy
RPP; *See* Reverse Peer to Peer
systems
RTL; *See* Replication Time Limit
RTTs; *See* Round trip times
RTT unfairness problem, 297,
310–311
Rural area connectivity
Kiosknet, 166
VANET applications,
231–232

S
SACK, 296
Saratoga, 262, 270–271, 290, 314
Satellite communications,
283–315; *See also* Space
communications
applications based on
end-to-end connectivity,
285
bundle layer DTN
architecture overview,
286–288
DTN concept development,
288
end-to-end reliability and
custody transfer option,
287–288
channel impairments and
possible solutions,

288–293
delay/disruption tolerant
networking alternative,
292–293
end-to-end TCP
enhancements and other
optimized protocols,
289–290
performance enhancing
proxies (protocol
accelerators), 290–292
PEPs (performance
enhancing processes),
284–285
research, 312–313
research and experimental
activities, 272
scenario 1: end-to-end
connectivity, continuous,
294–300
performance evaluation,
296–300
testbed and tools used in
performance evaluations,
294–296
scenario 2: end-to-end
connectivity, random
intermittent, 300–311
analysis of DTN behavior
in presence of disruption,
300–308
performance evaluation in
realistic case study of
railway communications,
308–311
TCP resilience to
disruption, 301–302
scenario 3: end-to-end
connectivity, scheduled
intermittent, 312
scenario 4: without
end-to-end connectivity,
312–313
Scalability
P2P architectures, 164, 165

VANETs, 244
Scalable MPP systems, 170
Scale Free Routing (SFR), 39–40
Scheduled intermittent
    connectivity
    bundle service, 10
    satellite communications, 285
Scheduled routing, 33
Scheduling, path planning, 195
SCPS-TP, 290
SDNVs (self-delimiting numeric
    values), bundle format,
    11–12
Second-hop processing and
    forwarding, bundle
    protocol, 18
Secure Multi-Layer Credit-based
    Incentive scheme; *See*
    SMART scheme
Security
    design goals, SMART
        scheme, 108
    MPP systems, 169–170
    open issues, 26–27
    open issues, VANET, 251–252
    P2P architectures, 164, 165
    SMART scheme, 104–105
    space communications, 274
    VANET research, 225
Segment losses, satellite
    communications,
    288–289
Selective Negative ACK, 290
Selective repeat ARQ, 271
Self-delimiting numeric values,
    11–12
Self-delimiting numeric values
    (SDNVs), 11–12
Selfishness, 103, 120; *See also*
    SMART scheme
Selfish nodes, 103
Semi-structured mWSN scenarios
    data collection, 204
    gMAP data collection
        performance, 211, 212

gMAP scenario classification,
    199
path planning of assist nodes,
    199, 200–203
    breakpoints, 200–201, 202
    optimal path, 203
    overlaps, 201–203
Sending window performance
    improvement, satellite
    communications,
    298–300
Sensor networks; *See*
    Mobility-assisted WSN,
    delay-tolerant
    monitoring
Sensor Network Tomography
    (SNT), 193
Sensor nodes (SN)
    downloading from, 233–234
    mWSN, 197
    collection efficiency, 210
    data collection, 204, 205
    data delivery techniques,
        195
    gMAP, 191
Service, bundle protocol, 9–11
Service container, RPP, 138,
    139–140
SETI@home project, 164
Shortest-path tour, mWSN path
    planning, 203
SI (swarm intelligence), 171,
    173–174, 181, 182
Signatures, aggregate, 115
Single copy schemes, 106
Single-path planning algorithm,
    202
Sink, data delivery techniques,
    195
Sink nodes, epidemic routing, 79,
    82
Smart forwarding, 47, 48
Smart replication, 48, 57, 62
SMART scheme, 102–124
    credit creation, 105–106

details of, 112–114
efficiency enhancement,
114–117
open issues, 123
overview of, 109–112
layer injection or nodular
tontine attack
prevention, 109–111
motivating nodes to submit
coins, 111–112
pairing technique, 109
performance evaluation,
117–123
cryptographic overhead,
117–118
simulation, 118–122
security properties, 104–105
selfishness issue, 102–105
system model and design
goals, 106–108
Smart Spraying, 60
SN; *See* Sensor nodes
SNAKs, 290
Snapshots, 43, 47
collection efficiency, 210
gMAP data collection
performance, 211, 212,
213
mWSN data collection, 204,
205
SNT (Sensor Network
Tomography), 193
Social networks, 44–45
Solaris, 272
SONC, 310–311
Source coding, 40
opportunistic routing
components, 34
Sources, RPP, 132
Space communications, 165–166,
261–275; *See also*
Satellite communications
bundle protocol (BP), 269
convergence layer protocol
(CLP), 269–271

DTN for space, 268–269
DTN overview, 267–268
open issues, 274–275
overview of, 263–264
performance problems,
264–266
research and experimental
activities on DTN,
271–273
transport protocol categories,
266–267
Space Communications Protocol
Standard-Transport
Protocols (SCPS-TP),
290
Space Networking Strategy Group
(SISG), 272
Space partitioning, eMAP
construction, 206
Space-time paths, DTNs as, 168
Sparse networks, 49
energy-aware routing
protocols, 70–74
intermittently connected
networks, 50–51
store-carry-forward
paradigm, 166
Spatial correlation
mobility, 54
WSN monitoring, 192
SPAWN protocol, 233–234, 242
Speed, VANET characteristics,
228
Spray and focus, 47, 48, 56, 58,
59, 61
Spray and wait, 21, 23–24, 38–39,
47, 48
applications with priorities,
61–63
incentive schemes, 107, 119
MPP, 181
performance comparison, 60
pocket switched networks, 58,
59
routing in sparse networks, 73

RPP, 134
SMART simulation, 119
Spraying, 52
    applications with priorities,
        62–63
    MPP, 174–175, 181
Stale position problem, VANETs,
    236, 237, 238
Stateless channel access, IEEE
    802.11p, 250–251
Statistics of encounter times, ER
    analytic model, 77
Status report, end-to-end DTN
    microanalysis, 306–308
Storage
    distributed storage systems,
        132, 138, 139
    DTN classification, 34
    MPP, 173
    node resources, 54, 55
    persistent, 4, 5
        bundle creation, 15
        bundle reception, 19
        custody transfer, 25,
            267–268, 285
        first hop processing and
            forwarding, 17
        second hop processing and
            forwarding, 18
        space communications, 269
    RPP functional blocks, 138
    RPP layered architecture, 133
    U-Hopper, 152
Store-carry-forward paradigm, 102
    energy aware routing, 70
    MPP DTNs, 166, 167
    SMART scenarios, 123
    in sparse networks, 71
    VANETs, 224–225, 239, 244
Store-forward paradigm, 70, 160,
    161, 239
    architecture, 5–6
    MPP systems, 182
Stretch effect
    ITR and, 178

MPP systems, 169, 182
Structured mobility
    gMAP scenario classification,
        198–199
    mWSN model, 197
Structured mWSN scenarios
    data collection, 204
    gMAP data collection
        performance, 211, 212
    mWSN gMAP, 212
    path planning of assist nodes,
        198–199
Submission refusal attack, 108,
    111–112
Super-peers, 165
Suppression-based approaches,
        WSN monitoring, 194
Surrey Satellite Technology Ltd.,
    314
Swarm intelligence (SI), 171,
    173–174, 181, 182
SWIM, 48, 79
Sympathy, 192, 193
SYN, 312
SYNACK, 312
Synchronization errors
    DTN performance, 273
    space communications, 274
Synchronous programming style,
    VANET, 231
System initialization, SMART,
    112
System model, SMART, 106–108

T
Tables
    structured MPP overlay
        networks, 175
    structured P2P systems, 163
    virtual ring routing, 179
Tapestry, 163
Target Delivery Probability
    (TDP), ERP network
    lifetime, 76

TATPA testbed, 294–296, 305, 309

Taxonomy of DTNs, 49–52
  application requirements, 55–56
  connectivity, 49–52
  mobility, 52–54
  node resources, 55–56

TCP; *See* Transmission control protocol

TCP-based CLP (TCPCL), 262, 269–270, 273

TCP-based DTN-2, 275

TCPCL, 262, 269–270, 273

TCP/IP, x, 3, 4, 10, 269

TCP-like functionality, LTP and, 271

TCP NewReno/SACK, 296–297

TCP splitting, 290, 291–292, 293

Temporal versus spatial correlation, mobility, 54

Terrestrial mobile networks, iv

T-function, RER, 86–91

Three-dimensional models, throwbox, 148

Throwboxes, RPP, 132
  functional blocks, 138, 139
  implementation, 153
  interaction controller, 142
  interactions with, 142–143
  interaction with user nodes, 142–143
  layered architecture, 133
  U-Hopper, 151

Throwbox nodes, RPP, 139, 144–148
  alternative solutions for data storage, 147
  DHT, 144, 146–147
  DHT alternatives/modifications, 147
  interaction diagram for sequence of operations with, 146

  location criteria, 147–148
  network interface, 144
  organization and modeling, 148
  throwbox primitives, 144, 146

Throwbox primitives
  functional blocks, 138, 139
  throwbox nodes, 144, 146

Time
  DTN performance, 272–273
  mWSN gMAP open issues, 215–216
  RER tradeoff function, 86–87, 88
  temporal versus spatial correlation in mobility, 54

Time-limited replication, 37

Timeouts, VANET characteristics, 229

Time-to-Live (TTL)
  ERP network lifetime, 74–76
  rewarding models, 107
  sink nodes, 79

T-nodes, U-Hopper, 151

Topological routing, 234, 235, 241

Topology
  DTN overview, 267–268
  gMAP scenario classification, 198–199
  mWSN unstructured scenarios, 211
  throwbox organization, 148
  VANET, 224

Topology-based distance prediction, 148

Topology based forwarding, VANETs, 245

Trace-based mobility, 58, 59

Tracking and data relay satellite system (TDRSS), 263, 264

Tradeoff function, RER, 86–91

Traditional routing table entry, 45

Traffic Adaptive Packet Relaying
        (TAPR), 241, 244
Traffic alerts, 241
Traffic load impact scenario,
        SMART simulation,
        120–121
Traffic monitoring, CarTel
        applications, 232
Traffic simulator, 227
Trajectory-based routing,
        VANETs, 234, 238–239,
        244
Transitional epidemic routing
        model, 77–79
Transition diagrams, Markov
        Chain Model, 78, 84, 85
Transmission by source, bundle
        protocol processing, 15,
        16
Transmission control protocol
        (TCP)
    bundle processing, x, 14, 15,
        17, 18
    performance problems in
        space, 264–266
    satellite communications,
        284, 288
    DTNperf_2 tool, 298–300
    end-to-end enhancements,
        289–290
    military applications, 314
    performance comparison,
        309
    railway, 308–311
    resilience to disruption,
        301–302
    single TCP connections on
        DTN hops, 300
    TATPA testbed, 295
    TCP/IP, 3, 4, 10
Transmission overhead, SMART
        scheme, 105, 115
Transmission range, VANET
        characteristics, 228
Transport protocols

end-to-end; *See* End-to-end
        connectivity in satellite
        communications
    space communications,
        266–267
Traveling salseman problems
        (TSP), 203
    gMAP construction, 191
    mWSN
        data collection
            performance, 211, 212
        path planning
            performance, 208
        with neighborhoods (TSPN),
            196
        path planning, 195–196
Tree-based routing, MPP, 175,
        176–178, 181
Tree structure, data, eMAP
        construction, 204
Triangulation, space partitioning,
        206
Trustworthiness
    destination-independent (DI)
        utility, 46
    VANET open issues, 252
TSP (traveling salesman
        problem); *See* Traveling
        salseman problems
Two-dimensional topology,
        throwbox, 148

U
UDP (user datagram protocol),
        262, 270–271, 290
UDP-based CLP, 270–271
UDP/IP, 269
UDP-like functionality, LTP and,
        271
UDP-Lite, 271, 290
U-Hopper, 151–153
UK-DMC LEO system, 272, 274,
        314
UMassDieselNet, 246

Unbounded networks, VANET
    characteristics, 229
Uncontrolled utility-based
    replication, 38
Unfairness problem, RTT, 297,
    310
United Kingdom, UK-DMC LEO
    system, 272, 274, 314
Unknown topology, gMAP
    scenario classification,
    199
Unpredictable end-to-end delays,
    165
Unreliable communications, 165
Unstructured mobility, mWSN
    model, 197
Unstructured mWSN scenarios
    data collection, 205
    gMAP data collection
        performance, 211, 212,
        213
User datagram protocol (UDP),
    262, 270–271, 290
User nodes, RPP, 132, 133, 138,
    139–144
    content manager and data,
        143
    interaction controller,
        140–143
        interaction with
            throwboxes, 142–143
        interaction with user
            nodes, 142
    interaction diagram, sequence
        of operations, 145
    network interface, 143–144
    service container, 139–140
Utility-based replication,
    intermittently connected
    networks, 38–39
Utility criteria, forwarding, 39
Utility functions, routing, 42–47,
    48
    additional considerations, 47

destination-dependent (DD)
    utility, 42–45
destination-independent (DI)
    utility, 45–46
examples of routing
    protocols, 47, 48
routing primitives, 47, 48
Utilized resources, RPP
    application requirements,
    150

V
VANETCode, 240, 241
VANETs; *See* Vehicular Ad Hoc
    Networks
Vehicle Assisted Data Delivery
    (VADD), 241, 245
Vehicle density, VANET
    characteristics, 228
Vehicle routing, path planning,
    195
Vehicular Ad Hoc Networks
    (VANETs), 224–253
    applications classification,
        230
    characteristics of, 226–229
    connectivity islands, 52
    factors affecting
        communications,
        224–225
    innovative applications,
        229–234
        cooperative downloading,
            233–234
        data collection from sensor
            nodes, 232–233
        miscellaneous, 234
        rural areas, low cost digital
            connectivity in, 231–232
    medium access control for,
        247, 249–251
        cooperative ARQ, 247, 249
        802.11p MAC, 249–251
    message dissemination,
        234–247, 248

forwarding schemes
   suitable for VANET,
   236–247, 248
MANET forwarding
   schemes and their
   limitations, 235–236
network characteristics,
   226–229
open issues, 251–253
   acknowledgments, 252
   addressing, 253
   emerging new application,
      253
   security, 251–252
research categories, 225
routing, 33
routing case studies, 61–62
Velocity, VANET characteristics,
   228
Viceroy, 163
Virtual bank, 106
Virtual currency, 103, 105
Virtual marketplace, VANET
   applications, 234
Virtual ring routing (VRR),
   178–179
VISSIM, 227
Voronoi triangulation, 206
Voxel grid, 206
VRR; *See* Virtual ring routing
VxWorks, 272

W
Wave BSS (WBSS), 250–251
W-copy selective query spraying
   (WSS), 174–175
Wireless Access in Vehicular
   Environment (WAVE),
   249
Wireless Sensor Networks (WSN),
   190–191; *See also*
   Mobility-assisted WSN,
   delay-tolerant
   monitoring